Android应用程序开发基础

蒲晓妮　赵 睿　主编

化学工业出版社
·北京·

内容简介

本书内容包含走进 Android 世界、Android 中的资源、用户界面设计基础、Activity、Fragment、数据存储、内容提供者、服务、广播、网络编程、媒体应用技术和综合项目——智慧党建等，书中对每一个项目都编写了详细的操作步骤和参考代码，学生参考教材就可以完成项目。为便于学生考取“移动应用开发”的“1+X”职业技能等级证书，书中融入了“移动应用开发”职业技能取证中级标准规定的对应考试内容。

本书配套了丰富的数字资源，教材中核心知识点和操作以二维码的形式提供了微视频。配套建设有国家高等教育智慧教育平台，省级职业教育在线精品课程“Android 应用程序开发基础”，课程网站中提供了微视频、电子课件、电子教案、源代码、题库及实训案例等教学资源。

本书适合作为高职高专院校、职业本科院校移动互联应用技术、移动应用开发等相关专业的教学用书，也可作为“1+X”职业技能取证培训的教材或移动应用开发兴趣爱好者自学的参考书。

图书在版编目（CIP）数据

Android 应用程序开发基础 / 蒲晓妮，赵睿主编. 一北京：化学工业出版社，2022.9
ISBN 978-7-122-41360-4

Ⅰ.①A… Ⅱ.①蒲… ②赵… Ⅲ.①移动终端-应用程序-程序设计-高等职业教育-教材 Ⅳ.①TN929.53

中国版本图书馆 CIP 数据核字（2022）第 074459 号

责任编辑：韩庆利
责任校对：杜杏然
装帧设计：王晓宇

出版发行：化学工业出版社
（北京市东城区青年湖南街 13 号 邮政编码 100011）
印 装：河北鑫兆源印刷有限公司
787mm×1092mm 1/16 印张 18 字数 477 千字
2022 年 9 月北京第 1 版第 1 次印刷

购书咨询：010-64518888
售后服务：010-64518899
网 址：http://www.cip.com.cn
凡购买本书，如有缺损质量问题，本社销售中心负责调换。

定 价：55.00 元

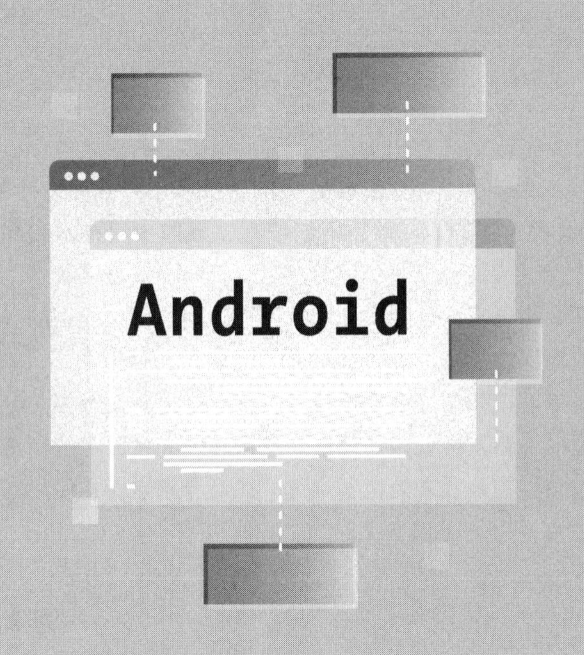

前言
PREFACE

随着大数据、人工智能、物联网及5G的应用普及，行业出现了融合趋势，智能终端在各种“智能”领域的应用得到了飞速发展。“移动应用开发”课程已成为大部分“智能”类专业的专业基础课程或专业课程。

本书的主要特色如下：

1.课程思政内容丰富

每章以二维码的形式提供了与专业内容相关的思政内容，培养学生的职业素养，提升综合素质。最后一章是“智慧党建”综合案例，通过项目实践实现了课程思政与专业学习的有效衔接，弘扬传统文化、增强民族自豪感。

2.新技术、新规范、新标准

随着Android操作系统快速发展，应用程序开发工具也在不断推出新版本。《Android应用程序开发基础》紧跟技术发展，教材内容体现新技术、新规范、新标准。本书采用Android10.0系统为技术标准，以Android Studio的2020.3.1版本为开发工具编写。

3.富媒体立体化教材

本书是新形态教材，教材中核心知识点和操作以二维码的形式提供了微视频，有利于“翻转课堂”教学活动的实施。

4.数字教学资源

本书配套建设有国家高等教育智慧教育平台，省级职业教育在线精品课程“Android应用程序开发基础”，课程网站中提供了微视频、电子课件、电子教案、源代码、题库及实训案例等教学资源。

5.强化实践教学指导

在本书中，对每一个项目都编写了详细的操作步骤和参考代码，学生参考教材就可以完成项目。

6.“1+X”职业技能取证

“1+X”职业技能取证是职业教育的特色，“移动应用开发”职业技能等级证书是最受欢迎的职业技能等级证书之一。本书中融入了“移动应用开发”职业技能取证中级标准规定的对应考试内容。

本书有 12 章，主要内容包含走进 Android 世界、Android 中的资源、用户界面设计基础、Activity、Fragment、数据存储、内容提供者、服务、广播、网络编程、媒体应用技术和综合项目——智慧党建。

本书由蒲晓妮和赵睿任主编，赵晓菲、张玉梅、马会军参加编写，在共同讨论提纲的基础上分别收集材料和编写，最后由蒲晓妮修改定稿。

本书适合作为职业本科院校和高职高专院校移动互联应用技术、移动应用开发等相关专业的教学用书，也可作为"1+X"职业技能取证培训的教材或移动应用开发兴趣爱好者自学的参考书。

由于编者理论水平和专业知识有限，教材中疏漏之处在所难免，恳请专家学者、使用本书的老师、同学和读者批评指正。

编者

目 录 Contents

第1章

走进 Android 世界

Android 是目前市场占有率最高的移动操作系统，本章将介绍 Android 的基础知识，学习搭建 Android 开发环境，通过一个简单的应用“Hello World”演示 Android Studio 的常用操作以及 Android 程序开发过程。

学习目标：

1. 掌握 Android Studio 环境的安装与配置；
2. 熟悉 Android Studio 开发环境；
3. 掌握 Android 简单应用的开发过程。

1.1 走进 Android

（1）Android

Android 简介

Android 是基于 Linux 内核的开源移动操作系统，是目前市场占有率最高的移动操作系统，主要应用于智能手机、平板电脑、智能电视、智能家居及可穿戴设备等。

Android 版本升级非常快，几乎每隔半年就会有一个新版本。自从 2008 年 Android 1.0 系统正式发布至今已经有 20 多个版本。在正式发行之前，最开始拥有两个内部测试版本，并且以著名的机器人名称来对其进行命名。它们分别是：阿童木（AndroidBeta）、发条机器人（Android 1.0）。从 Android 1.5 发布的时候开始，Android 命名规则变为以用甜点作为它们系统版本的代号的命名方法，每个版本代表的甜点的尺寸越变越大，然后按照字母顺序来进行命名：纸杯蛋糕（Cupcake，Android 1.5），甜甜圈（Donut，Android 1.6），松饼（Eclair，Android 2.0/2.1），冻酸奶（Froyo，Android 2.2），姜饼（Gingerbread，Android 2.3），蜂巢（Gingerbread，Android 3.0），冰激凌三明治（Ice Cream Sandwich，Android 4.0），果冻豆（Jelly Bean，Android 4.1 和 Android 4.2），奇巧（KitKat，Android 4.4），棒棒糖（Lollipop，Android 5.0），棉花糖（Marshmallow，Android 6.0），牛轧糖（Nougat，Android 7.0），奥利奥（Oreo，Android 8.0），派（Pie,Android 9.0）。从 Android 10 系统命名开始转换为版本号，就像 Windows 操作系统和 ios 系统一样，有 Android 10.0(Q)，Android 11.0（R），Android 12.0(S)等。

Android 的系统架构和其操作系统一样，采用了分层的架构。Android 分为四个层，从高层到低层分别是应用程序层、应用程序框架层、系统运行库层和 Linux 内核层，体系架构如图 1-1 所示。

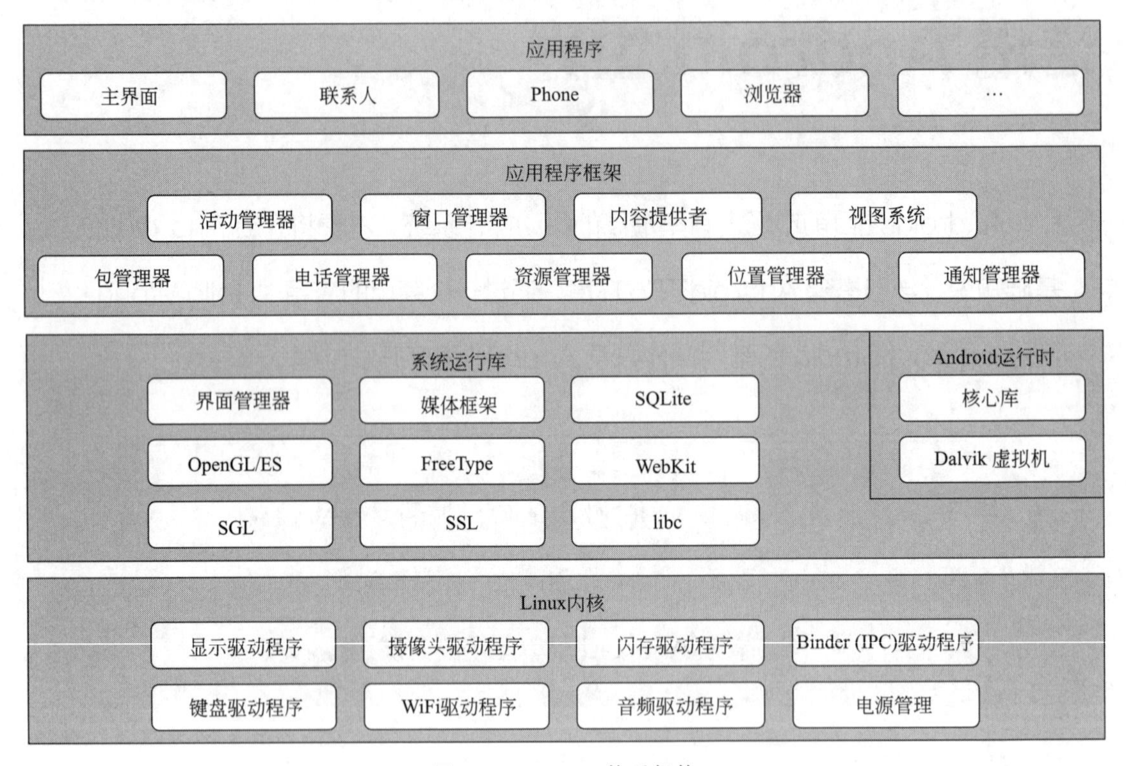

图 1-1　Android 体系架构

应用程序层：所有安装在手机上的应用和用户自己开发的应用都属于这一层，如电话、信息浏览器、联系人管理程序等。所有的应用程序都是使用 Java 语言编写的。

应用程序框架层：这一层提供开发应用程序时可能用到的各种 API 框架，开发人员可以使用这一层的 API 构建自己的应用程序，如使用电话管理器拦截电话，使用内容提供者访问系统联系人等。

系统运行库层：这一层包括 Android Runtime 和原生态的 C/C++库，这些库能被 Android 系统中不同的组件使用，它们通过 Android 应用程序框架为开发者提供服务。

Linux 内核层：Android 系统是基于 Linux 内核的，这一层为 Android 设备的各种硬件提供了底层的驱动，如电池管理、音频驱动、摄像头驱动等。

（2）Android Studio

Android Studio 是用于开发 Android 程序的开发平台，它提供了集成的 Android 开发工具，用于开发和调试 Android 应用。Android Studio 的开发环境和模式丰富便捷，能够支持多种语言，还可以为开发者提供测试工具和各种数据分析，是目前开发 Android 应用的主流工具。

1.2　搭建 Android 开发环境

Android Studio 的安装包的下载地址为：https://developer.android.google.cn/studio/index.html，

打开网址如图 1-2 所示。

图 1-2　Android Studio 下载页面

点击“Download Anroid Studio”,下载安装包。下载完成后，运行安装包弹出如图 1-3 所示的对话框。

图 1-3　“Welcome to Android Studio Setup”对话框

在图 1-3 中点击“Next”命令按钮，进入如图 1-4 所示的“Choose Components”窗口，选中界面中的选项。

单击图 1-4 中“Next”命令按钮，进入“Configuration Settings”窗口，如图 1-5 所示。

在图 1-5 中设置 Android Studio 的安装路径，可以使用默认路径，也可以选择“Browse”更换安装路径。然后点击“Next”命令按钮，进入如图 1-6 所示“Choose Start Menu Folder”窗口。

在图 1-6 中点击“Install”命令按钮，进入如图 1-7 所示的“Installing”窗口。

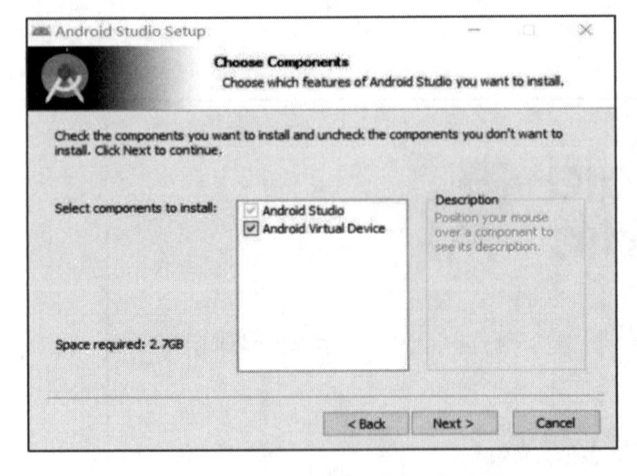

图 1-4 “Choose Components”窗口

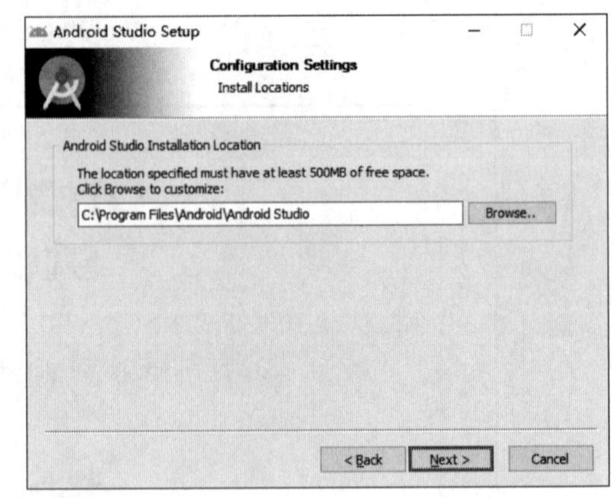

图 1-5 “Configuration Settings”窗口

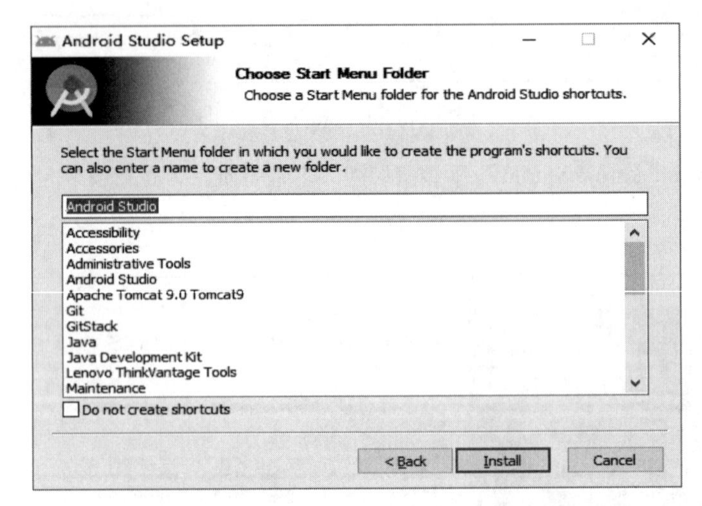

图 1-6 “Choose Start Menu Folder”窗口

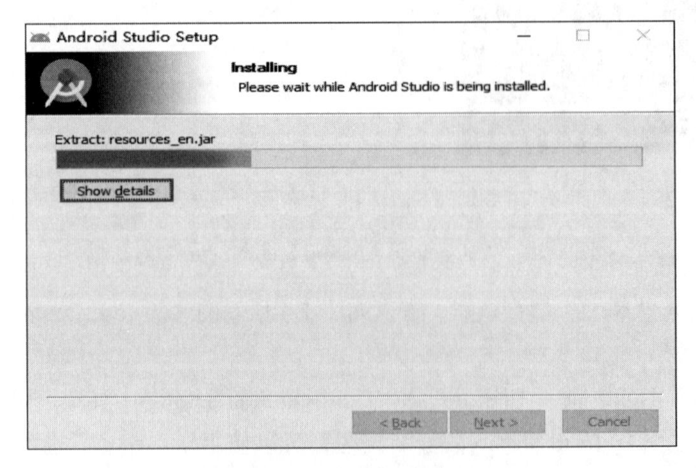

图 1-7 “Installing”窗口

安装完成后，单击“Next”命令按钮进入如图 1-8 所示“Completing Android Studio Setup”窗口。

图 1-8　“Completing Android Studio Setup”窗口

在图 1-8 中单击“Finish”命令按钮出现如图 1-9 所示 Android Studio 启动窗口。

图 1-9　Android Studio 启动窗口

图 1-9 的加载进度完成后，出现“Android Studio First Run”窗口，如图 1-10 所示。

第一次运行 Android Studio 的时候会弹出图 1-10 窗口，选择“Cancel”命令按钮，进入“SDK Components Setup”窗口，如图 1-11 所示。

在图 1-11 中选择“Next”命令按钮，进入“Verify Settings”窗口，如图 1-12 所示。

在图 1-12 窗口可以看到需要下载的 Android Studio 组件，选择默认“Finish”选项，会进入“Downloading Components”窗口下载组件，如图 1-13 所示。

在图 1-13 窗口中，显示了 SDK 等组件的下载进度。下载完成后，单击“Finish”，进入如图 1-14 所示“Welcome to Android Studio”窗口，现在即可开始 Android 开发之旅了。

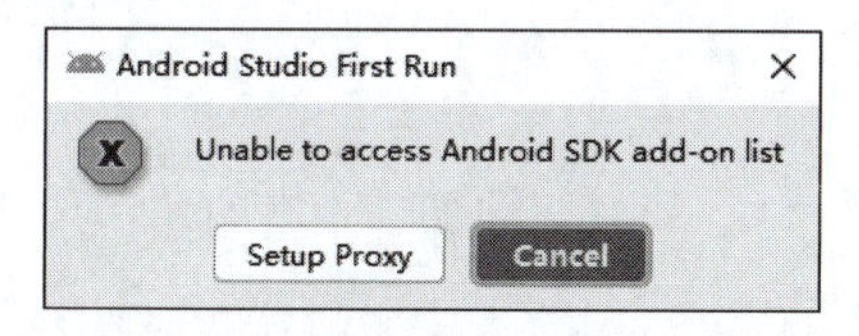

图 1-10　“Android Studio First Run”窗口

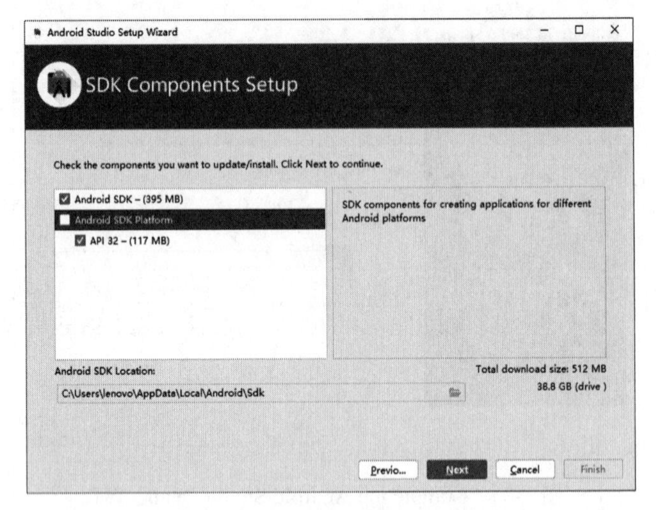

图 1-11 “SDK Components Setup”窗口

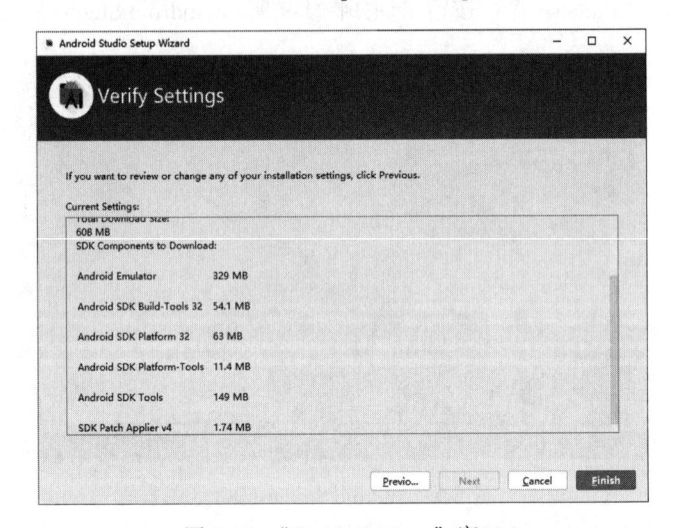

图 1-12 “Verify Settings”窗口

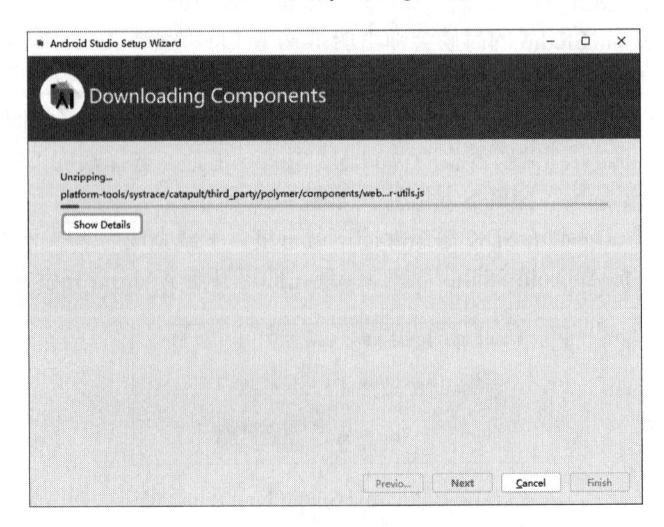

图 1-13 “Downloading Components”窗口

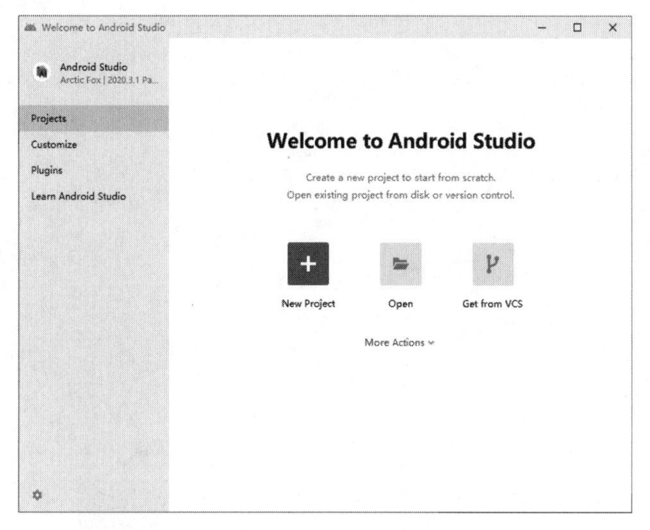

图 1-14　“Welcome to Android Studio”窗口

1.3　创建第一个应用程序

1.3.1　创建程序

启动 Android Studio 后，弹出如图 1-14 所示“Welcome to Android Studio”窗口。选择“New Project”，出现如图 1-15 所示的“New Project”对话框。

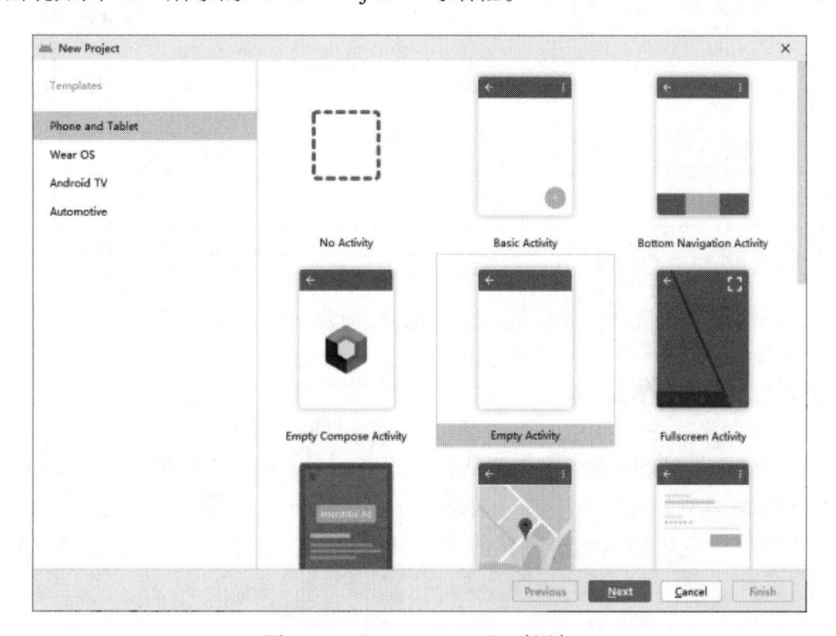

图 1-15　“New Project”对话框

在图 1-15 所示的“New Project”对话框中，默认项为“Empty Activity”选项，点击“Next”命令按钮，出现如图 1-16 所示“New Project”的 Empty Activity 对话框。

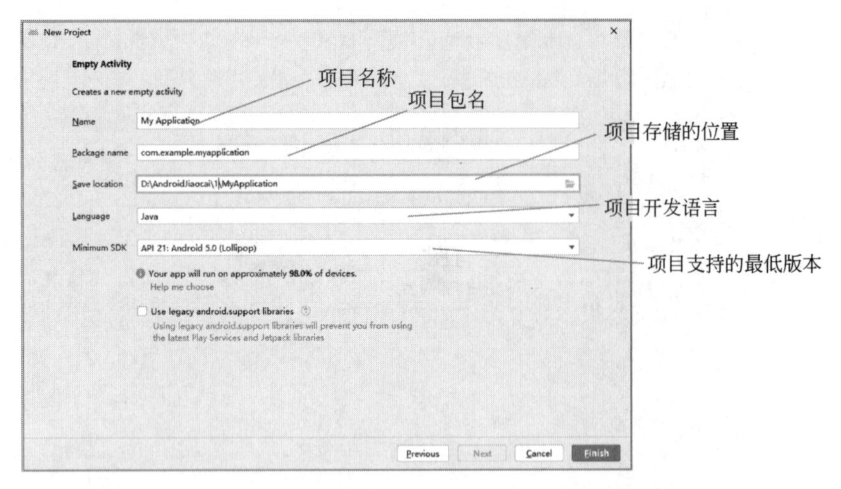

图 1-16 “New Project”的 Empty Activity 对话框

在图 1-16 中可以设置项目的名称，项目的包名，项目存储的位置以及开发项目所用的语言（本书中均选择 Java 语言）等。第一个项目采用默认的项目名称和包名。最后点击“Finish”命令按钮，项目创建完成，出现如图 1-17 所示的 Android Studio 集成开发环境。

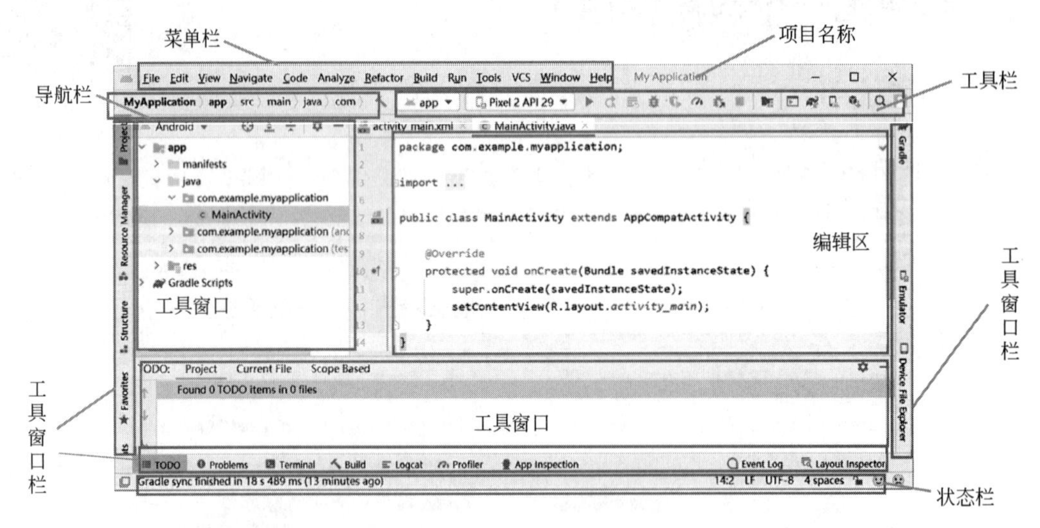

图 1-17 Android Studio 集成开发环境

1.3.2 Android 程序结构

完成了第一个应用程序的创建，我们来认识一下 Android 程序的结构。在 Android Studio 中，提供了多种程序结构类型。一般常用的结构是：Project，Android 和 Packages 。开发中一般使用的是 Android 结构，这里是显示最完整的项目文件目录，图 1-18 所示为选择 Project 时的程序结构。

其中：

.gradle 和.idea：目录下是 Android Studio 自动创建的一些文件；

app：项目内代码、资源均存放在这个目录下；

app/build：app 模块编译输出的文件；

app/libs：放置第三方 jar 包；
app/src：放置应用的主要文件目录；
app/src/androidTest：单元测试目录；
app/src/main：主要的项目目录和代码；
app/src/main/java：项目的源代码；
app/src/main/res：项目的资源文件（主要有图片资源、布局资源、颜色、字符串等资源）；
app/src/main/res/AndroidManifest.xml：项目的清单文件（名称、版本、SDK、权限等配置信息）；

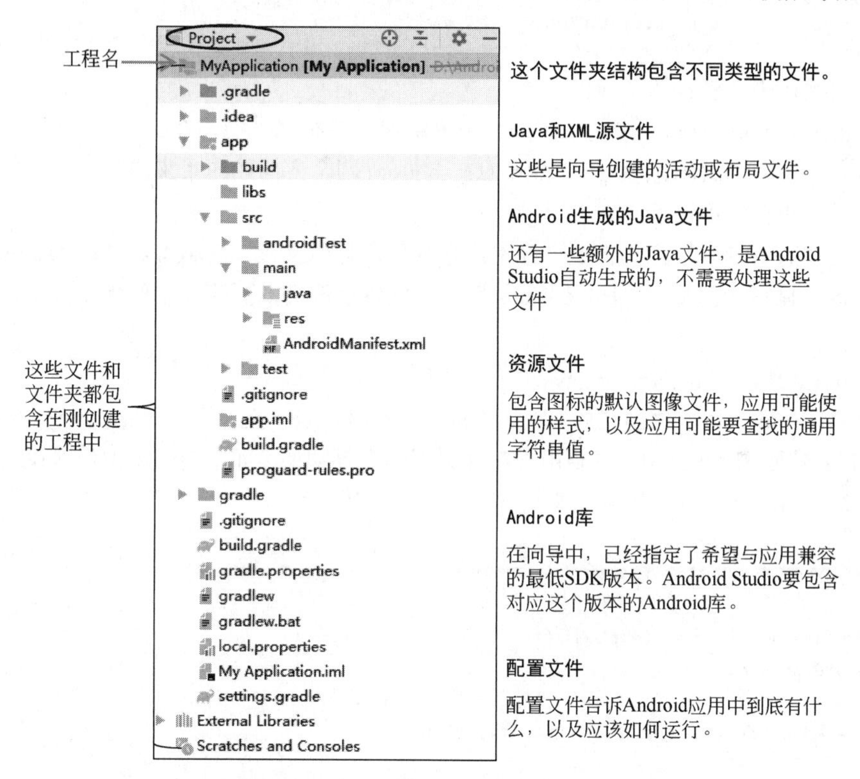

图 1-18　应用程序的“Project”结构图

Android Studio 为我们自动生成了几个关键文件。分别是 activity_main.xml、MainActivity.java 和 AndroidManifest.xml。

（1）布局文件 activity_main.xml

在 Android 应用程序中，界面是通过布局文件设定的，布局文件采用 XML 格式。每一个 Android 项目成功创建后，默认生成一个布局文件 activity_main.xml，用于指定应用界面是什么样的，该文件位于 res/layout 文件夹中，代码如下：

```
<?xml version="1.0" encoding="utf-8"?>
<androidx.constraintlayout.widget.ConstraintLayout xmlns:android="http://schemas. android.com/apk/
res/android"
    xmlns:app="http://schemas.android.com/apk/res-auto"
    xmlns:tools="http://schemas.android.com/tools"
    android:layout_width="match_parent"
    android:layout_height="match_parent"
    tools:context=".MainActivity">
```

```
    <TextView
        android:layout_width="wrap_content"
        android:layout_height="wrap_content"
        android:text="Hello World!"
        app:layout_constraintBottom_toBottomOf="parent"
        app:layout_constraintLeft_toLeftOf="parent"
        app:layout_constraintRight_toRightOf="parent"
        app:layout_constraintTop_toTopOf="parent" />

</androidx.constraintlayout.widget.ConstraintLayout>
```

上面这段代码主要包含两个元素：

第一个 ConstraintLayout，即约束布局，是系统默认的布局类型。在 ConstraintLayout 中的子控件，如 View 通过 start(left)、top、end(right)、bottom 四个方向的约束来决定控件的位置。后面将在布局管理中进行详细讲解。

第二个是“TextView”元素，这个元素用来为用户显示文本。它嵌套在“Constraint Layout”元素中，在这里显示示例文本“Hello World”。可以将示例文本更改成我们想要显示的文本，观察结果。

（2）Activity 文件 MainActivity.java

每一个 Android 项目成功创建后，默认生成的一个 Activity 文件即 MainActivity.java，该文件位于项目的 java/项目所在包中，主要用于实现界面的交互功能，即指定应用做什么以及如何与用户交互等。MainActivity.java 中自动产生的代码如下所示：

```
package com.example.myapplication;

import androidx.appcompat.app.AppCompatActivity;
import android.os.Bundle;

public class MainActivity extends AppCompatActivity {

    @Override
    protected void onCreate(Bundle savedInstanceState) {
        super.onCreate(savedInstanceState);
        setContentView(R.layout.activity_main);
    }
}
```

在上面的代码中 package com.example.myapplication 表示程序的包名；使用 import 导入了两个 MainActivity 中用到的 Android 包；定义类 MainActivity 继承自 AppCompatActivity 类，并且重写了父类的 onCreate()方法，在此方法中使用 setContentView(R.layout.*activity_main*)方法给当前类指布局为 activity_main 布局文件。

（3）清单文件 AndroidManifest.xml

AndroidManifest 文件是 Android Studio 项目的全局配置文件，是 Android 应用程序中最重要的文件之一，每个 Android 程序中都必须有 AndroidManifest.xml 这一个文件，这也是 Android 程序的入口文件，记录应用程序中所使用的各种组件。该文件提供了 Android 系统所需要的关于该应用程序的必要信息，即在该应用程序的任何代码运行之前系统所必须拥有的信息。另外当新添加一个Activity的时候，也需要在这个文件中进行注册，Activity只有注册之后才能被调用。创建项目自动生成的AndroidManifest.xml文件代码如下所示。

```
<?xml version="1.0" encoding="utf-8"?>
<manifest xmlns:android="http://schemas.android.com/apk/res/android"
    package="com.example.myapplication">

    <application
        android:allowBackup="true"
        android:icon="@mipmap/ic_launcher"
        android:label="@string/app_name"
        android:roundIcon="@mipmap/ic_launcher_round"
        android:supportsRtl="true"
        android:theme="@style/Theme.MyApplication">
        <activity
            android:name=".MainActivity"
            android:exported="true">
            <intent-filter>
                <action android:name="android.intent.action.MAIN" />
                <category android:name="android.intent.category.LAUNCHER" />
            </intent-filter>
        </activity>
    </application>

</manifest>
```

从上面的代码中可以看出xml文件的根元素为manifest，它里面包含了<application>标记，此标记用来配置应用程序的属性，如应用的图标、标签、主题等。其中属性icon引用了图标资源文件，属性label用来设置应用程序的标题，theme用来设置应用程序的样式主题。Activity标签，指定Activity的类名、标签等，一个应用程序可以包含多个Activity。所有的Activity只有在清单文件中注册了才能使用。Activity中有如下代码：

```
<activity android:name=".MainActivity" >
    <intent-filter>
        <action android:name="android.intent.action.MAIN" />
        <category android:name="android.intent.category.LAUNCHER" />
    </intent-filter>
</activity>
```

这段代码对MainActivity活动进行了注册，其中intent-filter标签中的两行代码表示整个项目是由MainActivity活动启动的。也就是说运行应用程序时，首先启动的是这个MainActivity，之后MainActivity使用setContentView()方法调用并显示activity_main.xml布局文件所设计的界面。

1.3.3　运行程序

Android 程序创建好以后，我们就要来运行这个程序了。它可以在手机上运行，也可以在Android Studio模拟器上运行。

（1）在模拟器上运行

要在模拟器上运行，我们首先需要创建 Android 模拟器，创建模拟器的过程如下所示。

在工具栏上选择“ADV Manager”选项，弹出如图 1-19 所示的“Your Virtual Devices”窗口。

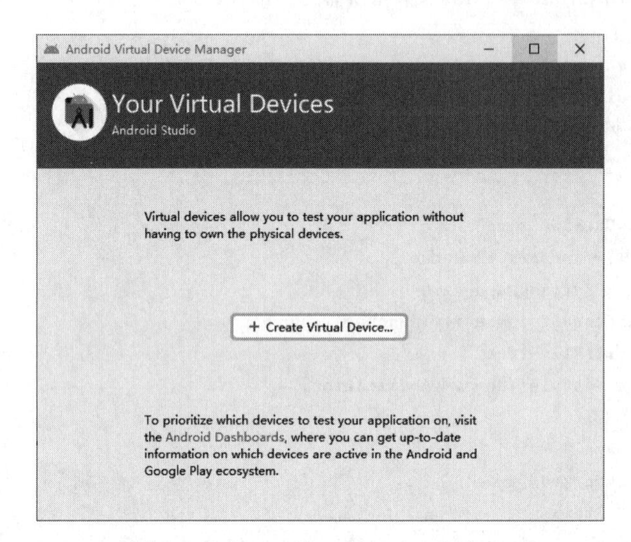

图 1-19 “Your Virtual Devices”窗口

在图 1-19 中选择创建模拟设备“Create Virtual Device”命令按钮，进入“Select Hardware”窗口，如图 1-20 所示。

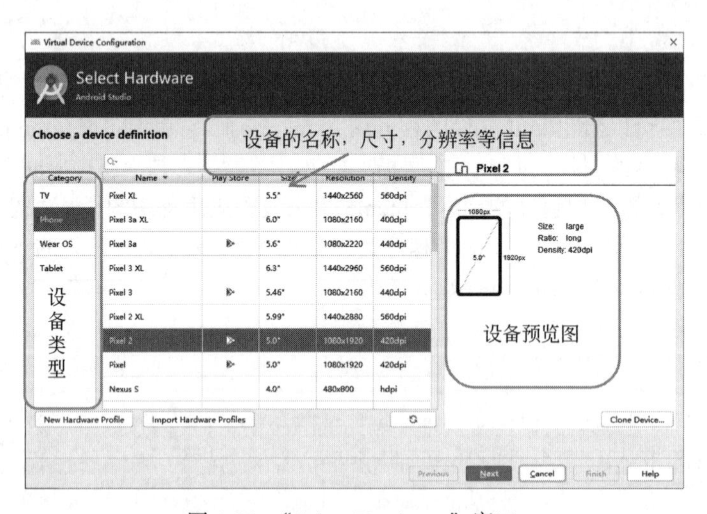

图 1-20 “Select Hardware”窗口

在图 1-20 中设备类型选择默认的“Phone”，设备名称选择“Pixel-2”，然后选择“Next”命令按钮，弹出如图 1-21 所示的“System Image”窗口，这里系统推荐了 Android 系统镜像，需要先点击“Download”，我们选择 Android 10.0 的系统版本进行下载，先选中 Q 的系统版本，单击“Download”下载后，下载完成后如图 1-22 所示。

在图 1-22 中选择“Next”，弹出如图 1-23 所示“Android Virtual Device（AVD）”窗口。

在图 1-23 中选择“Finish”，出现“Your Virtual Devices”窗口，如图 1-24 所示。

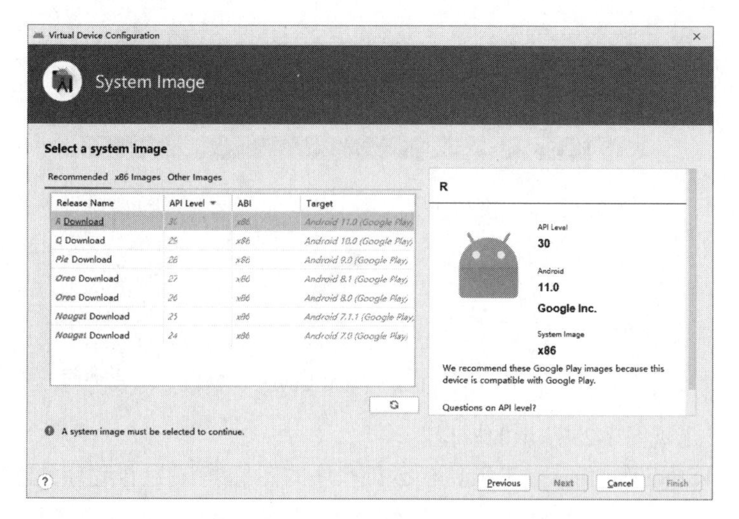

图 1-21　“System Image”窗口

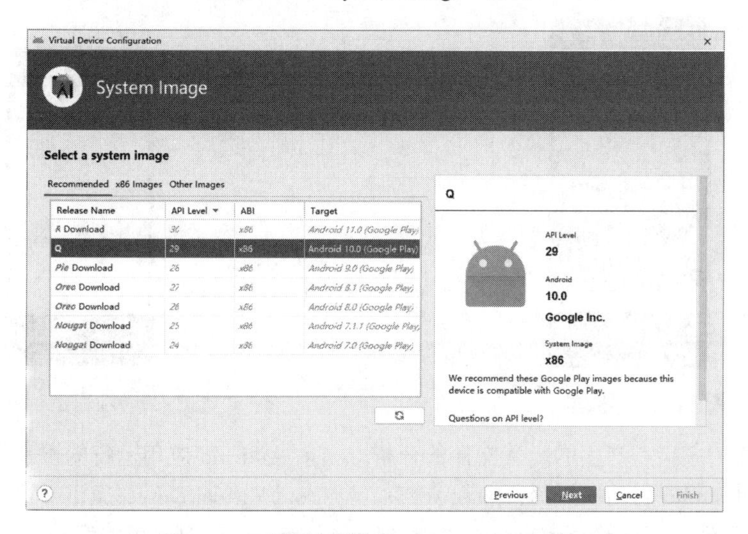

图 1-22　下载完成后“System Image”窗口

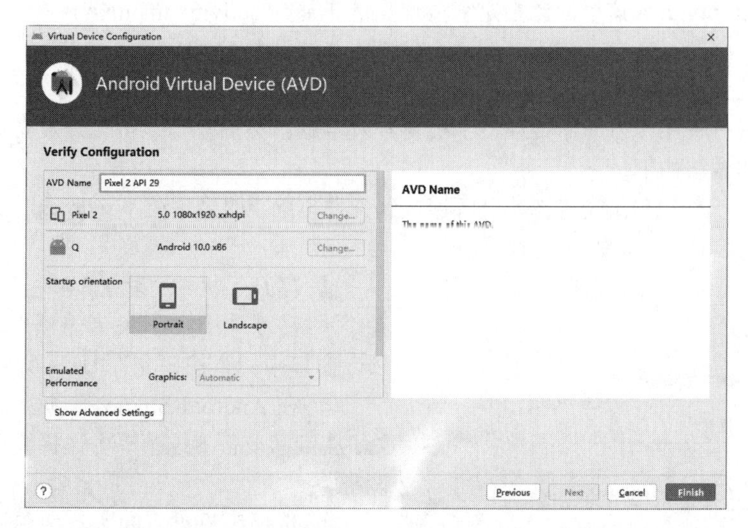

图 1-23　“Android Virtual Device（AVD）”窗口

运行应用程序

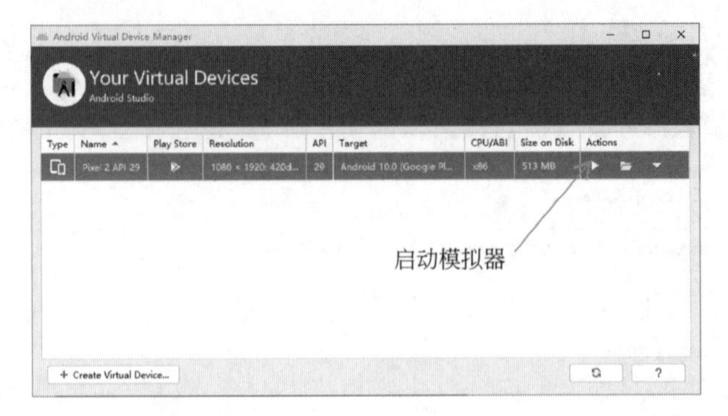

图 1-24 “Your Virtual Device”窗口

启动模拟器，出现图 1-25 所示的模拟器。

回到 Android Studio 界面，点击工具栏运行程序的图标▶，运行程序，运行效果如图 1-26 所示。

图 1-25 模拟器

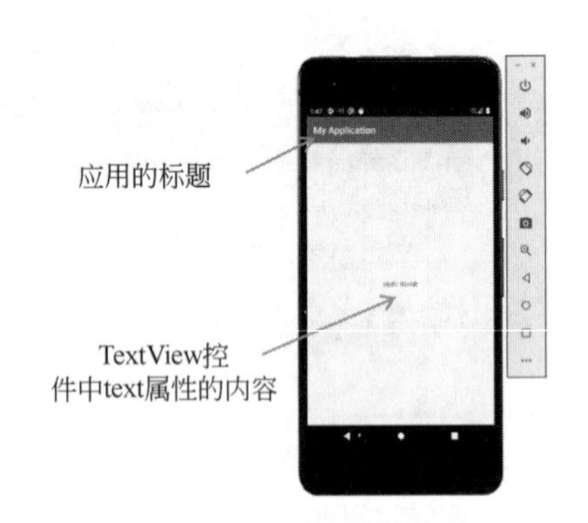

图 1-26 项目运行效果

通过对创建的应用所产生的默认文件的分析，当前 Android 应用运行的过程为:

① Android Studio 启动模拟器，加载 AVD（Android Virtual Device）；

② 由清单文件 AndroidManifest.xml 指定 MainActivity.java 为应用的启动活动;

③ MainActivity.java 这个活动指定它要使用的布局为 activity_main.xml;

④ activity_main.xml 布局告诉 Android 在屏幕上显示内容“Hello world!”。

（2）在手机上运行

在手机上运行程序，首先需要连接手机与开发电脑，然后打开手机“开发者选项”，再选择手机运行，具体做法如下;

① 需要使用一根 USB 线将移动设备连接到开发电脑，如果是在 Windows 上开发的，则可能需要为设备安装合适的 USB 驱动程序。

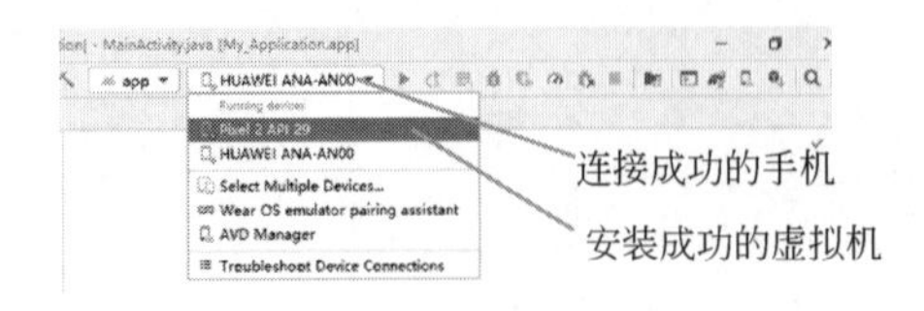

图 1-27 选择运行程序的设备窗口

② 打开手机中“设置”→“系统与更新”（或者“系统”）找到“开发者选项”，在“开发者选项”窗口中，查找并允许 USB 调试。

③ 在 Android Studio 工具栏中如图 1-27 位置处，从目标设备下拉菜单中选择要用来运行应用的设备。

④ 选择手机对应的设备名称后，点击工具栏

中的运行按钮▶，即可实现在手机上进行测试和运行 Android 项目。

1.3.4　生成 APK

通过前面的学习我们知道 Android 程序的调试既可以在 Android Studio 中的虚拟机上进行，也可以通过数据线连接手机在真机上进行。有时我们需要把 Android 程序打包成为一个 APK 文件，即 Android 程序安装包，只要拿到 APK 文件的 Android 手机就可以安装、调试。

导出 APK 安装包的过程如下所示:

① 在如图 1-28 所示的界面中，依次点击菜单项 Build→Generate Signed Bundle/APK...,弹出如图 1-29 所示对话框。

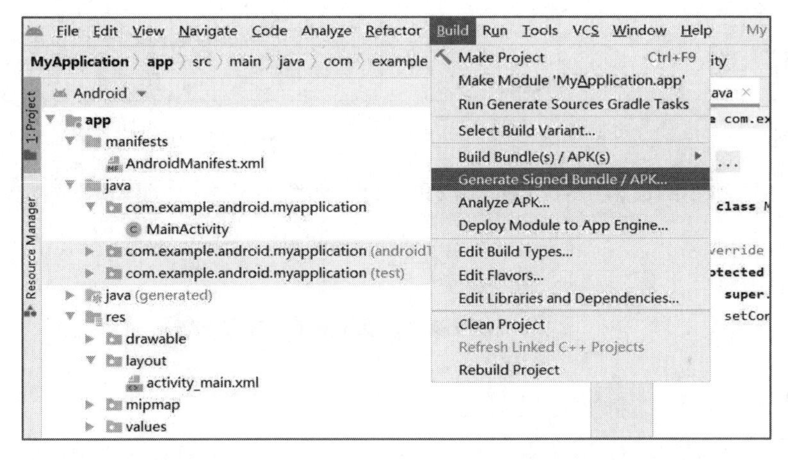

图 1-28　选择“Build”菜单项

② 在图 1-29 对话框中选择 APK 选项，单击 Next 按钮，进入如图 1-30 所示 APK 签名对话框。

③ 在图 1-30 所示的对话框中，系统默认的模块名为当前项目的名称，选择密钥文件的保存路径，如果原来有密钥文件，单击 Choose existing...按钮，在弹出的对话框中选择密钥文件；如果是第一次打包没有密钥文件，则单击 Create new...命令按钮,弹出如图 1-31 所示的密钥创建对话框。

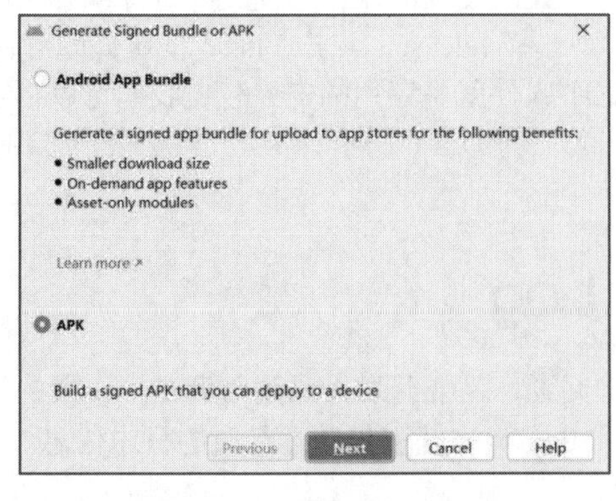

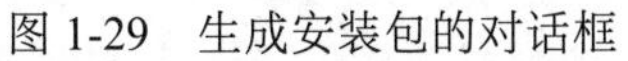
图 1-29　生成安装包的对话框

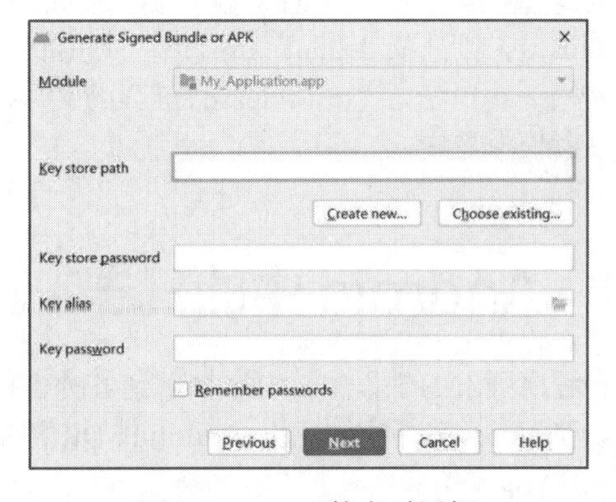

图 1-30　APK 签名对话框

④ 在图 1-31 所示的图中选择密钥文件保存路径，并在下方输入密钥文件名称，点击 OK 按钮返回密钥文件生成窗口，并依次填写密码 Password、确认密码 Confirm、别名 Alias、别名密码 Password、别名确认密码 Confirm、修改秘钥的有效期限 Validity，接下来的 Certificate 部分只有

First and Last Name(姓名)为必填项，其他选项可选填，填写完成后点击 OK 按钮返回到 APK 签名窗口，此时 Android Studio 自动把密码和别名都填上了，如图 1-32 所示。如果一开始选择已存在的密钥文件，则需要在这里手工输入密码和别名。

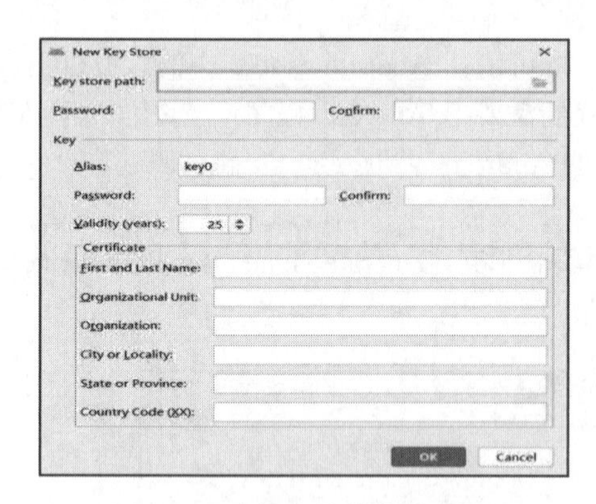

图 1-31　密钥文件的生成窗口

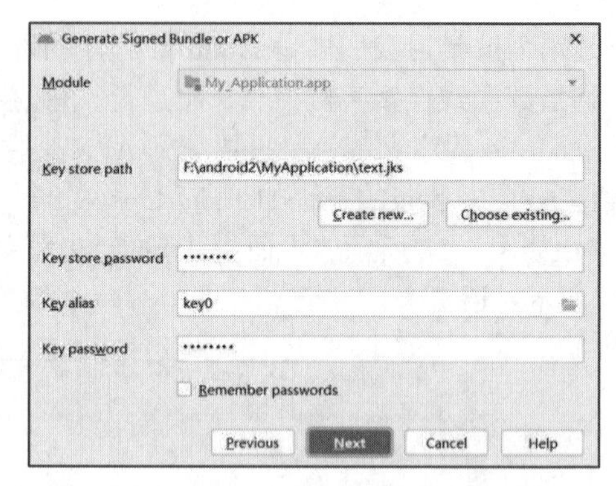

图 1-32　密钥文件对话框

⑤ 在图 1-32 所示的窗口中点击 Next 按钮，进入如图 1-33 所示的 APK 保存页面。

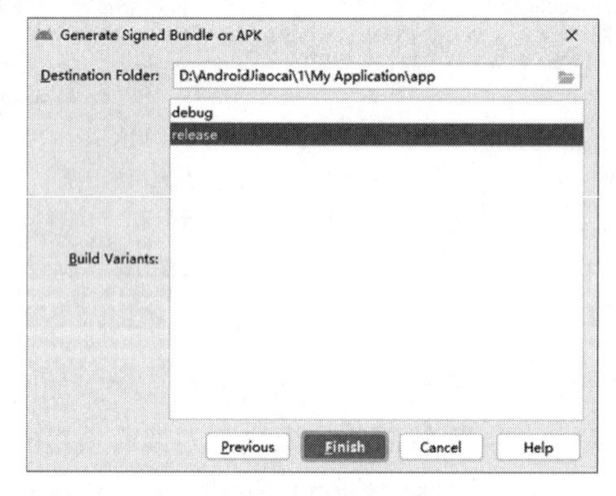

图 1-33　APK 保存页面

⑥ 在图 1-33 所示的 APK 保存页面中可以选择 APK 文件的保存路径，并选择编译类型，如果是调试用，则选择 debug；如果是正式发布用，则选择 release。点击 Finish 按钮，即可在 APK 保存路径下看到名为“test-release.apk”的文件，这个文件就是应用的安装包，我们可以将它装到手机或者模拟器中。

1.4　Android 中的日志工具 Log

在程序开发的过程中，经常会遇到各种各样的问题，需要开发人员耐心调试。一般的程序错误可以使用 Logcat 视图查看，Android SDK 提供了 Log 类来获取程序运行时的日志信息。

（1）Log 简介

Log(android.util.Log)是 Android 中的日志工具类，这个类中提供了如下 5 个方法来供我们打印日志，在今后的程序调试过程中会经常用到下面几个方法。

Log.v()方法：用于打印那些最为琐碎的、意义最小的日志信息。对应级别是 Verbose，是 Android 日志里面级别最低的一种。

Log.d()方法：用于打印一些调试信息，这些信息对调试程序和分析问题有帮助。对应级别为 Debug,比 Verbose 高一级。

Log.i()方法：用于打印一些比较重要的信息。例如用户行为数据、数据库数据等。这些信息可以帮助你分析用户行为。级别为 Info，比 Debug 高一级。

Log.w()方法：用于打印一些警告信息。提示程序这个地方可能会有潜在的风险，最好进行修复，级别为 Warn，比 Info 高一级。

Log.e()方法：用于打印程序中的错误信息。当有错误信息时就需要及时修改。对应级别为 error，比 warn 高一级。

日志等级越高，其所代表的日志信息对程序越重要，等级从建议级到中断级严重程度层层递进，我们在开发中灵活使用会有效提升效率。

（2）Log 方法的使用

打开我们创建的第一个项目，在 MainActivity 中的 onCreate()方法中添加一条日志打印的语句：

```
public class MainActivity extends AppCompatActivity {

    @Override
    protected void onCreate(Bundle savedInstanceState) {
        super.onCreate(savedInstanceState);
        setContentView(R.layout.activity_main);
        Log.d("AAA","HelloWorld");
    }
}
```

Log.d()方法有两个参数，一个参数为 tag，用来过滤信息；第二个参数为 msg，用来打印具体内容。

运行程序 ，选择底部的 Logcat 可以看到打印的日志信息，打印的日志很多，可以在图中搜索部分输入“AAA”，这样对日志信息进行过滤，只显示 tag 为“AAA”的信息。显示结果如图 1-34 所示：

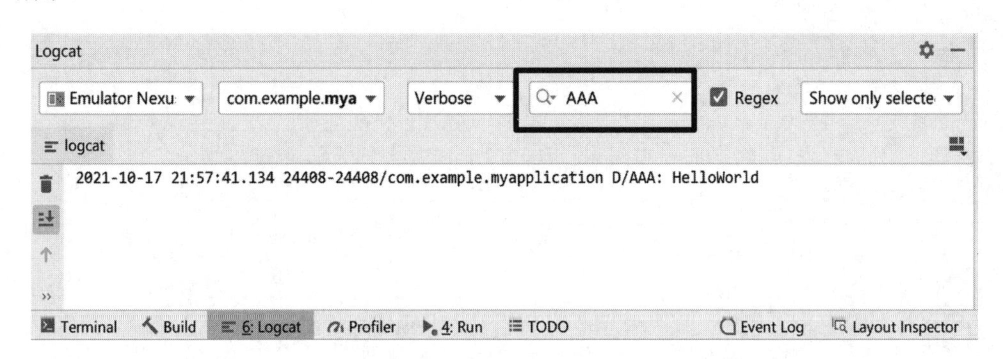

图 1-34　Logcat 中的打印信息

这里 Logcat 是 Android 中的命令工具，用于获取程序从启动到关闭的日志信息。在调试程序中加入 Log 可以有效提高开发效率，如果程序中使用了 Log 这 5 个方法的其他方法，在 Logcat 中会以不同的颜色显示，大家可以自己试一试，这 5 个方法的使用形式类似 Log.d()。

思考：

对 Android 应用开发的基础知识有了初步了解，找到清单文件（AndroidManifest.xml）、活动文件（MainActivity.java）和布局文件（Activity_main.xml）的位置。

观察 AndroidManifest.xml 文件中<application>标签中的 label 的属性值。

```
<application
    android:allowBackup="true"
    android:icon="@mipmap/ic_launcher"
    android:label="@string/app_name"
    android:roundIcon="@mipmap/ic_launcher_round"
    android:supportsRtl="true"
    android:theme="@style/Theme.MyApplication">
```

按住“Ctrl”键，双击“@string/app_name”，会打开 String.xml 文件，其内容为如下所示：

```
<resources>
    <string name="app_name">My Application</string>
</resources>
```

在<application>标签中 icon，roudnIcon，theme 这几个属性，其值的形式与 label 属性值类似，都引用了 Android 的资源，那么 Android 中的资源有哪些，又是如何来使用呢？这将是我们第 2 章要学习的内容。

【职业素养拓展1】

职业素养拓展

职业素养拓展内容详见文件“职业素养拓展”。

扫描二维码查看【职业素养拓展 1】

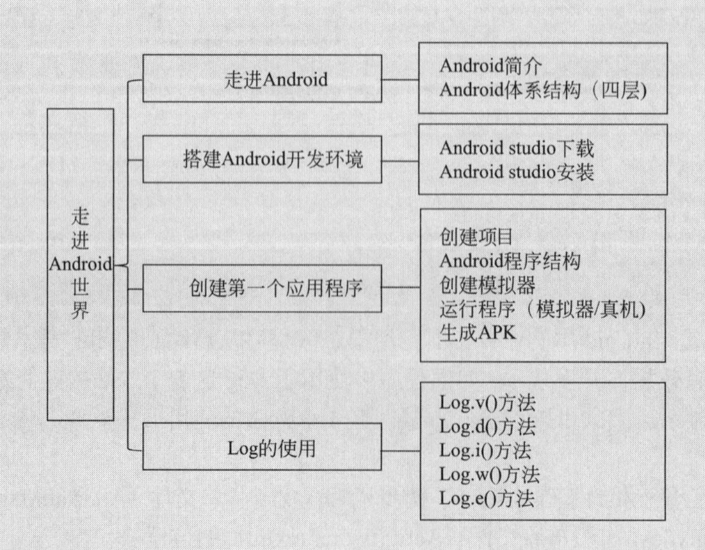

第 2 章 Android 中的资源

在第 1 章的学习中我们创建了第一个 Android 应用程序，了解了 Android 应用程序的基本结构，提出 Android 中资源的使用。

我们知道手机上比较热门的应用，选择不同国家的语言，它能适应很多不同的语言。还有当我们在应用程序的多处使用了同一个字符串，或是图片、颜色等时，如果应用程序需要修改时，逐一去修改工作量大又容易出错。针对这些情况 Android 提供了资源的使用。通过这章内容的学习要求学生了解 Android 中的资源类型，掌握常用资源的使用方法。

学习目标：

1. 了解 Android 资源的特点，在项目中存放各项资源的文件夹位置；
2. 了解各种资源对应的 xml 文件的书写格式，能够编制资源文件内容；
3. 重点掌握布局、字符串、颜色、尺寸、主题和样式等资源的应用；
4. 了解程序国际化的原理和实现方法。

2.1 资源概述

Android 中的资源主要指非 Java 代码部分，只要是不与业务流程相关的，用于界面部分的都可以用资源表示。Android 中资源的概念非常宽泛，常用的有布局、字符串、颜色、尺寸、图片、样式和主题等。

Android Studio 开发环境为每个 Android 的项目都会默认生成一个 res 目录，这个目录用于存储 Android 程序中的资源。常用资源存储位置如图 2-1 所示。

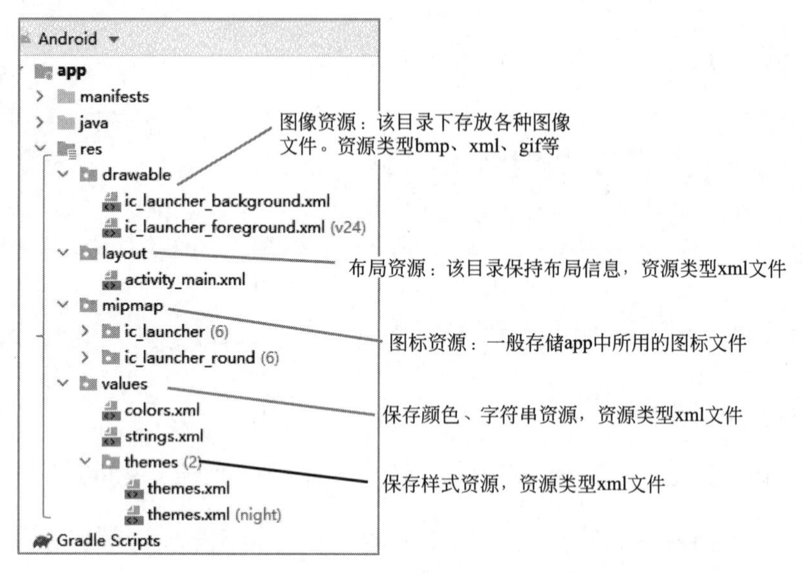

图 2-1　Android 中常用资源的存储位置

从图 2-1 中我们可以看到，res 目录下有固定名字的文件夹，用于保存各类资源文件。Android 提供了一个资源编译工具，它会按照事先约定的目录结构，把 res 目录下的文件自动编译，并生成 R.java 文件，应用程序可以通过 R.java 对资源文件采用“R.type.name”的方式进行引用，R 常用的资源引用类别（type）有：R.layout，R.drawable，R.string，R.color，R.style，R.demin，R.array，等等。对于资源文件使用常用的有两种形式，一种是在 xml 文件中使用，还有一种是在 Java 文件中使用 getResources()方法获取资源。

2.2　布局资源

布局资源用于定义在屏幕上要显示的内容。简单说就是一个用户界面，或者是界面的一个组成部分的内容。存储在应用程序的/res/layout/目录下。我们在创建第一个应用程序的时候系统会自动创建了一个布局文件 activity_main.xml，根据需要还可以添加布局。

（1）新建布局文件

新建布局文件常用方法有两种。

方法一：选中“res”目录，单击右键选择“New”→“Layout Resource File”，弹出图 2-2 所示的“New Resource File”界面。

选择图 2-2 中的“OK”命令按钮，在 res/layout 目录下会创建一个新的布局文件 mylayout，默认为约束布局。

注意：布局文件名只能由小写字母、数字、下划线组成，而且必须以小写字母开头。

方法二：　选中“res”目录，单击右键选择“New”→“XML”→“Layout XML File”，弹出如图 2-3 所示的“Create a new XML Layout file”界面。

单击“Finish”命令按钮，在 res/layout/目录下会新建出了一个 layout.xml 布局文件。

（2）编辑布局文件

布局文件的编辑有两种方法。

方法一：在 Design 模式下进行设计。如图 2-4 所示为布局文件的 Design 模式界面。

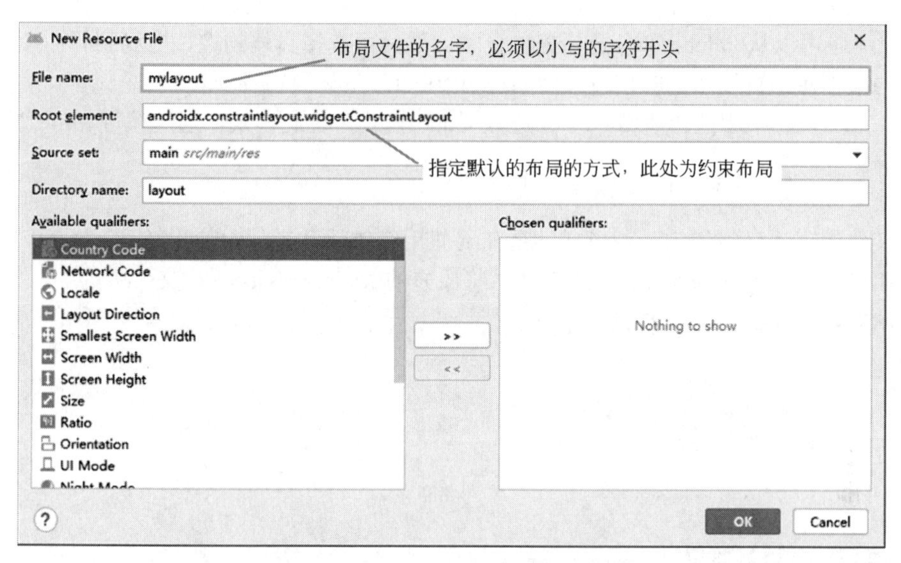

图 2-2　“New Resource File”界面

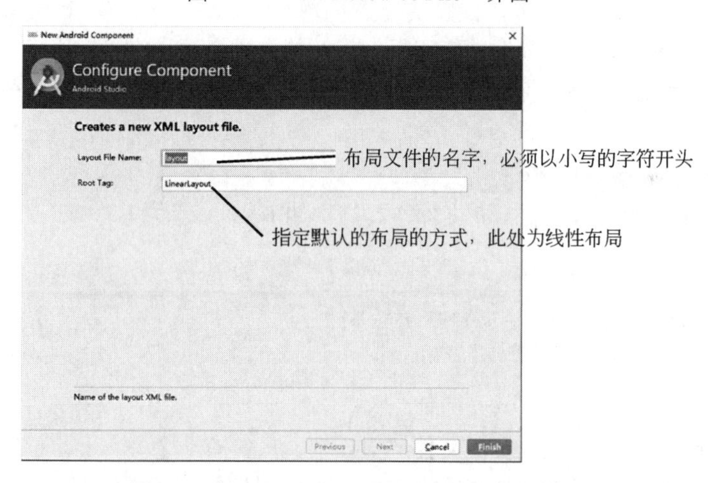

图 2-3　“Create a new XML Layout file”界面

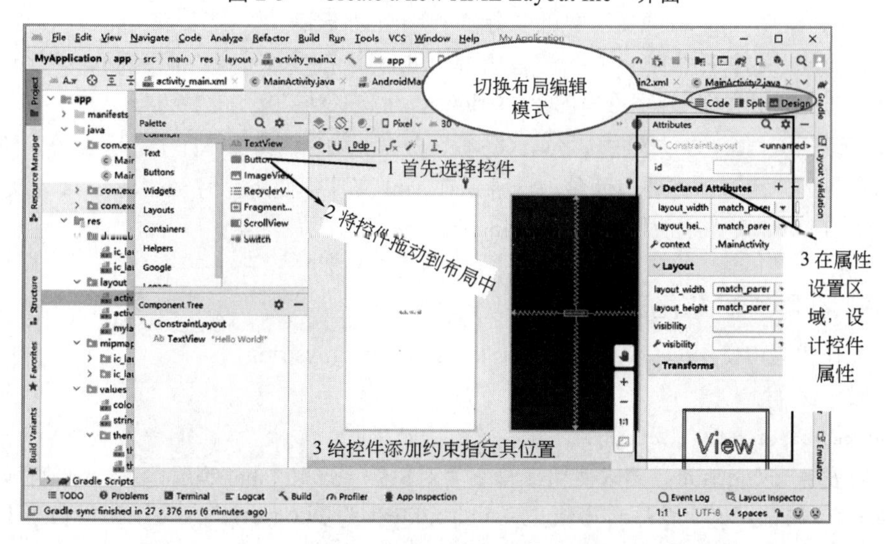

图 2-4　Android Studio 集成环境布局文件的 Design 模式界面

在图 2-4 中可以从 Palette 部分选择控件分类，然后再选择具体的控件拖曳到界面中，调整控件位置，为其指定垂直和水平的位置，然后在 Attributes 下设置控件属性。

方法二：在图 2-4 中选择“Code”，进入 Code 模式，在此模式下通过编写 xml 代码来实现界面的布局。

(3) 使用布局文件

布局文件的应用最常用的一种形式就是在要使用它的活动文件中使用 SetContentView()方法来加载。如果要使用新建的 layout.xml 文件，将原先的 MainActivity.java 文件中的

```
setContentView(R.layout.activity_main);
```

更改为:

```
setContentView(R.layout.layout);
```

即可更改布局。

2.3 字符串资源

字符串资源是 Android 应用开发中使用最为广泛的一种资源，定义字符串资源的 strings.xml 文件位于/res/values 目录下。我们在清单文件 Android Manifest.xml 中使用了字符串资源。代码如下:

```
android:label="@string/app_name"
```

打开定义字符串资源的 strings.xml 文件，其结构如下:

```
<resources>
    <string name="app_name">My Application</string>
</resources>
```

其中:

<resources></resources>标记是资源文件的根元素，只能有一个。

<string></string>标记定义的就是字符串资源。name 是字符串资源的名称，两个标签中间的内容就是字符串的内容。一个资源文件中可以有多个<string>标记。

在 strings.xml 文件中添加一个字符串资源，也就是在<resources></resources>标记内添加<string name=“字符串名称”>字符串内容</string>。示例代码如下:

```
<string name="mystring">Welcome to you!!!</string>
```

字符串资源的使用方法主要有两种，一种是在 xml 文件中使用，另一种是在 Java 文件使用。

方法一： 在 xml 文件中使用@string/stringname

示例代码:

```
android:text="@string/mystring"
```

方法二：在代码中使用 getResource().getString(R.string.mystring)

示例代码:

```
getResources().getString(R.string.mystirng);
```

在实际开发 Android Studio 要求使用字符串为对象的 text 和 label 等属性进行赋值，如果直接给 text 或者 label 属性定义一个字符串值，如 android:text="Hello world",则会给出警告“Hardcoded Text”。

2.4　颜色资源

在 Android 中使用颜色资源可以设置对象的颜色，定义颜色资源的 color.xml 文件位于 /res/values 目录中。颜色资源四种形式：#AARRGGBB、#ARGB、#RRGGBB、#RGB。

其中：A 表示透明度，R、G、B 代表的是三原色， RGB 值越大，颜色越深，都为 0，表示黑色，都为 f 表示白色。打开默认的 color.xml 文件看到如下代码：

```
<?xml version="1.0" encoding="utf-8"?>
<resources>
    <color name="colorPrimary">#008577</color>
    <color name="colorPrimaryDark">#00574B</color>
    <color name="colorAccent">#D81B60</color>
</resources>
```

根据应用开发的需要，可以在<resources></resources>标签内添加<color>标签，定义自己所需的颜色，代码如下所示：

```
    <color name="myred">#D53535</color>
```

颜色资源使用方法有两种，一种是在 xml 文件中使用，另一种是在 Java 文件中使用。

（1）在 xml 文件中使用

通过@color/colorname 实现，示例代码：

```
    android:textColor="@color/myred"
```

（2）在 Java 文件中使用

getResource().getColor(R.color.colorname) ，示例代码：

```
    getResources().getColor(R.color.myred);
```

2.5　尺寸资源

Android 支持的尺寸资源比较丰富，可以适应不同手机分辨率的需求。在不同的应用场合下需要选择不同的尺寸单位。尺寸资源通常定义在 res/values/dimens.xml 文件中，Android Studio 中没有默认的 dimen.xml 文件，需要手动创建。这里给大家介绍尺寸资源中常用的两个单位 dp 和 sp。

dp：独立设备像素，常用来设置对象大小、边距等。可根据设备屏幕大小自动调整，不同的设备上显示的大小不一样。

sp：可伸缩像素，是 Android 开发中推荐的字体大小的单位，可以根据用户手机上字体大小的首选项进行缩放。

（1）创建尺寸资源文件

选择“res/values”文件夹，单击鼠标右键选择“New”→“XML” →“Values XML File”，文件命名为 dimen。打开 dimen.xml 文件，编写如下代码：

```
<?xml version="1.0" encoding="utf-8"?>
<resources>
    <dimen name="textsize">30sp</dimen>
    <dimen name="topmargin">100dp</dimen>
</resources>
```

（2）应用尺寸资源

尺寸资源的使用方法有两种，一种是在 xml 文件中使用，另一种是在 Java 文件中使用。

① 在 xml 文件中使用

例如，在布局文件中设置控件字体大小，使用如下代码：

```
android:textSize="@dimen/textsize"
```

② 在 Java 文件中使用

使用方法：getResources().getDimension()。示例代码：

```
getResources().getDimension(R.dimen.textsize)
```

2.6 图片资源

在新建一个项目的时候，Android Studio 会自动创建“/res/drawable/”和“/res/mipmap/”两个目录，Android 中的图片资源存放在 drawable 和 mipmap 文件夹中。一般把有固定的尺寸、不需要更改的图片，放在 drawable 文件夹中；把需要缩放变化的处理的图片，放在 mipmap 文件夹中。对于图片资源使用主要是通过 xml 文件和 Java 文件。例如在 drawable 或是 mipmap 文件夹中存放 myimage.png 这个文件。

（1）在 xml 文件中

通过@drawable/imagename 或者是@mipmap/myimage 来使用，示例代码：

```
android:background="@drawable/myimage"
```

（2）在 Java 文件中

使用方法：getResources().getDrawable(R.drawable.*imagename*) 或者是 getResources().getDrawable(R.mipmap.imagename)。示例代码：

```
getResources().getDrawable(R.drawable.myimage);
```

2.7 样式和主题

（1）样式

样式（style）是包含一种或多种控件的属性集合，可以指定高度、填充、字体颜色、字号、背景色等许多属性。例如设置按钮、文本框等控件的高、宽、字体大小、颜色等。通过样式可以将设计与内容分离。样式定义在“/res/values/”目录的 themes.xml 文件中。

① 样式的定义　可以直接打开“/res/values/themes”下的 themes.xml 文件，在其中添加一个<style>标记，并且给这个 style 设置 name 属性，添加 item 项目，示例代码如下：

```
<style name="textStyle">
        <item name="android:layout_width">wrap_content</item>
        <item name="android:layout_height">wrap_content</item>
        <item name="android:textColor">#673AB7</item>
        <item name="android:textSize">30sp</item>
</style>
```

说明：定义样式是通过<item>标签的“name”属性值实现的，“name”的属性值可以是高、宽、字体颜色、字体大小等。

还可以通过新建一个“styles”文件：首先在“/res/values/”目录下创建一个“Resource File”文件，名为 styles.xml，添加一个 style 标签，styles 中的代码如下：

```
<?xml version="1.0" encoding="utf-8"?>
<resources>
    <style name="textStyle">
        <item name="android:layout_width">wrap_content</item>
        <item name="android:layout_height">wrap_content</item>
        <item name="android:textColor">#673AB7</item>
        <item name="android:textSize">30sp</item>
    </style>
</resources>
```

② 样式的应用　在 xml 文件中控件的 style 属性上添加引用即可。如果要使用定义好的样式“textStyle”，示例代码如下：

```
<TextView
……
style="@style/textStyle"
……/>
```

这样就可以一次性设置当前控件的高、宽、字体颜色和字体大小。

(2) 主题

主题（theme）指应用到整个 Activity 和 Application 的样式，当设置好主题后，Activity 或整个程序中的视图都将使用主题中的属性，当主题和样式中的属性发生冲突时，样式的优先级要高于主题。主题定义在“/res/values/themes”目录的 themes.xml 文件中。

① 在 themes.xml 中定义主题　Android 项目创建成功后，在 themes.xml 文件中有如下代码：

```
<resources xmlns:tools="http://schemas.android.com/tools">
    <!-- Base application theme. -->
    <style name="Theme.MyApplication" parent="Theme.MaterialComponents.DayNight.DarkActionBar">
        <!-- Primary brand color. -->
        <item name="colorPrimary">@color/purple_500</item>
        <item name="colorPrimaryVariant">@color/purple_700</item>
        <item name="colorOnPrimary">@color/white</item>
        <!-- Secondary brand color. -->
        <item name="colorSecondary">@color/teal_200</item>
        <item name="colorSecondaryVariant">@color/teal_700</item>
        <item name="colorOnSecondary">@color/black</item>
        <!-- Status bar color. -->
        <item name="android:statusBarColor"tools:targetApi="l">?attr/colorPrimaryVariant</item>
        <!-- Customize your theme here. -->
    </style>
</resources>
```

这里设置了项目的默认主题，其中 name 是主题的名称，parent 指明 Android 系统提供的父主题。@color/colorPrimary 引用了在 colors.xml 中定义的颜色资源，也可以将自定义主题“AppTheme”作为新的主题的父主题。示例代码：

```
<style name="myTheme" parent="Theme.MyApplication">
    <item name="colorPrimary">#F00</item>
</style>
```

② 主题的应用　对于主题的应用有两种方式，一种是在 AndroidManifest.xml 清单文件中使用，另一种是在 Java 代码中使用。

a. 在 AndroidManifest.xml 文件中使用。打开 AndroidManifest.xml 清单文件，在<application>标签中看到的代码：android:theme="@style/AppTheme"就是对主题的使用，如果使用自定义的主题 myTheme，则代码如下：

```
android:theme="@style/myTheme"
```

b. 在 Java 代码中设置主题。示例代码：

```
setTheme(R.style.Appthem);
```

2.8 国际化与资源自适应

有的应用程序存在多个语言的版本，Android 为了让应用程序自适应手机不同的语言环境，比如在英文环境中显示英文，在中文环境中显示中文，需要对界面的字符串进行国际化处理，也就是为应用程序提供不同语言的相应字符串信息，开发者需要做的是为各种语言的字符串资源建立国际化目录，然后将相应的同名资源文件放到这些目录中。

在 Android Studio 中国际化处理的 strings.xml 文件都存储在“/res/values/strings”目录下，例如这个文件夹中存储有英文和中文的字符串资源文件，则分别为 strings.xml 和 strings.xml(zh-rCN)。这样程序在使用时，系统会根据环境的语言设置自动选择英文或中文的字符串定义作为对应的显示内容。

创建文件的过程，选择“/res/values/”，单击右键，“New”→“Android Resources File”，出现如图 2-5 所示“New Resources File”界面。

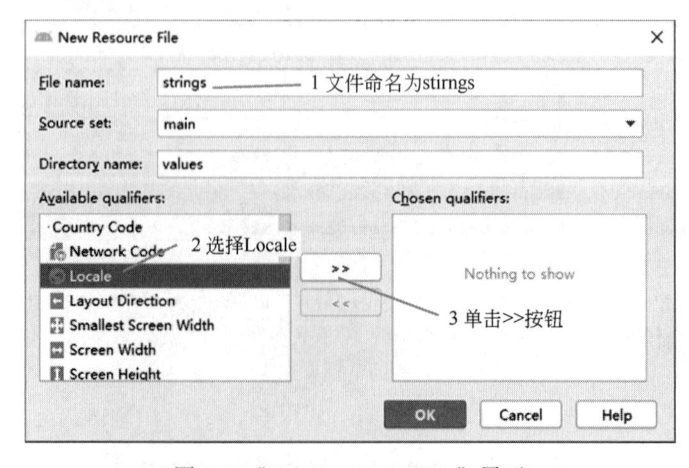

图 2-5 “New Resources File”界面

在图 2-5 中输入文件名“strings”,选择 Avaliable quelifieres 中的“Local”，然后单击“>>”命令按钮，点击“OK”，出现图 2-6 所示界面。

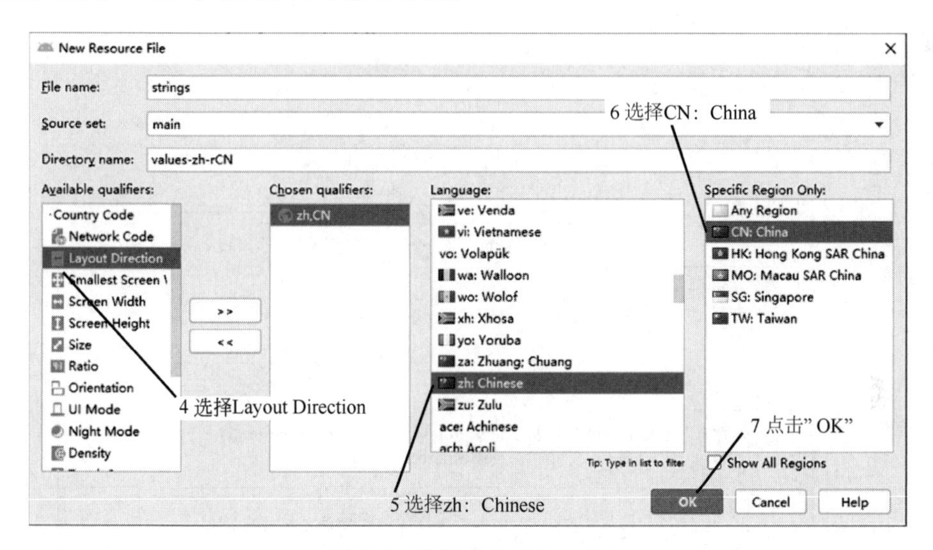

图 2-6　选择“zh:Chinese”

在图 2-6 中选择“Layout Direction”，然后在“Language”下找到 “zh:Chinese”选中，然后在“Specific Region Only”选择“CN: China”,这时将在 Directory name 后自动出现“values-zh-rCN”,选择“OK”命令按钮，在 res/values/目录下会创建出一个新 strings 文件夹，并创建一个 strings(zh-rCN)，如图 2-7 所示。

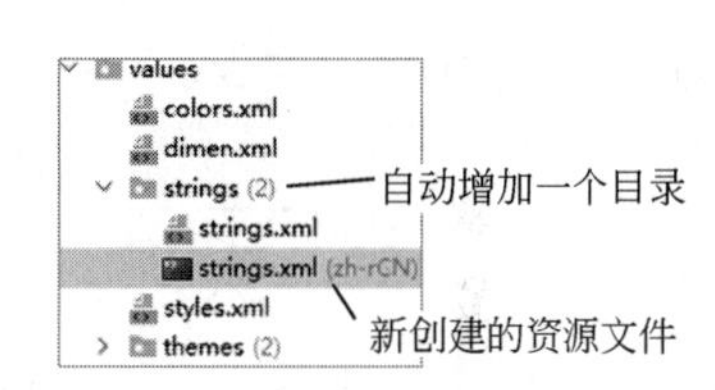

图 2-7　stirngs 目录

将 strings.xml 文件下的代码:

```
<resources>
    <string name="app_name">My Application</string>
</resources>
```

复制到 strings.xml(zh)文件中，用于中文表示对应的字符串资源。

```
<resources>
    <string name="app_name">我的应用</string>
</resources>
```

程序运行时，显示中文还是英文是根据设备设置的语言来决定的。其他的字符串资源也可以采用同样的方式处理语言和地区的国际化。

职业素养拓展

职业素养拓展内容详见文件“职业素养拓展”。

扫描二维码查看【职业素养拓展 2】

【职业素养拓展2】

本章小结

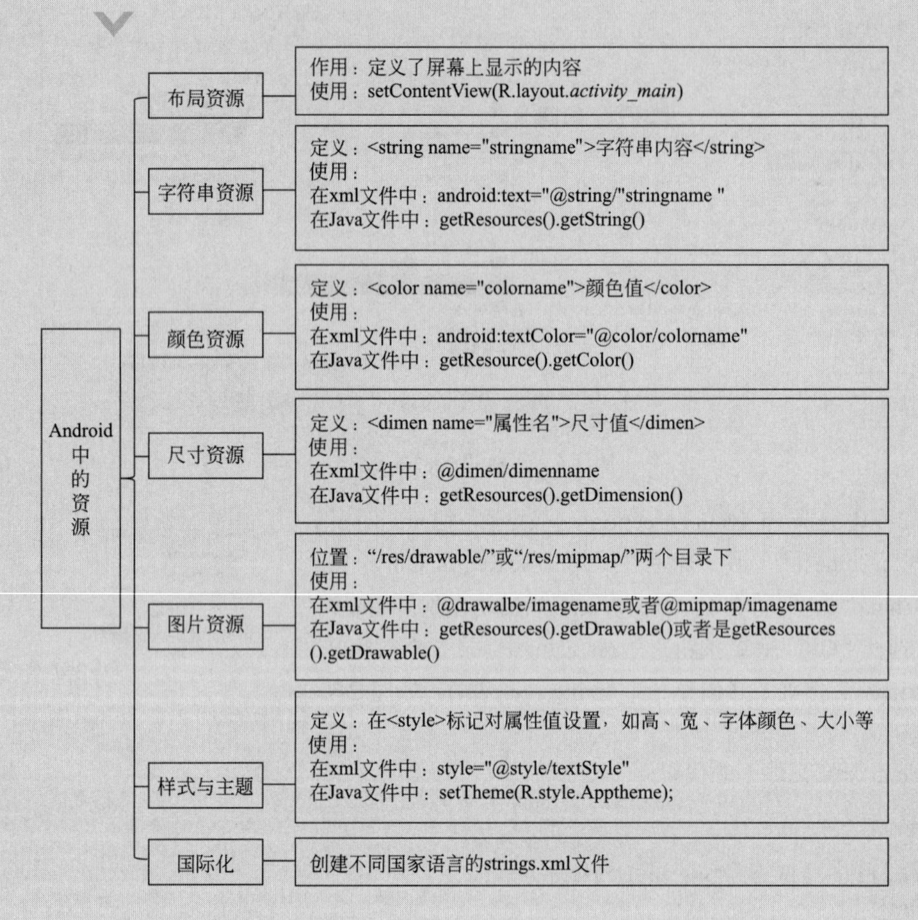
Android中的资源
布局资源
作用：定义了屏幕上显示的内容
使用：setContentView(R.layout.activity_main)
字符串资源
定义：<string name="stringname">字符串内容</string>
使用：
在xml文件中：android:text="@string/"stringname "
在Java文件中：getResources().getString()
颜色资源
定义：<color name="colorname">颜色值</color>
使用：
在xml文件中：android:textColor="@color/colorname"
在Java文件中：getResource().getColor()
尺寸资源
定义：<dimen name="属性名">尺寸值</dimen>
使用：
在xml文件中：@dimen/dimenname
在Java文件中：getResources().getDimension()
图片资源
位置："/res/drawable/"或"/res/mipmap/"两个目录下
使用：
在xml文件中：@drawalbe/imagename或者@mipmap/imagename
在Java文件中：getResources().getDrawable()或者是getResources().getDrawable()
样式与主题
定义：在<style>标记对属性值设置，如高、宽、字体颜色、大小等
使用：
在xml文件中：style="@style/textStyle"
在Java文件中：setTheme(R.style.Apptheme);
国际化
创建不同国家语言的strings.xml文件

第 3 章

用户界面设计基础

Android 开发离不了界面，而只要涉及界面，就要用到布局和放置在布局中的控件。本章主要介绍 Android 开发中常用的布局、基本控件和对话框。通过学习要求掌握布局和基本控件的使用，能够设计实现美观大方的应用界面。

学习目标：

1. 熟悉 Android 开发中常用的控件；
2. 掌握 Android 常用的控件的使用；
3. 能够在实际应用中灵活使用所学控件。

3.1 常用布局

漂亮的界面是由控件组成的，如何组织这些控件使其按一定的规则摆放在界面上呢？这就需要借助布局来实现了。布局可以说是一种容器，在其上可以放置多个控件，并且让这些控件按照一定的规则摆放在合适的位置，构成美观的用户界面。常用的布局有 ConstraintLayout 约束布局，LinearLayout 线性布局，RelativeLayout 相对布局，FrameLayout 帧布局，TableLayout 表格布局。本节将给大家讲解 Android 中常用的布局。

3.1.1 ConstraintLayout 约束布局

ConstraintLayout 约束布局是 Android Studio2.2 中新使用的一种布局方式。从 Android Studio 2.3 起，官方的模板默认使用 ConstraintLayout。约束布局非常适合使用可视化的方式来编写界面，通过拖曳将控件放置在布局中，然后设置控件的垂直和水平的相对位置，来实现用户界面的设计。布局文件是 xml 文件，使用可视化设计界面的时候会自动生成 xml 代码。

新建一个项目，选择 activity_main.xml 文件，将布局文件切换到“Design”模式，可见如图 3-1 所示界面，主操作区域内有两个类似于手机屏幕的界面，左边的是预览界面，右边的是蓝图界面。这两部分都可以用于布局编辑工作，区别是左边部分主要用于预览最终的界面效果，右边部分主要用于观察界面内各个控件的约束情况。左侧 palette 部分是常用控件，可直接拖入到白色区域生成一个控件。右侧显示的是属性窗口，可以根据需要显示或者隐藏。在属性窗口中有“layout”下面的“Constraint Widget”部分，在这部分可以添加或者删除约束条件，设置外边距、约束偏差和设置高度/宽度模式。

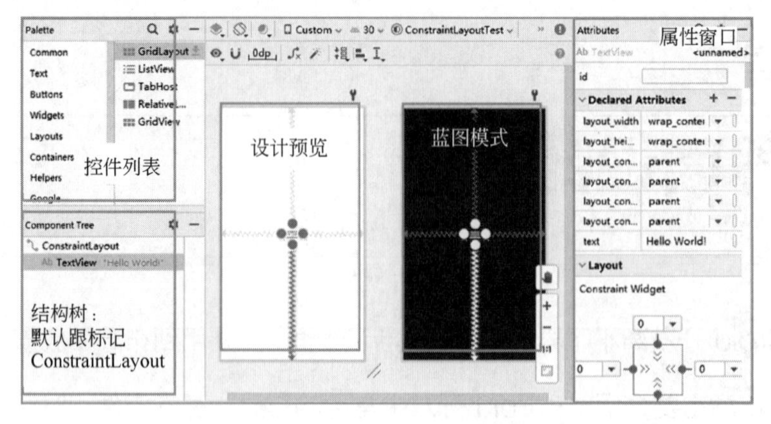

图 3-1 “Design”布局编辑界面

在图 3-1 所示的结构树中可以看到根标记为 ConstraintLayout，下一级有一个控件 TextView；在布局预览中可以看到界面中间有一个 TextView 控件，从蓝图模式可以看到这个控件与整个界面的上下左右都有连线，这是什么意思呢。我们切换到布局的“Code”来看生成的布局代码。activity_main 的代码如下所示。

```
<?xml version="1.0" encoding="utf-8"?>
<androidx.constraintlayout.widget.ConstraintLayout
    xmlns:android="http://schemas.android.com/apk/res/android"
    xmlns:tools="http://schemas.android.com/tools"
    xmlns:app="http://schemas.android.com/apk/res-auto"
    android:layout_width="match_parent"
    android:layout_height="match_parent"
    tools:context=".MainActivity">

    <TextView
        android:layout_width="wrap_content"
        android:layout_height="wrap_content"
        android:text="Hello World!"
        app:layout_constraintBottom_toBottomOf="parent"
        app:layout_constraintLeft_toLeftOf="parent"
        app:layout_constraintRight_toRightOf="parent"
        app:layout_constraintTop_toTopOf="parent" />

  </androidx.constraintlayout.widget.ConstraintLayout>
```

从代码可以看出 activity_main 默认使用的就是约束布局，此布局的默认根元素是 androidx.constraintlayout.widget.ConstraintLayout。默认布局中有一个控件 TextView:

```
android:layout_width="wrap_content"设置 TextView 宽度为刚好容纳它的内容
android:layout_height="wrap_content"设置 TextView 高度为刚好容纳它的内容
android:text="Hello World!" 设置 TextView 显示的文本为"Hello World!"
```

后面有四行代码:

```
app:layout_constraintBottom_toBottomOf="parent"
app:layout_constraintLeft_toLeftOf="parent"
app:layout_constraintRight_toRightOf="parent"
app:layout_constraintTop_toTopOf="parent"
```

其中:

app:layout_constraintBottom_toBottomOf="parent"，它的作用是将此对象的底部约束到其父容器的底部。

app:layout_constraintTop_toTopOf="parent"，它的作用是将此对象的顶部约束到其父容器的顶部。

当 Bottom 和 Top 这两个属性的值都为 parent 时，则此控件垂直居中显示在父布局中。

app:layout_constraintLeft_toLeftOf="parent", 它的作用是将此对象的左侧约束到其父容器的左侧。

app:layout_constraintRight_toRightOf="parent", 它的作用是将此对象的右侧约束到其父容器的右侧。

如果 Left 和 Right 属性都以父容器为约束的话，则此控件水平居中显示在父布局中。

ConstraintLayout 是使用约束的方式来指定各个控件的位置和关系的, 每个对象至少需要一个水平和一个垂直约束。

将布局设计切换到“Design”模式，在布局视图中向上，向左拖动 TextView 控件。然后再切换到“Code”模式, 观察代码变化。发现增加了两个表示控件在水平和垂直方向上偏移量的属性, 代码如下:

```
app:layout_constraintHorizontal_bias="0.2" 此控件在水平方向的偏移
app:layout_constraintVertical_bias="0.3"  此控件在垂直方向的偏移
```

如果控件位于父布局的中心，其值都为 0.5，没有实质的意义。

我们从 Palette 中拖动 Button 控件到界面，观察图 3-2 所示界面。

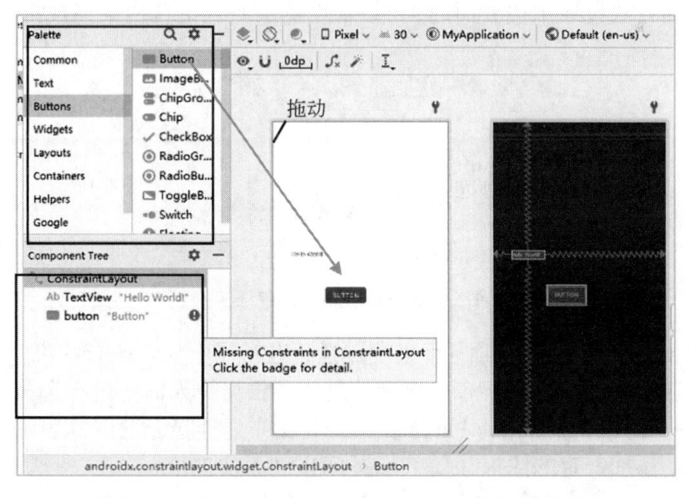

图 3-2　ConstraintLayout 布局“Design”模式图

在“Component Tree”中可以看到，button 后面有一个红色❶图标提示，提示这个 view 没有被约束。我们运行程序时可以发现，没有添加约束的控件运行时会在窗口的左上角出现提示。这就提出必须对控件指定竖直和水平的约束，才能确定控件的位置。对于控件的位置约束主要由以下属性来实现。

layout_constraintLeft_toLeftOf 当前控件的左边与另一个控件的左边对齐。例如有如下代码:

layout_constraintLeft_toLeftOf="@+id/textView"当前控件与 id 为 textView 的控件左对齐。

layout_constraintRight_toRightOf 当前控件的右边与另一个控件的右边对齐。

当 Button 控件同时对 TextView 控件设置左对齐和右对齐两个属性时，则 Button 控件与 TextView 控件水平中心对齐。位置关系如图 3-3 所示。

按钮的左边与文本的左边对齐　　按钮的右边与文本的右边对齐

按钮的左右边与文本的左右边同时对齐

图 3-3　ConstraintLayout 布局中 2 个控件的约束

layout_constraintTop_toTopOf：当前控件的顶部与另一控件的顶部对齐
layout_constraintBottom_toBottomOf：当前控件的底部与另一控件的底部对齐
当一个控件的顶部和底部与另外一个控件同时对齐时，则这两个控件中心将垂直对齐。
除了上面属性外，常用的约束定位的属性还有以下几个：
layout_constraintTop_toBottomOf：当前控件的顶部与另一控件的底部对齐
layout_constraintBottom_toTopOf：当前控件的底部与另一控件的顶部对齐
layout_constraintLeft_toRightOf：当前控件的左边与另一控件的右边对齐
layout_constraintRight_toLeftOf：当前控件的右边与另一控件的左边对齐
与上面几个属性类似的还有：
layout_constraintStart_toStartOf：当前控件的起始边与另一控件的起始边对齐
layout_constraintStart_toEndOf：当前控件的起始边与另一控件的尾部对齐
layout_constraintEnd_toEndOf：当前控件的尾部与另一控件的尾部对齐
layout_constraintEnd_toStartOf：当前控件的尾部与另一控件的起始边对齐
在使用约束布局方式设计用户界面时，可以借助约束布局工具栏来完成布局设计，如图 3-4。

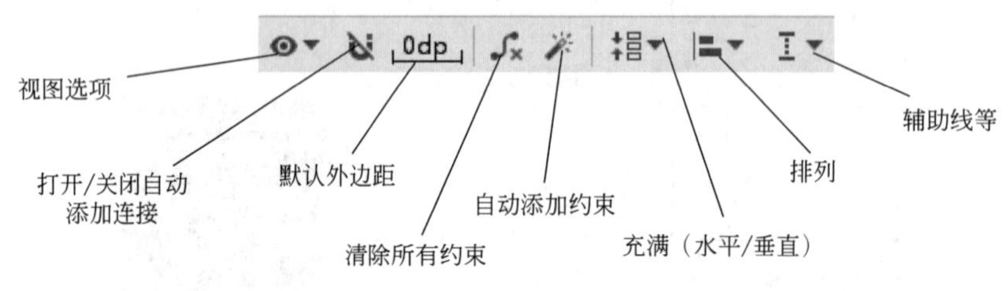

图 3-4　ConstraintLayout 布局工具栏

① ：View Options 视图选项，用来设置是否在布局上显示所有约束，显示边距，未选择视图变淡等，可以通过选择与否观察布局显示的变化，根据个人需要进行选择。

② ：Enable autoConnection to Parent，自动与父布局连接，此图标表示在布局中拖曳控件到父布局边上会自动添加一定的约束，表示关闭自动添加连接。

③ 0dp：default Margins，可以为布局中的每个视图设置默认外边距。

④ ：clear all Constraints，清除布局中所有的约束。

⑤ ：infer Constraints，自动创建约束添加，将布局中的每个视图移动到希望的位置后，点击此图标，会为每个视图添加约束，不过有时也可能根据需要进行一些调整。

⑥ ：pack，一个控件，选择“Expand Horizontally”时设置当前控件充满水平可用空间，选择“Expand Vertically”时当前控件充满垂直可用空间；针对多个控件可以设置整体的水平垂直的填充和等间距分布等。

⑦ ：align，排列，当选择“Vertically”或者是“Horizontally”时表示当前控件在可用空间垂直或者是水平居中，当选择“Vertically in Parent”或者是“Horizontally in Parent”表示在父容器中垂直或者水平居中。

⑧ ：Guideline 分类中提供了在 ConstraintLayout 布局中的工具类，用于辅助布局。选择

倒三角，可以根据需要添加垂直或水平的辅助线，垂直或是水平的 Barrier 或者 Group，Set of Constraints,Layer 或者 Flow，完成多种要求的布局。

对于约束布局的使用，一般都需要鼠标在布局中拖动控件，与在 Attributes 属性窗口设置属性值和同时在 XML 布局文件中设置结合应用。由于此布局的很多用法都是通过对控件动态拖曳设置相关属性完成的，在此只做简单介绍，我们可以通过多练习来熟悉其用法，本书的多数案例都是使用约束布局进行界面设计，在后面的学习中会逐步熟悉。

3.1.2 LinearLayout 线性布局

线性布局

LinearLayout 线性布局是一种简单常用的布局形式，它会将包含的控件在水平或者垂直单一的方向排列。如果一组控件设置为水平排列，那么它只在布局中显示一行，如果设置为垂直排列，则只显示一列。具体的实现是通过 orientation 属性来设置的。属性值有 vertical 和 horizontal。其中：

```
android:orientation="vertical"  表示布局中的控件垂直排列
android:orientation="horizontal"  表示布局中的控件水平排列
```

新建一个项目，我们将布局文件 activity_main.xml 进行修改，代码如下：

```
<?xml version="1.0" encoding="utf-8"?>
<LinearLayout xmlns:android="http://schemas.android.com/apk/res/android"
    xmlns:app="http://schemas.android.com/apk/res-auto"
    xmlns:tools="http://schemas.android.com/tools"
    android:layout_width="match_parent"
    android:layout_height="match_parent"
    android:orientation="vertical"
    tools:context=".MainActivity">

    <Button
        android:id="@+id/button1"
        android:layout_width="wrap_content"
        android:layout_height="wrap_content"
        android:text="Button1" />

    <Button
        android:id="@+id/button2"
        android:layout_width="wrap_content"
        android:layout_height="wrap_content"
        android:text="Button2" />

    <Button
        android:id="@+id/button3"
        android:layout_width="wrap_content"
        android:layout_height="wrap_content"
        android:text="Button3" />
</LinearLayout>
```

在代码中使用 LinearLayout 布局，设置其属性 android:orientation="vertical"，即布局中的控件垂直显示。在布局中放置了三个命令按钮，每个命令按钮的宽高都是 wrap_content。运行效果如图 3-5 所示。

如果需要控件水平排列，只需要将上面代码中的 android:orientation="vertical" 修改为 android:orientation="horizontal"即可，运行效果如图 3-6 所示。

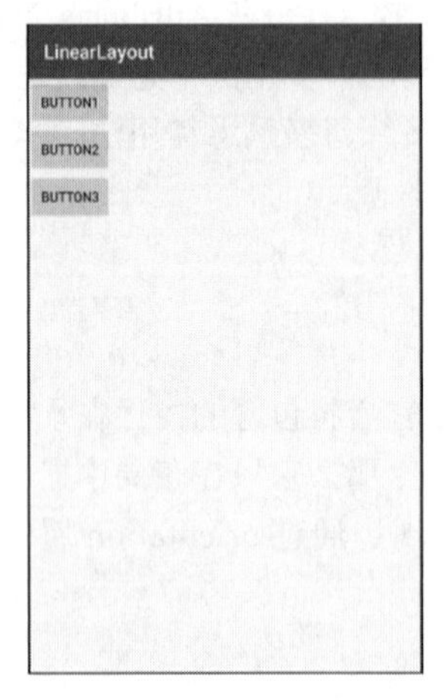

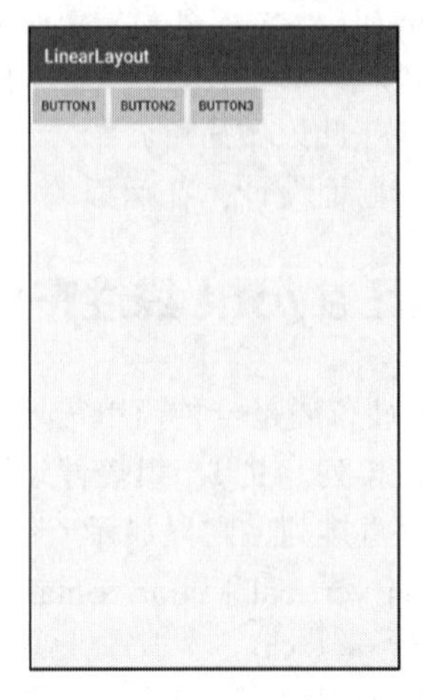

图 3-5　LinearLayout 垂直排列

图 3-6　LinearLayout 水平排列

注意:

① 如果在一个 LinearLayout 中没有指定 android:orientation 属性值，则默认的是 horizontal。

② 对于 LinearLayout 布局在水平方向，如果上面的例子将 BUTTON1 的宽度指定为 match_parent，运行时 BUTTON1 一个按钮将占满整个水平位置，只显示 BUTOTN1 一个按钮。如果只将 BUTTON3 的宽度指定为 match_parent，BUTTON1 和 BUTTON2 的宽度根据内容显示，BUTTON3 将水平占满剩余可用的空间。如果需要三个按钮等分水平空间，则可以使用 LinearLayout 布局 android:layout_weight 属性。代码如下所示:

```
<?xml version="1.0" encoding="utf-8"?>
<LinearLayout xmlns:android="http://schemas.android.com/apk/res/android"
    xmlns:app="http://schemas.android.com/apk/res-auto"
    xmlns:tools="http://schemas.android.com/tools"
    android:layout_width="match_parent"
    android:layout_height="match_parent"
    android:orientation="horizontal"
    tools:context=".MainActivity">

    <Button
        android:id="@+id/button1"
        android:layout_width="0dp"
        android:layout_height="wrap_content"
        android:layout_weight="1"
        android:text="Button1" />

    <Button
        android:id="@+id/button2"
        android:layout_width="0dp"
        android:layout_height="wrap_content"
        android:layout_weight="1"
        android:text="Button2" />

    <Button
        android:id="@+id/button3"
```

```
        android:layout_width="0dp"
        android:layout_height="wrap_content"
        android:layout_weight="1"
        android:text="Button3" />
</LinearLayout>
```

运行程序效果如图 3-7 所示，这里将三个命令按钮的 android:layout_weight 属性的值同时设置为 1 就可以平分屏幕宽度。其原理是系统先计算在 LinearLayout 布局下所有控件的 android:layout_weight 值的总和，然后根据每个控件所占额比来分配其所占的大小。对于这个例子，三个按钮的 android:layout_weight 的值都为 1，其和为 3，则每个按钮宽度分别占屏幕宽度的 1/3。如果修改第三个按钮 android: layout_weight="2"，则第一个和第二个按钮的宽度各占屏幕宽度的 1/4，第三个按钮宽度占屏幕宽度的 2/4。

图 3-7　layout_weight 属性应用

注意：当使用 android:layout_weight 属性时控件的宽度是根据其值占比来决定，所以指定宽度为 0dp，即 android: layout_width="0dp"。

在 LinearLayout 的应用中有两个比较形似的属性 android: layout_gravity 和 android:gravity。其中：

android:gravity 用于指定文字在控件中的对齐方式。

android:layout_gravity 用于指定控件在布局中的对齐方式。示例代码如下所示：

```
<?xml version="1.0" encoding="utf-8"?>
<LinearLayout xmlns:android="http://schemas.android.com/apk/res/android"
    xmlns:app="http://schemas.android.com/apk/res-auto"
    xmlns:tools="http://schemas.android.com/tools"
    android:layout_width="match_parent"
    android:layout_height="match_parent"
    android:orientation="horizontal"
    tools:context=".MainActivity">

    <Button
        android:id="@+id/button1"
        android:layout_width="wrap_content"
        android:layout_height="wrap_content"
        android:layout_gravity="top"
        android:gravity="center_vertical"
        android:text="Button1" />

    <Button
        android:id="@+id/button2"
        android:layout_width="wrap_content"
        android:layout_height="wrap_content"
        android:layout_gravity="center"
        android:text="Button2" />

    <Button
        android:id="@+id/button3"
        android:layout_width="wrap_content"
```

```
        android:layout_height="wrap_content"
        android:layout_gravity="bottom"
        android:text="Button3" />
</LinearLayout>
```

对于 LinearLayout，当设置 android:orientation="horizontal"时，其中控件的 android:layout_gravity 只能指定垂直方向的排列方向，如指定第一个 Button 在布局中的对齐方式为 top，第二个 Button 指定为 center，第三个 Button 指定为 bottom，运行程序，效果如图 3-8(a)所示。

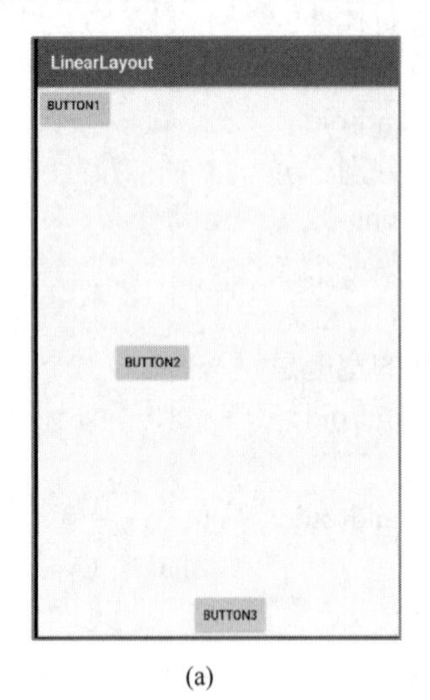

(a)

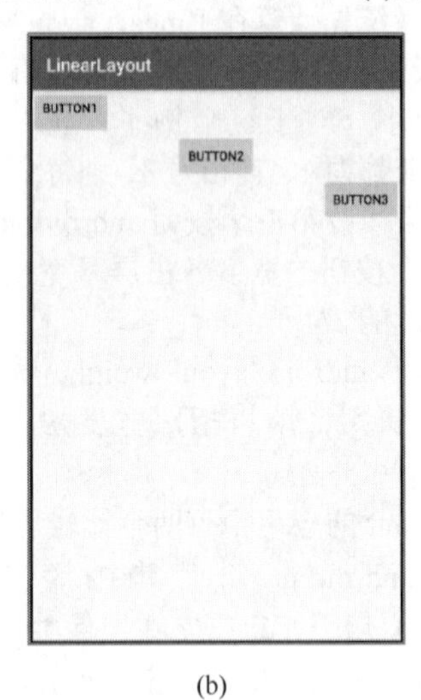

(b)

图 3-8 layout_gravity 属性应用

当 LinearLayout 中设置 android:orientation="vertical"时，其中控件的 android:layout_gravity 只能指定水平方向的排列方向，如指定第一个 Button 对齐方式为 left，第二个 Button 对齐方式指定为 center，第三个 Button 对齐方式指定为 right，运行程序，效果如图 3-8(b)所示。

3.1.3 RelativeLayout 相对布局

相对布局

RelativeLayout 也称为相对布局，也是一种常用的布局。它允许通过指定显示对象相对于父容器或其他兄弟控件的相对位置来完成布局。例如指定一个控件位于父容器的顶部、底部、左端、右端，或者是一个控件可以在另一个控件的左边、右边、上边、下边等。

（1）控件相对于父容器定位

对于控件在父容器的相对位置的应用，新建一个项目，修改 activity_main 布局文件，代码如下所示。

```
<?xml version="1.0" encoding="utf-8"?>
<RelativeLayout xmlns:android="http://schemas.android.com/apk/res/android"
    android:layout_width="match_parent"
    android:layout_height="match_parent">
    <Button
        android:id="@+id/button1"
        android:layout_width="wrap_content"
        android:layout_height="wrap_content"
        android:text="Button1"
        android:layout_centerInParent="true"/>
```

```
    <Button
        android:id="@+id/button2"
        android:layout_width="wrap_content"
        android:layout_height="wrap_content"
        android:text="Button2"
        android:layout_alignParentLeft="true" />
    <Button
        android:id="@+id/button3"
        android:layout_width="wrap_content"
        android:layout_height="wrap_content"
        android:text="Button3"
        android:layout_alignParentRight="true"/>
    <Button
        android:id="@+id/button4"
        android:layout_width="wrap_content"
        android:layout_height="wrap_content"
        android:text="Button4"
        android:layout_alignParentBottom="true"/>
    <Button
        android:id="@+id/button5"
        android:layout_width="wrap_content"
        android:layout_height="wrap_content"
        android:text="Button5"
        android:layout_alignParentRight="true"
        android:layout_alignParentBottom="true"/>
</RelativeLayout>
```

在这个 RelativeLayout 布局中，我们使用属性 android:layout_centerInParent="true"指定 button1 父容器居中，button2 使用属性 android:layout_alignParentLeft="true"指定其与父容器的左边对齐，button3 使用属性 android:layout_alignParentRight="true"指定其与父容器的右边对齐，button4 使用属性 android:layout_alignParentBottom="true"指定其与父容器的左下角对齐，button5 设置属性 android:layout_alignParentBottom="true"和 android:layout_alignParentBottom="true"使其与父容器的右下角对齐。运行程序，效果如图 3-9 所示。

图 3-9　相对于父容器定位

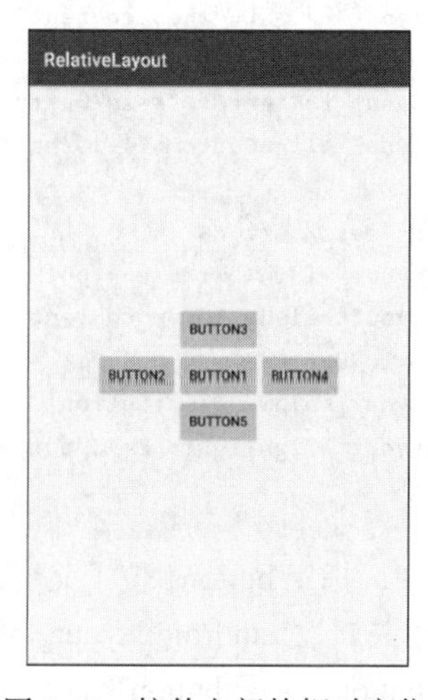

图 3-10　控件之间的相对定位

除了使用的这几个属性，设置控件与父容器对齐方式还有：

android:layout_centerHorizontal 相对于父容器水平居中；

android:layout_centerVertical　相对于父容器垂直居中。

(2) 控件与控件之间的定位

常用的相对位置有一个控件位于指定控件的左方、右方、上方、下方和一个控件与指定控件的上下左右边缘对齐,等等。修改 activity_main.xml，代码如下所示。

```
<?xml version="1.0" encoding="utf-8"?>
<RelativeLayout xmlns:android="http://schemas.android.com/apk/res/android"
    android:layout_width="match_parent"
    android:layout_height="match_parent">
    <Button
        android:id="@+id/button1"
        android:layout_width="wrap_content"
        android:layout_height="wrap_content"
        android:text="Button1"
        android:layout_centerInParent="true"/>
    <Button
        android:id="@+id/button2"
        android:layout_width="wrap_content"
        android:layout_height="wrap_content"
        android:text="Button2"
        android:layout_toLeftOf="@+id/button1"
        android:layout_alignTop="@+id/button1"/>
    <Button
        android:id="@+id/button3"
        android:layout_width="wrap_content"
        android:layout_height="wrap_content"
        android:text="Button3"
        android:layout_above="@+id/button1"
        android:layout_alignLeft="@+id/button1"/>
    <Button
        android:id="@+id/button4"
        android:layout_width="wrap_content"
        android:layout_height="wrap_content"
        android:text="Button4"
        android:layout_toRightOf="@+id/button1"
        android:layout_alignBottom="@+id/button1"/>
    <Button
        android:id="@+id/button5"
        android:layout_width="wrap_content"
        android:layout_height="wrap_content"
        android:text="Button5"
        android:layout_below="@+id/button1"
        android:layout_alignRight="@+id/button1"/>
</RelativeLayout>
```

在上面代码中，指定 button1 位于父容器中央，button2 位于 button1 的左侧，与 button1 上边缘对齐。使用的相关属性: android:layout_toLeftOf 和 android:layout_alignTop 这两个属性的值为另一个控件的 id。button3 位于 button1 的上方，与其左边缘对齐，使用的属性 android:layout_above 和 android:layout_alignLeft 。button4 位于 button1 的右方，button4 的底部与 button1 的底部对齐，使用的属性为 android:layout_toRightOf 和 android:layout_alignBottom。Button5 位于 button1 的下方，button5 的右边缘与 button1 的右边缘对齐，相关的属性 android:layout_below 和

android:layout_alignRight。运行程序，效果如图 3-10 所示。

3.1.4　FrameLayout 帧布局

FrameLayout 也称为帧布局，最大的特点是所有控件都默认显示在屏幕左上角，并按照先后放入的顺序重叠摆放，先放入的控件将会在最底层，后放入的控件显示在最顶层。帧布局适用图层设计。

新建一个项目，修改 activity_main 代码如下所示。

```
<?xml version="1.0" encoding="utf-8"?>
<FrameLayout xmlns:android="http://schemas.android.com/apk/res/android"
    android:layout_width="match_parent"
    android:layout_height="match_parent">
    <Button
        android:id="@+id/button1"
        android:layout_width="156dp"
        android:layout_height="100dp"
        android:text="Button1"
        android:background="#A7A2A2"/>
    <Button
        android:id="@+id/button2"
        android:layout_width="wrap_content"
        android:layout_height="wrap_content"
        android:text="Button2" />
</FrameLayout>
```

从运行效果图 3-11(a)可以看到，在 FrameLayout 布局中的控件都显示在布局的左上角，BUTTON2 添加在 BUTTON1 之后，所以显示在 BUTTON1 的上方。当然可以对 Frame Layout 布局中的控件通过指定 android:layout_gravity="center"属性指定控件的位置，代码如下所示，进行设置后运行效果如图 3-11(b)所示。

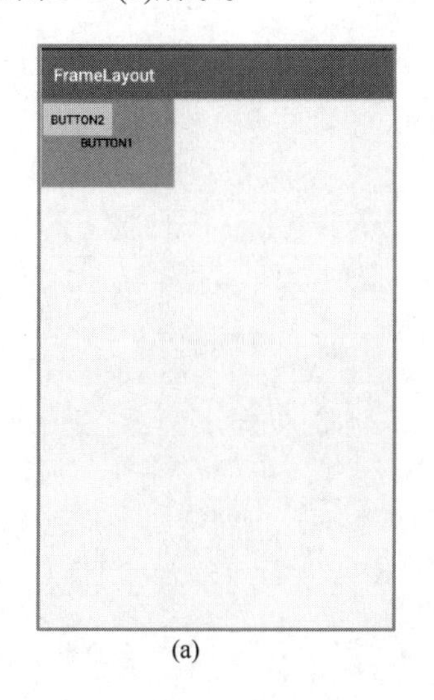

(a)

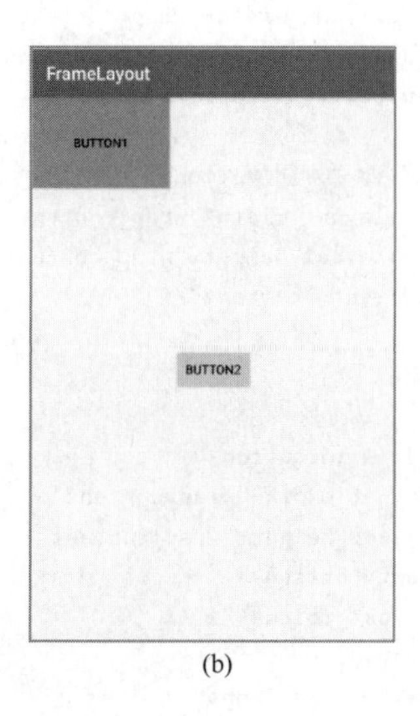

(b)

图 3-11　FrameLayout 应用

```
<FrameLayout xmlns:android="http://schemas.android.com/apk/res/android"
    android:layout_width="match_parent"
    android:layout_height="match_parent">
    <Button
        android:id="@+id/button1"
        android:layout_width="156dp"
        android:layout_height="100dp"
        android:text="Button1"
        android:background="#A7A2A2"/>
    <Button
        android:id="@+id/button22"
        android:layout_width="wrap_content"
        android:layout_height="wrap_content"
        android:text="Button2"
        android:layout_gravity="center"/>
</FrameLayout>
```

3.1.5 TableLayout 表格布局

TableLayout 又称为表格布局，它是通过 TableRow 设置行，列数由 TableRow 中的子控件个数最多的值决定，直接在 TableLayout 中添加子控件会占据整个一行。

创建一个新的项目，将 activity_main.xml 中的代码修改为如下所示的代码:

```
<?xml version="1.0" encoding="utf-8"?>
<TableLayout xmlns:android="http://schemas.android.com/apk/res/android"
    android:layout_width="match_parent"
    android:layout_height="match_parent"
    android:stretchColumns="1">
    <TableRow>
        <Button
            android:id="@+id/button1"
            android:layout_width="wrap_content"
            android:layout_height="wrap_content"
            android:text="button1"/>
        <Button
            android:id="@+id/button2"
            android:layout_width="wrap_content"
            android:layout_height="wrap_content"
            android:text="button2"/>
    </TableRow>
    <TableRow>
    <Button
        android:id="@+id/button3"
        android:layout_width="wrap_content"
        android:layout_height="wrap_content"
        android:text="button3"
        android:layout_column="1"/>
    <Button
        android:id="@+id/button4"
        android:layout_width="wrap_content"
        android:layout_height="wrap_content"
```

```
        android:text="button4"/>
    <Button
        android:id="@+id/button5"
        android:layout_width="wrap_content"
        android:layout_height="wrap_content"
        android:text="button5" />
    </TableRow>
    <TableRow>
        <Button
            android:id="@+id/button6"
            android:layout_width="wrap_content"
            android:layout_height="wrap_content"
            android:text="button6"
            android:layout_span="2"/>
        <Button
            android:id="@+id/button7"
            android:layout_width="wrap_content"
            android:layout_height="wrap_content"
            android:text="button7"/>
    </TableRow>
    <TableRow>
        <Button
            android:id="@+id/button8"
            android:layout_width="wrap_content"
            android:layout_height="wrap_content"
            android:layout_column="3"
            android:text="button8"/>
    </TableRow>
</TableLayout>
```

运行效果如图 3-12 所示，分析代码对照图，我们可以看到此表格有 4 行 4 列。在 TableLayout 布局中设置了 android:stretchColumns="1"，用来设置表格的第 2 列可被拉伸（列号从 0 开始），第二行中指定 android: layout_ column="1"表示此行的第一个控件显示的位置在第 2 列。第三行指定 android:layout_span="2"，表示此控件占两列。第四行 BUTTON8 显示在第 4 列。

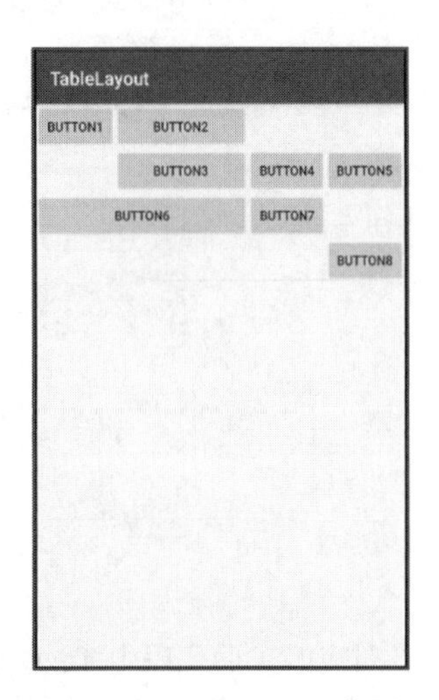

图 3-12 TableLayout 应用

设置界面布局主要通过相对应的属性来完成。对于我们提到的布局，它们有一些通用的属性:

① android:id 属性用于指定当前布局的唯一标识，用法与控件的 id 相同。

② android:layout width 属性和 android:layout height 属性，用于指定布局的宽和高，我们之前使用的都是默认的，在布局中也可以将其设为“match_parent”，“wrap_content”或者是具体的值如“200dp”。

③ android:layout_margin 属性，用于设置布局与屏幕边距，或者是其周围控件的距离，可以整体设置，也可以使用 android:layout_marginTop，android:layout_marginBottom，android: layout_marginLeft 和 android:layout_ marginRight 分别对上下左右的边距进行设置。

④ android:padding 属性，用于设置布局内的控件与布局的距离，使用 android:padding 同时设置上下左右的距离，也可以使用 android:paddingTop，android:paddingBottom，android:paddingLeft 和 android:paddingRight 分别设置上下左右的距离。

⑤ android:background 属性，用于设置布局的背景，其值可以直接引用颜色资源，如 android:background="@color/teal_700"；可以为一个3位或者6位的颜色值，如android:background="#f00"或者 android:background="#ff0000"；还可以为一个 4 位或者 8 位的颜色值，如 android:background="#af00"或者 android:background="#a0ff0000",在这种情况下，4 位值的首位和 6 位值的前两位标识透明度。对于布局的背景还可以使用已有的图片来设置，将背景图片复制到 res/drawble 目录下，设置 android:background="@drawable/beijing"。

3.2 基础控件及应用

美观的用户界面是 Android 应用给用户的第一印象，这些界面如何来实现，需要哪些控件呢？在这一节我们学习 Android 界面设计中常用的控件。在布局中添加控件，可以在“Design”设计模式下直接拖到布局界面中，也可以在布局文件的“Code”模式下的 xml 文件中通过代码添加。

3.2.1 TextView 控件

TextView 控件主要用于显示静态的文本，可以显示单行，或者是多行文本。文本的内容由 text 属性的值决定。创建一个项目，在“Design”视图下，查看布局文件 activity_main 中 TextView 控件的属性。在布局中选择 TextView 控件，观察 Attributes 属性窗口，如图 3-13 所示。对于控件的显示设置可以在属性窗口来完成。

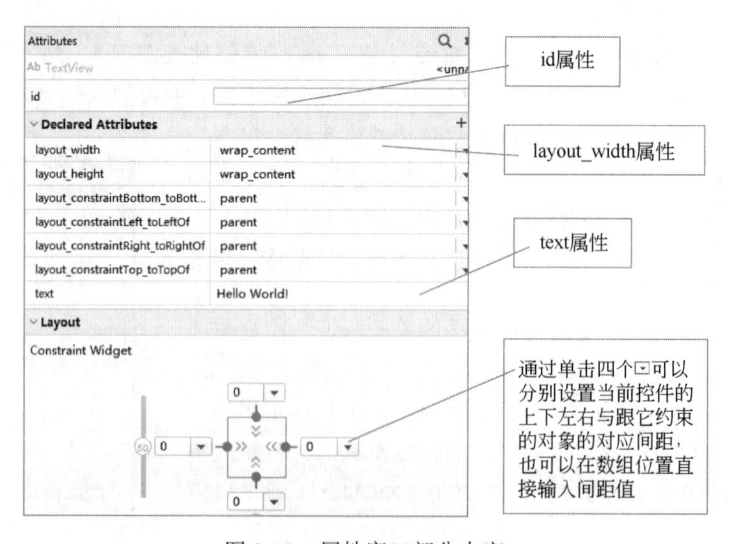

图 3-13　属性窗口部分内容

对应的布局代码中 TextView 控件相关代码如下所示。

```
<TextView
    android:layout_width="wrap_content"
    android:layout_height="wrap_content"
    android:text="Hello World!"
```

```
app:layout_constraintBottom_toBottomOf="parent"
app:layout_constraintLeft_toLeftOf="parent"
app:layout_constraintRight_toRightOf="parent"
app:layout_constraintTop_toTopOf="parent" />
```

上面的代码是在布局中添加了一个 TextView 控件，及其默认的属性设置。除了代码中所示的控件属性，TextView 常用属性及其含义如下。

① android:id，用于设置当前对象的唯一标识。这一属性对 Android 中的所用对象都适用。在 xml 文件中它的属性值是通过 android:id="@+id/属性名称"定义的。为布局指定 id 属性后，在 R.java 文件中，会自动生成对应的 int 值。在 Java 代码中通过使用 findViewById()方法传入该 int 值来获取该对象。

② android:layout_width 和 Android:layout_height，设置控件的宽高，其值有三种形式：

- wrap_content，内容包裹，即控件宽或高以显示 text 属性的全部内容为准。
- match_parent，适应父容器，即控件的宽或高充满其所在父容器。
- 具体的值，如 android:layout_width="200dp" 设置 TextView 的宽为 200dp。

③ android:textSize，设置文本字体的大小，可以是具体的大小或是使用尺寸资源。

```
如：android:textSize="30sp"。
```

④ android:textColor，设置文本颜色,其值可以是“#”加三位或是四位的十六进制数，也可以使用颜色资源。

```
如：android:textColor="#CA1212"。
```

⑤ android:background，设置背景颜色。其值可以是已经存在的图片文件，或是颜色值。

```
示例：android:background="@color/colorAccent"  背景使用颜色资源
android:background="@drawable/beijing"  背景为 drawable 文件夹中的图片 beijing.png
```

一个控件的属性可以在属性窗口进行设置，也可以直接在布局文件 code 模式中编辑，在属性窗口设置结束会自动生成对应的代码。

3.2.2　EditText 控件

输入框 EditText 用于输入信息，可以是普通字符串，也可以是指定格式的文本，如密码、电话号码和 Email 地址等。EditText 继承自 TextView 类，大部分属性与 TextView 类似。通过实例来学习 EditText 的使用。实现如图 3-14 所示的运行界面，操作过程如下所示。

第一步：新建项目

新建一个项目，设置名字为 EditText_Test，项目所在的包名为 com.example.edittext。

第二步：设计用户界面

修改界面中自动创建的 TextView，text 属性改为“用户名”，从 palette 窗口的 Text 控件组中拖一个 PlainText 控件到界面，修改其 id 为 etname，设置 hint 属性为“请输入用户名”，要求其文字居中，页面布局预览效果如图 3-15 所示。

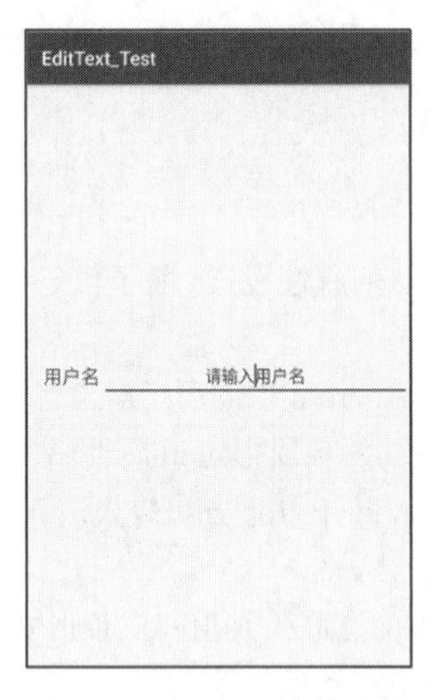

图 3-14　EditText 控件示例

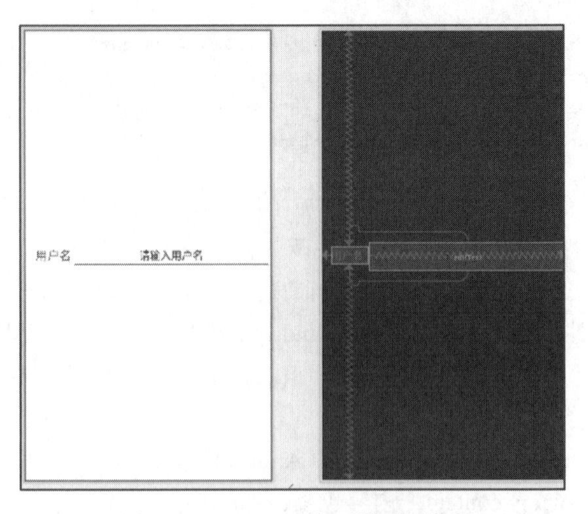

图 3-15　EditText 应用界面预览图

activity_main 参考代码如下所示:

```
<?xml version="1.0" encoding="utf-8"?>
    <androidx.constraintlayout.widget.ConstraintLayout xmlns:android="http://schemas. android.com/
apk/res/android"
    xmlns:app="http://schemas.android.com/apk/res-auto"
    xmlns:tools="http://schemas.android.com/tools"
    android:layout_width="match_parent"
    android:layout_height="match_parent"
    tools:context=".MainActivity">

    <TextView
        android:id="@+id/textView"
        android:layout_width="wrap_content"
        android:layout_height="wrap_content"
        android:layout_marginStart="10dp"
        android:text="用户名"
        android:textSize="20dp"
        app:layout_constraintBottom_toBottomOf="parent"
        app:layout_constraintStart_toStartOf="parent"
        app:layout_constraintTop_toTopOf="parent" />

    <EditText
        android:id="@+id/editText"
        android:layout_width="0dp"
        android:layout_height="41dp"
        android:layout_marginStart="8dp"
        android:gravity="center"
        android:hint="请输入用户名"
        android:textColorHint="#f00"
        app:layout_constraintBottom_toBottomOf="@+id/textView"
        app:layout_constraintEnd_toEndOf="parent"
        app:layout_constraintStart_toEndOf="@+id/textView"
        app:layout_constraintTop_toTopOf="@+id/textView" />
</androidx.constraintlayout.widget.ConstraintLayout>
```

在布局文件中指定了TextView控件的位置和相关属性，添加了一个EditText控件。使EditText控件顶部和底部都与TextView对齐，也就是它们的水平中心在同一水平线。EditText控件除了与TextView一样有id属性，表示高和宽的layout_width和layout_height属性，文本属性text，以及textsize等属性外，还有以下常用的属性。

① android:hint属性，用于显示提示信息，如有android:hint="请输入用户名"，当程序运行出现的editText控件中显示“请输入用户名”，开始输入信息时提示自动消失。

② android:minLines属性，表示输入框的最小输入行数。如android:minLines="1"设置单行输入。

③ android:inputType 属性，用于设置输入文本的类型。常用的值有“phone”(电话号码)、“number”（数字）、“textPassword”（文本型密码）、“numberPassword”（数字密码）等。

④ android:enable属性，输入框是否可以编辑。如有android:enabled="false"表示此输入框不可编辑。

⑤ android:gravity属性，用于指定提示内容对齐方式。如本例中代码：android:gravity="center"表示居中对齐。

⑥ android:textColorHint属性，用于指定提示文本的颜色。

第三步：运行程序

运行程序，效果如图3-14所示。当在EditText中输入文本时，提示信息“请输入用户名”消失。

3.2.3 Button控件

Button控件是我们常说的命令按钮，它继承自TextView类，既可以显示文本，又可以显示图片，常用来响应用户的点击。

下面我们通过一个应用介绍Button的使用。要求在默认的布局中添加一个Button，单击Button在TextView控件中显示“您点击了命令按钮”，运行效果如图3-16所示，操作步骤如下。

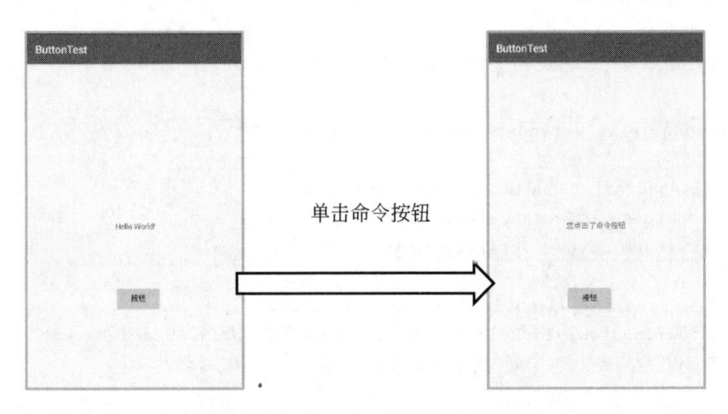

图3-16　Button应用运行效果

第一步：新建项目

创建一个项目，名为ButtonTest，包名为com.example.buttontest。

第二步：用户界面设计

在项目默认的用户界面中添加一个 Button 命令按钮，添加约束使其在 TextView 控件下面 100dp 处，修改布局文件 activity_main.xml，代码如下所示。

```
<?xml version="1.0" encoding="utf-8"?>
<androidx.constraintlayout.widget.ConstraintLayout
    xmlns:android="http://schemas.android.com/apk/res/android"  …>
    <TextView
        android:id="@+id/textView1"
        android:layout_width="wrap_content"
        android:layout_height="wrap_content"
        android:text="Hello World!"
        app:layout_constraintBottom_toBottomOf="parent"
        app:layout_constraintLeft_toLeftOf="parent"
        app:layout_constraintRight_toRightOf="parent"
        app:layout_constraintTop_toTopOf="parent" />
    <Button
        android:layout_width="wrap_content"
        android:layout_height="wrap_content"
        android:text="按钮"
        android:onClick="buttonClick"
        android:layout_marginTop="100dp"
        app:layout_constraintStart_toStartOf="@+id/textView1"
        app:layout_constraintTop_toBottomOf="@+id/textView1" />
</androidx.constraintlayout.widget.ConstraintLayout>
```

布局文件已经存在，TextView 添加 id 属性 textView1，添加一个 Button 控件，其 android:text 属性设置为“按钮”，使用 android:layout_marginTop 属性指定 Button 按钮与 TextView 控件间距为 100dp，并且与 TextView1 左对齐，在其下方。指定设置 onClick 属性为“buttonClick”，我们可以在 Activity 中定义专门的方法[buttonClick()方法]来实现 Button 控件的点击事件。

第三步：实现界面交互

在 MainActivity.java 中实现交互，具体代码如下所示。

```
public class MainActivity extends AppCompatActivity {
    @Override
    protected void onCreate(Bundle savedInstanceState) {
        super.onCreate(savedInstanceState);
        setContentView(R.layout.activity_main);
    }
    public void buttonClick(View view) {
        TextView tv=findViewById(R.id.textView1);//使用 findViewById()方法得到 TextView 的实例
          tv.setText("您点击了命令按钮");//设置布局中 TextView1 的文本内容
     }
}
```

在上面的代码中 MainActivity 类中增加了一个 buttonClick()方法，就是单击命令按钮需要执行的逻辑。在此方法中声明一个 TextView 对象 tv，使用 findViewById()得到 TextView 对象的实例（布局文件中 id 为 textView1 的 TextView 控件），在 Java 文件中要使用布局上的某个控件，需要先使用 findViewById()方法初始化，然后才能使用。当单击命令“按钮”时，设置 TextView 对象 tv 上显示的内容为“您点击了命令按钮”。这里使用到了 setText()方法，setText()方法的作用是设置对象上显示的文本内容。如：

```
tv.setText("您点击了命令按钮");
```

作用是设置布局上的 TextView 上显示的文本为括号中的字符串“您点击了命令按钮”。

对应的还有 getText()方法，它的作用是获取对象文本内容。如获取 tv 中的文本赋值给 String 类型变量 str，写法如下所示：

```
String str=tv.getText().toString();
```

第四步：运行程序

运行程序，运行效果如图 3-16 所示，当单击命令“按钮”时，界面中“Hello World!”变为了“您点击了命令按钮”。

对于 Button 控件最常用的就是单击事件，命令按钮 Button 控件的单击事件的实现方法有三种。

第一种方法 设置 onClick 属性，然后在 Java 文件中实现对应的方法，也就是上面实现按钮单击所用的方法。主要分两步：

① 在布局中添加 Button 控件，设置 onClick 属性。

我们在布局文件 activity_main.xml 文件中加入 Button 控件，具体代码如下：

```
<Button
  …
      android:onClick="buttonClick"
   …/>
```

② 在 Java 文件中实现单击事件。

可以直接在 MainActivity.Java 文件中实现以 onClick 的属性值命名的方法，也可以在布局文件 activity_main.xml 中将鼠标放置在 onClick 属性值中，按“Alt+Enter”键，选择提示“Create‘button Click（View）’in‘MainActivity’”即可打开 MainActivity.java 文件，并且增加如下所示代码框架：

```
   public void buttonClick(View view) {
}
```

在 buttonClick 方法体中添加待处理的逻辑就好。

第二种方法 使用内部匿名类的方法实现单击事件。

使用内部匿名类的方法实现 Button 点击事件，首先需要在布局文件中为 Button 控件设置 id 属性，例如 android:id="@+id/button1"。

在 MainActivity 的 onCreate()方法中编写如下代码，可以实现上面展示的效果。

```
public class MainActivity extends AppCompatActivity {
    private Button btn;
    @Override
    protected void onCreate(Bundle savedInstanceState) {
        super.onCreate(savedInstanceState);
        setContentView(R.layout.activity_main);
        btn = findViewById(R.id.button1);
        btn.setOnClickListener(new View.OnClickListener() {
            @Override
```

```
            public void onClick(View v) {
                TextView tv = findViewById(R.id.textView1);
                tv.setText("您点击了命令按钮");
            }
        });
    }
}
```

在这种方法中，编写按钮对象 btn 的 setOnClickListener()方法，此方法的参数是一个内部匿名类，如果监听到按钮被点击，程序将会执行内部类中的 onClick()方法中的代码逻辑。

第三种方法 在当前 Activity 中实现 OnClickListener 接口。

这种方法常用在命令按钮较多的情况下，具体操作为：

先在 Activity 中实例化 Button 对象 btn，然后调用 setOnClickListener()方法，方法的参数为 this，即为 btn.setOnClickListener(this)；选择“this”按“Alt+Enter”键，在弹出的快捷菜单中选择“Make 'MainActivity' implement 'android.view.View.setOnClickListener'”，重写 onClick()方法，MainActivity.java 代码如下所示：

```
public class MainActivity extends AppCompatActivity implements View.OnClickListener {
    private Button button;
    @Override
    protected void onCreate(Bundle savedInstanceState) {
        super.onCreate(savedInstanceState);
        setContentView(R.layout.activity_main);
        button = findViewById(R.id.button);
        button.setOnClickListener(this);
    }

    @Override
    public void onClick(View v) {
        TextView tv = findViewById(R.id.textView1);
        tv.setText("您点击了命令按钮");//设置布局中 TextView1 文本内容
    }
}
```

使用这种方法，MainActivity 通过实现 View.OnClickListener 接口中的 onClick()方法来设置点击事件，需要注意的是在实现 onClick()方法之前，必须调用 Button 控件的 setOnClickListener()方法设置点击监听事件，否则，Button 控件的点击不会生效。然后在 onClick()方法中编写点击按钮要执行的代码即可。这种方法一般用在按钮比较多的情况，如果有多个命令按钮，可以通过 getId()方法获取点击按钮的 id，然后在 case 语句中选择执行的逻辑，修改上面的代码如下所示：

```
    @Override
        public void onClick(View v) {
            switch (v.getId()) {
                case R.id.button1:
                    tv.setText("您点击了命令按钮 1");//设置布局中 TextView1 文本内容
                    break;
                case R.id.button2:
                    tv.setText("您点击了命令按钮 2");//设置布局中 TextView1 文本内容
                    break;
                ......
```

```
        }
    }
}
```

3.2.4　ImageView 控件

ImageView

ImageView 是用于在界面上显示图片的控件，它继承自 View 类，使用此控件可以从各种来源加载图像（如资源库或网络），并提供缩放、裁剪、着色（渲染）等功能，从而美化用户界面。

ImageView 控件的资源存放在 res/drawable 目录下，常用的属性如下。

① android:background，用来设置 ImageView 控件的背景图片。

② android:src，用来设置 ImageView 控件需要显示的图片资源。

background 和 src 都可以显示图片，backgroud 用来显示控件的背景图片，src 用来显示控件的前景图片。区别在于使用 background，图片会根据控件大小进行伸缩，使用 src 会以原图的大小显示。

下面通过实现如图 3-17 所示效果，演示 ImageView 的使用方法。

分析　实现图 3-17 效果，程序运行 ImageView 控件显示一幅图，单击命令按钮更换 ImageView 中的图片，这就要在代码中设置 ImageView 显示的图片，实现步骤如下所示。

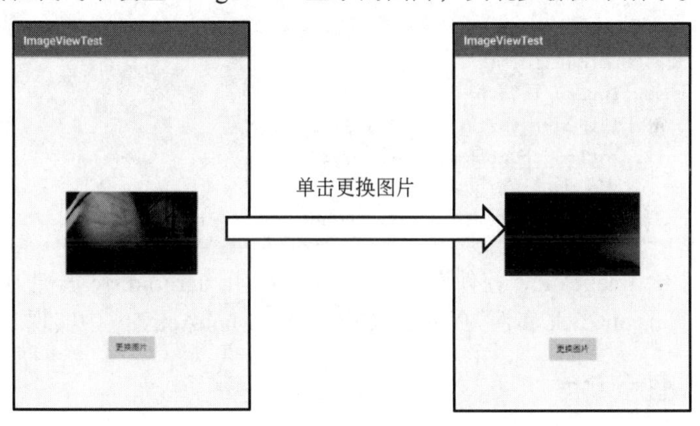

图 3-17　ImageView 应用

第一步：新建项目

创建一个项目，名为 ImageView，项目包名为 com.example.imageview。

第二步：准备图片

将准备好的图片 tu1.png 和 tu2.png 复制到 res/drawable 目录下。

第三步：设计用户界面

在默认的布局文件中，删除 TextView 控件，添加 ImageView 控件，设置 android:src 属性为要显示的图片。添加一个命令按钮，位于 ImageView 控件的下方，与 ImageView 控件居中对齐。

activity_main.xml 参考代码，如下所示。

```
<?xml version="1.0" encoding="utf-8"?>
<androidx.constraintlayout.widget.ConstraintLayout xmlns:android="http://schemas.android. com/apk/
```

```
res/android"
    xmlns:app="http://schemas.android.com/apk/res-auto"
    xmlns:tools="http://schemas.android.com/tools"
    android:layout_width="match_parent"
    android:layout_height="match_parent"
    tools:context=".MainActivity">

    <ImageView
        android:id="@+id/iv1"
        android:layout_width="wrap_content"
        android:layout_height="wrap_content"
        android:src="@drawable/tu1"
        app:layout_constraintBottom_toBottomOf="parent"
        app:layout_constraintLeft_toLeftOf="parent"
        app:layout_constraintRight_toRightOf="parent"
        app:layout_constraintTop_toTopOf="parent" />

    <Button
        android:layout_width="wrap_content"
        android:layout_height="wrap_content"
        android:layout_marginTop="100dp"
        android:text="更换图片"
        android:onClick="buttonClick"
        app:layout_constraintEnd_toEndOf="@+id/iv1"
        app:layout_constraintStart_toStartOf="@+id/iv1"
        app:layout_constraintTop_toBottomOf="@+id/iv1" />

</androidx.constraintlayout.widget.ConstraintLayout>
```

在上面的代码中给 ImageView 控件设置 id 为 iv1，使用 android:src 属性为其添加图片 tu1;我们给 Button 控件设置了 onClick 属性为“buttonClick”, 在 MainActivity 中实现 buttonClick()方法。

第四步：实现图片切换

除了在布局文件中静态地设置显示的图片，还可以在代码文件中动态地修改 ImageView 控件中显示的图片。修改 MainActivity 代码如下所示：

```
public class MainActivity extends AppCompatActivity {
    private ImageView imageView;
    @Override
    protected void onCreate(Bundle savedInstanceState) {
        super.onCreate(savedInstanceState);
        setContentView(R.layout.activity_main);
        imageView = findViewById(R.id.iv1);
    }
    public void buttonClick(View view) {//按钮单击事件
        imageView.setImageResource(R.drawable.tu2);
    }
}
```

在上面的代码中我们先在 onCreate()方法中初始化 ImageView 控件，然后实现了按钮的单击事件 buttonClick()方法，在此方法中通过 setImageResource()方法指定 ImageView 控件显示 drawable

文件夹下的图片资源，这也是我们常用的动态设置 ImageView 控件图片的方法。

第五步：运行程序

运行程序，运行效果如图 3-17 所示，单击命令按钮，ImageView 显示的图片发生变化。

3.2.5　用户登录功能的设计与实现

我们已经学习了常用布局和基础控件，下面以实现如图 3-18 所示的运行效果对 Android 基础控件综合应用。

图 3-18　用户登录运行效果

分析　界面上有一个 ImageView 控件用来显示一个图片，两个 TexView 控件分别用来显示用户名和密码，两个 EditText 控件，提示内容分别为“请输入用户名”和“请输入用户密码”，两个命令按钮“登录”和“取消”。并且对按钮进行判定，当用户名或密码为空弹出消息框提示，如果用户名和密码输入正确，提示“欢迎进入本系统”，否则提示“输入有误”；点击“取消”关闭运行界面。项目实施过程如下所示。

第一步：新建项目

新建项目，名为 UserLogin，项目包名为 com.example.userlogin。

第二步：准备工作

将准备好的图片 studio.png 复制到 res/drawble/目录下。

第三步：用户界面设计

在 activity_main 布局文件中，指定布局背景颜色，拖曳控件 ImageView、TextView、EditText、Button 命令按钮到布局，给 ImageView 添加图片，修改两个 TextView 的 text 属性分别为“用户名”和“密码”，两个输入框设置 hint 属性分别为“请输入用户名”和“请输入用户密码”，等等，为所有对象添加约束，预览效果如图 3-19 所示界面。

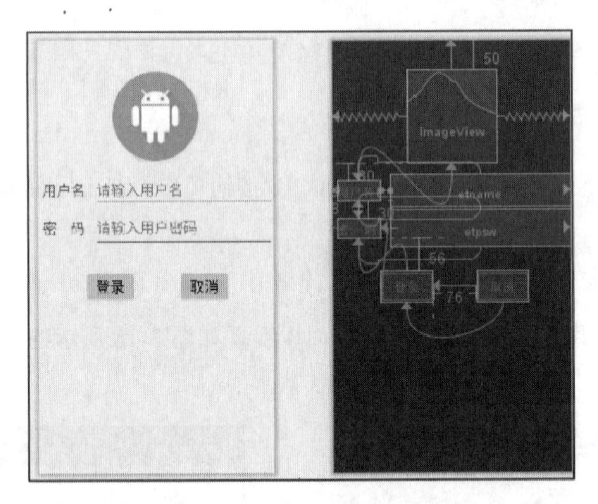

图 3-19　用户登录界面设计效果

布局文件 activity_main 的参考代码如下：

```
<?xml version="1.0" encoding="utf-8"?>
<androidx.constraintlayout.widget.ConstraintLayout xmlns:android="http://schemas.android.com/apk/
res/android"
    xmlns:app="http://schemas.android.com/apk/res-auto"
    xmlns:tools="http://schemas.android.com/tools"
    android:id="@+id/relativeLayout"
    android:layout_width="match_parent"
    android:layout_height="match_parent"
    android:background="#F5F5F5"
    tools:context=".MainActivity">

    <ImageView
        android:id="@+id/imageView"
        android:layout_width="wrap_content"
        android:layout_height="wrap_content"
        android:layout_marginTop="50dp"
        android:src="@drawable/android_img"
        app:layout_constraintEnd_toEndOf="parent"
        app:layout_constraintStart_toStartOf="parent"
        app:layout_constraintTop_toTopOf="parent" />

    <TextView
        android:id="@+id/textView"
        android:layout_width="wrap_content"
        android:layout_height="wrap_content"
        android:layout_marginStart="8dp"
        android:layout_marginTop="30dp"
        android:text="用户名"
        android:textSize="25sp"
        app:layout_constraintStart_toStartOf="parent"
        app:layout_constraintTop_toBottomOf="@+id/imageView" />

    <EditText
        android:id="@+id/etname"
        android:layout_width="0dp"
```

```
        android:layout_height="55dp"
        android:layout_marginStart="16dp"
        android:hint="请输入用户名"
        android:inputType="textPersonName"
        android:textSize="25sp"
        app:layout_constraintBottom_toBottomOf="@+id/textView"
        app:layout_constraintEnd_toEndOf="parent"
        app:layout_constraintStart_toEndOf="@+id/textView"
        app:layout_constraintTop_toTopOf="@+id/textView" />

    <TextView
        android:id="@+id/textView2"
        android:layout_width="wrap_content"
        android:layout_height="wrap_content"
        android:layout_marginTop="30dp"
        android:text="密    码"
        android:textSize="25sp"
        app:layout_constraintStart_toStartOf="@+id/textView"
        app:layout_constraintTop_toBottomOf="@+id/textView" />

    <EditText
        android:id="@+id/etpsw"
        android:layout_width="0dp"
        android:layout_height="64dp"
        android:hint="请输入用户密码"
        android:inputType="textPassword"
        android:textSize="25sp"
        app:layout_constraintBottom_toBottomOf="@+id/textView2"
        app:layout_constraintEnd_toEndOf="parent"
        app:layout_constraintStart_toStartOf="@+id/etname"
        app:layout_constraintTop_toTopOf="@+id/textView2" />

    <Button
        android:id="@+id/btlogin"
        android:layout_width="wrap_content"
        android:layout_height="wrap_content"
        android:layout_marginTop="56dp"
        android:text="登录"
        android:textSize="25sp"
        app:layout_constraintStart_toEndOf="@+id/textView2"
        app:layout_constraintTop_toBottomOf="@+id/textView2" />

    <Button
        android:id="@+id/btexit"
        android:layout_width="wrap_content"
        android:layout_height="wrap_content"
        android:layout_marginStart="76dp"
        android:text="取消"
        android:textSize="25sp"
        app:layout_constraintBottom_toBottomOf="@+id/btlogin"
        app:layout_constraintStart_toEndOf="@+id/btlogin" />

</androidx.constraintlayout.widget.ConstraintLayout>
```

在布局代码中我们看到，密码输入的 EditText 控件，输入密码时不能显示输入的内容，所以指定密码输入的控件 etpsw 的 android:inputType="textPassword"，即输入文本型密码，显示“.”。

第四步：实现用户交互

在 MainActivity 中声明并初始化所用对象，获取用户名和密码的内容，判定用户名和密码不能为空，如果用户名和密码正确给出提示，否则给出错误的提示。MainActivity 中的具体代码如下：

```
public class MainActivity extends AppCompatActivity implements View.OnClickListener {
    private EditText et_name,et_psw;
    private Button bt_login, bt_exit;
    @Override
    protected void onCreate(Bundle savedInstanceState) {
        super.onCreate(savedInstanceState);
        setContentView(R.layout.activity_main);
        et_name = findViewById(R.id.etname);
        et_psw = findViewById(R.id.etpsw);
        bt_login = findViewById(R.id.btlogin);
        bt_exit = findViewById(R.id.btexit);
        bt_login.setOnClickListener(this);
        bt_exit.setOnClickListener(this);
    }
    @Override
    public void onClick(View v) {
        switch (v.getId()){
          case R.id.btlogin:
             String strname,strpsw;
             strname=et_name.getText().toString();
             strpsw=et_psw.getText().toString();
             if(strname.equals("")||strpsw.equals("")){//判定用户名和密码是否为空
                 Toast.makeText(this,"用户名或密码不能为空",Toast.LENGTH_LONG).show();
             }else if(strname.equals("admin")&&strpsw.equals("123456")){
                 Toast.makeText(this,"欢迎进入本系统！！！",Toast.LENGTH_LONG).show();
                     }else{
                   Toast.makeText(this,"输入有误",Toast.LENGTH_LONG).show();
              }
              break;
          case R.id.btexit:
             Toast.makeText(this,"再见",Toast.LENGTH_LONG).show();
             finish();//关闭当前界面
             break;
        }
    }
}
```

在 MainActivty 中，我们定义 EditText 和 Button 对象，使用 findViewById()方法获取到布局文件 activity_main 中定义的元素 etname、etpsw、btlogin、btexit，得到 EditText 和 Button 的实例。按钮的单击事件是通过实现 MainActivity 的 OnClickListener 接口来完成的。在 onClick()方法中使用 v.getId()获取布局中控件的 id，使用 swicth 语句根据 v.getId()获取的值，选择执行控件的单击事件。单击事件中使用 getText().toString()方法获取 EditText 对象中的内容并转换成字符串赋值给了字符串变量，单击命令按钮会弹出信息提醒，使用 Toast 实现。

Toast 是 Android 中的一种简易的消息提示框。它不会获得焦点，无法被用户单击。Toast 显示的时间有限，会根据用户设置的显示时间自动消失。Toast 的一般用法格式：

```
Toast.makeText(Context context,CharSequence text,int duration);
```

使用 Toast 的静态方法 makeText()实现，其中第一个参数是 Context，是 Toast 显示在哪个上下文，通常是当前 Activity，这里使用 this，代表当前的 Activity 即 MainActivity；第二个参数是要显示的文本内容；第三个参数是显示的时长，有两个常量可以使用 Toast.LENGTH_LONG 和 Toast.LENGTH_SHORT 分别表示长时间显示和短时间显示。

在取消按钮中使用了 finish()方法，它的作用是销毁当前活动，效果和按下 Back 键是一样的。

第五步：运行程序

运行程序，效果如图 3-18 所示，输入用户名和密码如果正确提示“欢迎进入本系统！！！”如果不正确，则提示“输入有误”，如果用户名或者密码没有输入，则提示“用户名或密码不能为空”。这里我们只是简单获取输入框中的数据和判定内容的简单操作，真正在实际项目中还需进一步完善。

3.2.6　RadioButton 和 RadioGroup 控件

RadioButton

RadioButton 单选按钮，常用于多个选项中单选的情况，例如性别、职业、单选题等场景。RadioButton 有两种状态，“选中”和“未选中”，使用 Android:checked 属性来指定。单选按钮在使用时必须与 RadioGroup 结合。下面通过实现如图 3-20 所示的运行效果，来学习 RadioButton 和 RadioGroup 的使用。

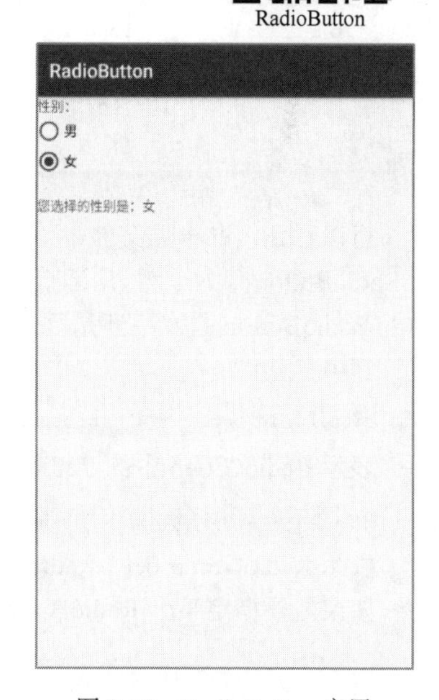

图 3-20　RadioButton 应用

分析　界面显示性别单选按钮，当点击“男”显示“您选择的性别是：男”，如果点击了“女”，则显示“您选择的性别是：女”。这里使用 RadioGroup 和 RadioButton 控件来完成，实现过程如下。

第一步：创建工程

创建一个项目 RadioButton，项目包名为 com.example.radiobutton。

第二步：设计用户界面

修改 activity_main.xml，代码如下：

```
<?xml version="1.0" encoding="utf-8"?>
<RelativeLayout xmlns:android="http://schemas.android.com/apk/res/android"
    xmlns:app="http://schemas.android.com/apk/res-auto"
    xmlns:tools="http://schemas.android.com/tools"
    android:layout_width="match_parent"
    android:layout_height="match_parent"
    tools:context=".MainActivity">
    <TextView
```

```
        android:id="@+id/tv1"
        android:layout_width="wrap_content"
        android:layout_height="wrap_content"
        android:text=" 性别: "/>
    <RadioGroup
        android:id="@+id/rgsex"
        android:layout_width="match_parent"
        android:layout_height="wrap_content"
        android:layout_below="@+id/tv1">
        <RadioButton
            android:id="@+id/rbname"
            android:layout_width="wrap_content"
            android:layout_height="wrap_content"
            android:text="男"/>
        <RadioButton
            android:id="@+id/rbfemale"
            android:layout_width="wrap_content"
            android:layout_height="wrap_content"
            android:text="女"/>
    </RadioGroup>
    <TextView
        android:id="@+id/tvresult"
        android:layout_width="wrap_content"
        android:layout_height="wrap_content"
        android:layout_below="@+id/rgsex"
        android:layout_alignLeft="@+id/rgsex"
        android:layout_marginTop="20dp"/>

</RelativeLayout>
```

可以看到，在布局文件中使用 TextView 控件对单选按钮做说明，使用 RadioGroup 控件将两个 RadioButton 分为一组。从运行效果图 3-20 可以看到两个 RadioButton 按钮垂直排列。RadioGroup 中的 RadioButton 是水平还是垂直排列，可以通过设置 RadioGroup 的 android:orientation 属性来实现，其中：

```
android:orientation="horizontal"
```

表示 RadioGroup 中的 RadioButton 单选按钮水平排列；

```
android:orientation="vertical"
```

表示 RadioGroup 中的 RadioButton 单选按钮垂直排列，这也是默认的排列方式。

如果需要指定某个 RadioButton 为默认选中项，需设置它的 android:checked 属性 true。

第三步：实现单选功能

在 MainActivity 中实现对所选 RadioButton 的响应。MainActivity 类中代码如下：

```
public class MainActivity extends AppCompatActivity {

    private RadioGroup radioGroup;
    private TextView tvresult;
    @Override
    protected void onCreate(Bundle savedInstanceState) {
        super.onCreate(savedInstanceState);
```

```
        setContentView(R.layout.activity_main);
        radioGroup = findViewById(R.id.rgsex);
        tvresult = findViewById(R.id.tvresult);
        radioGroup.setOnCheckedChangeListener(new RadioGroup.OnCheckedChangeListener() {
            @Override
            public void onCheckedChanged(RadioGroup group, int checkedId) {
                if(checkedId==R.id.rbname){
                    tvresult.setText("您选择的性别是：男");
            }else{
                    tvresult.setText("您选择的性别是：女");
                }
            }
        });
    }
}
```

对于判定单选按钮组中的单选按钮是否按下，需要使用 RadioGroup 对象的 setOnChecked-ChangeListener()方法来监听布局中的 RadioButton 选中状态是否发生改变，通过实现 onCheckedChanged()方法获取被点击的控件，以及点击此控件要执行的操作。

第四步：运行程序

运行程序，如果选择“男”单选按钮，则显示 “您选择的性别是：男”,如果选择“女”单选按钮，则显示“您选择的性别是：女”，效果如图 3-20 所示。

如果需要在程序运行开始就有单选按钮处于选中状态，则可以在布局中设置 RadioButton 的属性 android:checked="true"，在程序运行中如果当某种情况发生需要设置单选按钮为选中状态，需要使用 setChecked(true)方法，例如 radioButton1.setChecked(true)将 radioButton1 设置为了选中状态。

3.2.7 CheckBox 控件

CheckBox 复选框继承自 CompoundButton 类，具备选中和非选中两种状态，与 RadioButton 控件类似，只是不用进行分组，可以进行多选。下面通过实现图 3-21 所示的效果，来学习 CheckBox 的使用。

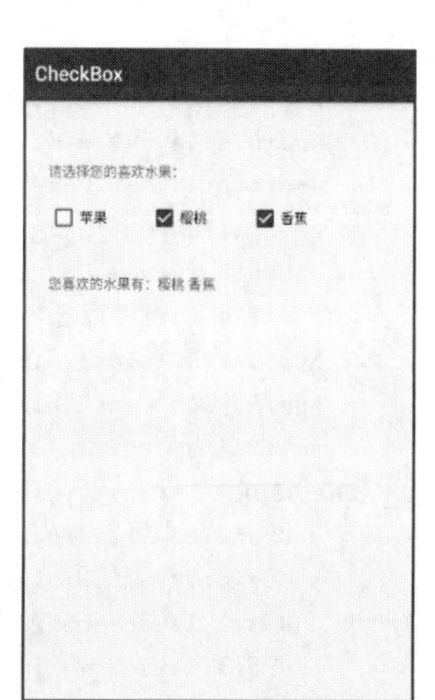

图 3-21 CheckBox 使用

分析 根据运行效果，这里使用了 CheckBox 控件，我们要解决的问题是根据复选框 CheckBox 是否选中来决定显示的内容，即当复选框选中则添加选项，取消选中则要删除当前复选框的内容。实现过程如下所示。

第一步：新建项目

创建一个项目 CkeckBox；项目包名为 com.example.checkbox。

第二步：设计用户界面

CheckBox

在 activity_main 布局中默认有一个 TextView,调整其位置，拖入三个复选框和一个 Text View，用来设置选项内容和显示选择结构，界面使用默认的布局 ConstraintLayout，调整控件位置，为其添加约束，参考界面如图 3-22 所示。

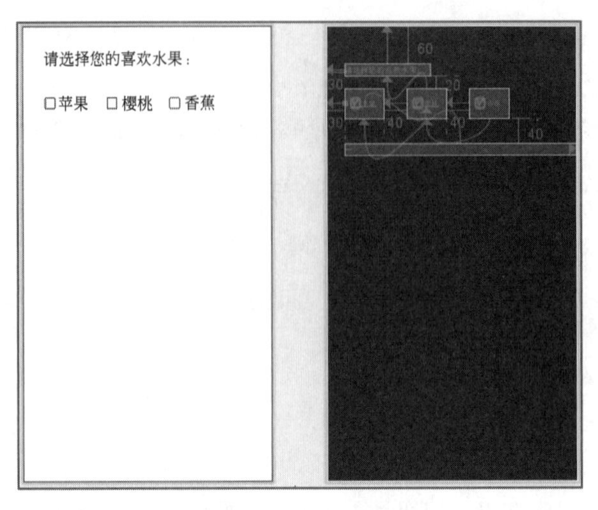

图 3-22 界面布局效果图

布局界面 activity_main.xml 的参考代码如下所示：

```
<?xml version="1.0" encoding="utf-8"?>
<androidx.constraintlayout.widget.ConstraintLayout  xmlns:android="http://schemas.android.com/apk/
res/android"
    xmlns:app="http://schemas.android.com/apk/res-auto"
    xmlns:tools="http://schemas.android.com/tools"
    android:layout_width="match_parent"
    android:layout_height="match_parent"
    tools:context=".MainActivity">

    <TextView
        android:id="@+id/textView"
        android:layout_width="wrap_content"
        android:layout_height="wrap_content"
        android:layout_marginStart="30dp"
        android:layout_marginTop="60dp"
        android:text="请选择您的喜欢水果："
        app:layout_constraintStart_toStartOf="parent"
        app:layout_constraintTop_toTopOf="parent" />

    <CheckBox
        android:id="@+id/cbapple"
        android:layout_width="wrap_content"
        android:layout_height="wrap_content"
        android:layout_marginStart="30dp"
        android:text="苹果 "
        app:layout_constraintBaseline_toBaselineOf="@+id/cbcherry"
        app:layout_constraintStart_toStartOf="parent" />

    <CheckBox
        android:id="@+id/cbcherry"
        android:layout_width="wrap_content"
        android:layout_height="wrap_content"
        android:layout_marginStart="40dp"
        android:layout_marginTop="20dp"
```

```
        android:text="樱桃 "
        app:layout_constraintStart_toEndOf="@+id/cbapple"
        app:layout_constraintTop_toBottomOf="@+id/textView" />

    <CheckBox
        android:id="@+id/cbbanana"
        android:layout_width="wrap_content"
        android:layout_height="wrap_content"
        android:layout_marginStart="40dp"
        android:text="香蕉 "
        app:layout_constraintBottom_toBottomOf="@+id/cbcherry"
        app:layout_constraintStart_toEndOf="@+id/cbcherry" />

    <TextView
        android:id="@+id/tvresult"
        android:layout_width="0dp"
        android:layout_height="wrap_content"
        android:layout_marginTop="40dp"
        android:layout_marginEnd="1dp"
        app:layout_constraintEnd_toEndOf="parent"
        app:layout_constraintStart_toStartOf="@+id/cbapple"
        app:layout_constraintTop_toBottomOf="@+id/cbapple" />
</androidx.constraintlayout.widget.ConstraintLayout>
```

第三步：实现用户名交互

在 MainActivity 中选择喜欢的水果显示在下面选择结果中。实现代码如下：

```
public class MainActivity extends AppCompatActivity implements CompoundButton.OnCheckedChangeListener {
    private CheckBox cbapple, cbcherry, cbbanana;
    private TextView tvresult;
    private String str = "您喜欢的水果有：";
    private String strfruits;

    @Override
    protected void onCreate(Bundle savedInstanceState) {
        super.onCreate(savedInstanceState);
        setContentView(R.layout.activity_main);
        tvresult = findViewById(R.id.tvresult);
        cbapple = findViewById(R.id.cbapple);
        cbcherry = findViewById(R.id.cbcherry);
        cbbanana = findViewById(R.id.cbbanana);
        cbapple.setOnCheckedChangeListener(this);
        cbcherry.setOnCheckedChangeListener(this);
        cbbanana.setOnCheckedChangeListener(this);
    }

    @Override
    public void onCheckedChanged(CompoundButton buttonView, boolean isChecked) {
        String strfruit = buttonView.getText().toString();
        if (isChecked) {
            if (!strfruits.contains(strfruit)) {
                    strfruits = strfruits + strfruit;
```

```
                tvresult.setText(str + strfruits);
            }
        } else {
            if (strfruits.contains(strfruit)) {
                strfruits = strfruits.replace(strfruit, "");
                tvresult.setText(str + strfruits);
            }
        }
    }
}
```

在类 MainActivity 中实现接口 OnCheckedChangeListener，从上面的代码中可以看到，在MainActivity 的 onCreate()方法中首先对控件使用 findViewById()方法进行初始化，然后为每个复选框添加 setOnCheckedChangeListener()监听，实现 onCheckedChanged()方法，在这个方法中有两个参数，第一个参数指的是所点击的 CheckBox，第二个参数是 boolean 型，表示当前 CheckBox 状态，选中为 true，否则为 false。

第四步：运行程序

运行程序，当表示某种水果的复选框被选中，则水果显示在下面，如果取消选中则从喜欢的水果中删除，运行效果如图 3-21 所示。

3.2.8 Spinner 控件

Spinner 下拉列表框用于提供一系列的列表项供用户选择，使用 Spinner 可以方便用户使用，还可以节省手机的屏幕空间。下面通过如图 3-23 所示实例讲解 Spinner 控件的使用。

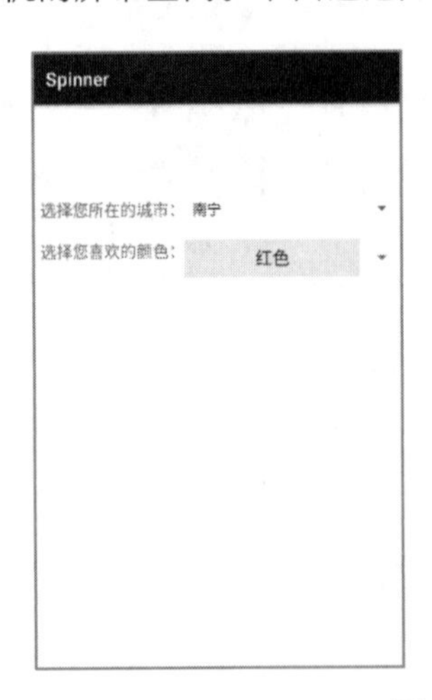

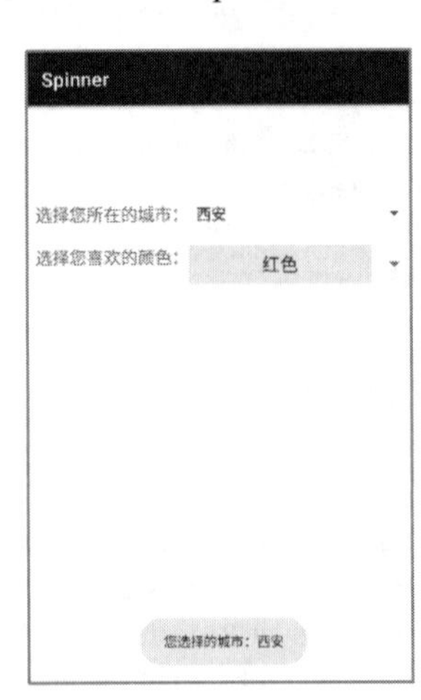

图 3-23　Spinner 控件

分析　如图 3-23 运行所示界面上使用了两个 Spinner 控件，一个显示样式选择了系统默认的，另一个是自定义的样式，实现这个实例的过程如下。

第一步：新建项目

创建一个项目 Spinner，项目包名为 com.example.spinner。

第二步：在字符串资源文件中添加数据

打开 res/values 目录下的 strings.xml 文件，在<resouces></resources>标签中添加一个<string-array></ string-array >标记，为第一个 Spinner 控件创建数据源，strings 的代码如下所示。

```
<resources>
    <string name="app_name">Spinner</string>
    <string-array name="arrCity">
        <item>南宁</item>
        <item>上海</item>
        <item>西安</item>
        <item>深圳</item>
    </string-array>
    <string name="strprompt">选择您喜欢的颜色</string>
</resources>
```

第三步：创建 Spinner 对应的布局文件

第二个 Spinner 控件使用了自定义布局，所以在 res/layout 目录下创建一个布局文件 color_layout.xml，为第二个 Spinner 指定样式代码如下。

```
<?xml version="1.0" encoding="utf-8"?>
<LinearLayout xmlns:android="http://schemas.android.com/apk/res/android"
    android:layout_width="match_parent"
    android:layout_height="match_parent"
    android:orientation="vertical">
    <TextView
        android:id="@+id/text1"
        android:layout_width="match_parent"
        android:layout_height="40dp"
        android:gravity="center"
        android:textSize="19sp"
        android:textColor="#00f"
        android:background="#E8F7E7"        />
</LinearLayout>
```

第四步：用户界面设计

修改布局文件 activity_main 整体布局为 RelativeLayout 相对布局，其参考代码如下。

```
<?xml version="1.0" encoding="utf-8"?>
<RelativeLayout xmlns:android="http://schemas.android.com/apk/res/android"
    xmlns:app="http://schemas.android.com/apk/res-auto"
    xmlns:tools="http://schemas.android.com/tools"
    android:layout_width="match_parent"
    android:layout_height="match_parent"
    tools:context=".MainActivity">
    <TextView
        android:id="@+id/tv"
        android:layout_width="wrap_content"
```

```
        android:layout_height="wrap_content"
        android:text=" 选择您所在的城市: "
        android:layout_marginTop="100dp"
        android:textSize="18sp" />
    <Spinner
        android:id="@+id/spCity"
        android:layout_width="match_parent"
        android:layout_height="wrap_content"
        android:entries="@array/arrCity"
        android:layout_alignTop="@+id/tv"
        android:layout_toRightOf="@+id/tv"/>
    <TextView
        android:id="@+id/tv2"
        android:layout_width="wrap_content"
        android:layout_height="wrap_content"
        android:text=" 选择您喜欢的颜色: "
        android:textSize="18sp"
        android:layout_marginTop="20dp"
        android:layout_below="@+id/tv"  />
    <Spinner
        android:id="@+id/spColor"
        android:layout_width="match_parent"
        android:layout_height="wrap_content"
        android:layout_alignTop="@+id/tv2"
        android:layout_toRightOf="@+id/tv2"/>
</RelativeLayout>
```

在此布局文件中有两个 TextView 分别用来对选择城市和颜色的 Spinner 控件进行说明。在第一个 Spinner 控件中指定 android:entries 属性，这个属性的作用是指定此控件中显示的内容为在 string.xml 中定义的数组资源"@array/arrCity"中的内容。第二个 Spinner 控件 spColor 显示的内容通过代码添加。

第五步：实现界面功能

在 MainActivity 中，主要有两个功能需要实现，一个是为 spColor 添加选择项，另一个是选择 Spinner 控件中的选项，并给出响应。MainActivity 的代码如下所示。

```
public class MainActivity extends AppCompatActivity {
    private Spinner spCity, spColor;

    @Override
    protected void onCreate(Bundle savedInstanceState) {
        super.onCreate(savedInstanceState);
        setContentView(R.layout.activity_main);
        spCity = findViewById(R.id.spCity);
        spColor = findViewById(R.id.spColor);
        ArrayList<String> arrColor = new ArrayList<>();
        arrColor.add("红色");
        arrColor.add("绿色");
        arrColor.add("黄色");
        arrColor.add("蓝色");
        ArrayAdapter<String> arrayAdapter = new ArrayAdapter(this, R.layout.layout, R.id.text1,
```

```
arrColor);
        spColor.setAdapter(arrayAdapter);
        spCity.setSelection(0, true);//禁止 OnItemSelectedListener 默认自动调用一次
         spCity.setOnItemSelectedListener(new AdapterView.OnItemSelectedListener() {
            @Override
            public void onItemSelected(AdapterView<?> parent, View view, int position, long id) {
              Toast.makeText(MainActivity.this, "您选择的城市: " + spCity.getSelectedItem().toString(),
Toast.LENGTH_LONG).show();
            }

            @Override
            public void onNothingSelected(AdapterView<?> parent) {

            }
        });
    }
}
```

在 MainActivity 中定义 ArrayList 对象 arrColor 用来存放要添加到 spColor 控件的列表项。定义下拉列表内容适配器 arrayAdapter，ArrayAdapter()方法原型如下所示：

```
ArrayAdapter(Context context, int resource,int textViewResourceId, List<T> objects)
```

ArrayAdapter()有四个参数，第一个参数是上下文 context，第二个参数是下拉列表的显示样式，使用的是自定义的布局文件 layout.xml，第三个参数是布局中的控件 id,本项目是自定义布局 layout 中的 TextView 控件 text1，第四个参数是所需的数据 arrColor。对 spColor 控件调用 setAdapter()方法设置适配器，将构建好的适配器对象传给 spColor，这样 spColor 和数据之间的关联就建立完成了。对于第一个 Spinner 对象 spCity 调用 setSelection()方法禁止 OnItemSelectedListener 默认自动调用一次。spCity 使用 setOnItemSelectedListener()方法对 spCity 注册监听器，当选择一个列表项时就弹出所选内容，其中 getSelectedItem()方法获取所选的列表项。

第六步：运行程序

运行程序，效果如图 3-23 所示，可以看出两个 Spinner 对象的显示的样式不一样，一个是选择表示城市的 Spinner 对象，城市列表显示的是控件本身的样式，而且数据也是在界面中设置好的；颜色 Spinner 对象是自定义的适配器。点击城市 Spinner 下拉按钮，选择某一个城市会弹出想要的选项。

3.2.9 用户注册

在很多应用中，常用到用户注册的功能，本节通过实现如图 3-24 所示的运行效果学习用户注册功能。

分析 本案例用户界面使用了文本框显示注册信息提示，用到了输入框、单选按钮、复选框和 Spinner 控件，使用命令按钮“注册”将用户注册信息显示在下方的文本框中。

用户注册界面实现过程如下：

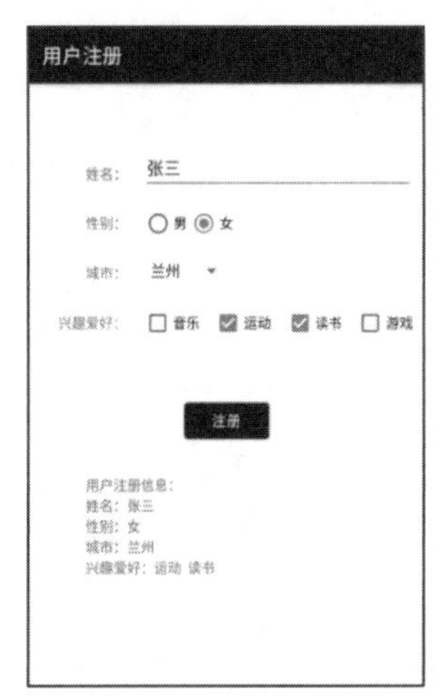

图 3-24 用户注册运行效果

第一步：创建项目

创建一个新的项目 UserRegister，包名为 com.example. userregister。

第二步：准备城市数据

注册界面中用到 Spinner 控件，选择所在城市，这里我们在资源文件 res/values/目录下的 strings.xml 中添加一个<string-array>名为 mycity，定义代码如下所示：

```
<string-array name="myCity">
    <item>北京</item>
    <item>兰州</item>
    <item>西安</item>
    <item>深圳</item>
</string-array>
```

第三步：设计用户界面

选择 activty_main 布局文件的 Design 模式下，在界面上投入四个 TextView 文本控件，分别用来显示“姓名：　”“性别：　”“城市：　”和“兴趣爱好：　”，先指定“姓名：”文本的位置，然后使得其他三个命令按钮与其右对齐，设置每个文本距上一个控件有相同的值。添加 EditText 控件用来输入用户名，在属性窗口中设置 id 为“uname”，hint 属性为“请输入用户名”，设置“姓名：”的位置；添加一个 RadioGroup 控件里面放置两个 RadioButton 控件用来设置性别，id 分别为 rbboy 和 rbgirl，对应的文本 text 内容为“男”和“女”,设置“男”单选按钮的 idchecked 属性值为“true”;添加一个 Spinner 控件 id 为“spcity”，entries 属性值为在 strings.xml 中定义的 myCity;然后添加“音乐”“运动”“读书”和“游戏”四个 CheckBox 控件，设置 id 为“cbmusic”“cbsports”“cbbook”和“cbgame”;添加一个命令按钮，text 属性值为“注册”，onClick 属性值设为 registerClick，即在 MainActivity 中必须实现 registerClick()方法；最后添加一个 TextView 控件用来显示注册信息，其 id 属性设置为 tvresiter，text 属性设置为“用户注册信息”，为所有控件都添加约束，设计结构如图 3-25 所示。

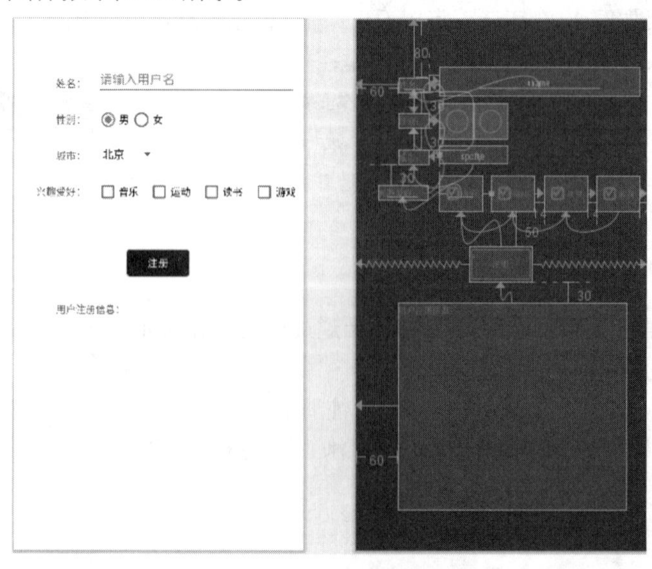

图 3-25　用户注册界面设计图

activity_main 布局文件中设计的代码如下所示：

```
<?xml version="1.0" encoding="utf-8"?>
<androidx.constraintlayout.widget.ConstraintLayout xmlns:android="http://schemas.android. com/apk
/res/android"
    xmlns:app="http://schemas.android.com/apk/res-auto"
    xmlns:tools="http://schemas.android.com/tools"
    android:layout_width="match_parent"
    android:layout_height="match_parent"
    tools:context=".MainActivity">

    <TextView
        android:id="@+id/textView"
        android:layout_width="wrap_content"
        android:layout_height="wrap_content"
        android:layout_marginStart="60dp"
        android:layout_marginTop="80dp"
        android:text="姓名："
        app:layout_constraintStart_toStartOf="parent"
        app:layout_constraintTop_toTopOf="parent" />

    <TextView
        android:id="@+id/textView2"
        android:layout_width="wrap_content"
        android:layout_height="wrap_content"
        android:layout_marginTop="30dp"
        android:text="性别："
        app:layout_constraintEnd_toEndOf="@+id/textView"
        app:layout_constraintTop_toBottomOf="@+id/textView" />

    <TextView
        android:id="@+id/textView3"
        android:layout_width="wrap_content"
        android:layout_height="wrap_content"
        android:layout_marginTop="30dp"
        android:text="城市："
        app:layout_constraintEnd_toEndOf="@+id/textView2"
        app:layout_constraintTop_toBottomOf="@+id/textView2" />

    <TextView
        android:id="@+id/textView4"
        android:layout_width="wrap_content"
        android:layout_height="wrap_content"
        android:layout_marginTop="30dp"
        android:text="兴趣爱好："
        app:layout_constraintEnd_toEndOf="@+id/textView3"
        app:layout_constraintTop_toBottomOf="@+id/textView3" />

    <EditText
        android:id="@+id/uname"
        android:layout_width="282dp"
        android:layout_height="40dp"
        android:layout_marginStart="16dp"
```

```
        android:ems="10"
        android:hint="请输入用户名"
        android:inputType="textPersonName"
        app:layout_constraintBaseline_toBaselineOf="@+id/textView"
        app:layout_constraintStart_toEndOf="@+id/textView" />

    <RadioGroup
        android:id="@+id/radioGroup"
        android:layout_width="wrap_content"
        android:layout_height="wrap_content"
        android:orientation="horizontal"
        app:layout_constraintBottom_toBottomOf="@+id/textView2"
        app:layout_constraintStart_toStartOf="@+id/uname"
        app:layout_constraintTop_toTopOf="@+id/textView2">

        <RadioButton
            android:id="@+id/rbboy"
            android:layout_width="wrap_content"
            android:layout_height="wrap_content"
            android:checked="true"
            android:text="男" />

        <RadioButton
            android:id="@+id/rbgirl"
            android:layout_width="wrap_content"
            android:layout_height="wrap_content"
            android:text="女" />
    </RadioGroup>

    <Spinner
        android:id="@+id/spcitye"
        android:layout_width="0dp"
        android:layout_height="wrap_content"
        app:layout_constraintBottom_toBottomOf="@+id/textView3"
        app:layout_constraintStart_toStartOf="@+id/radioGroup"
        android:entries="@array/myCity"/>

    <CheckBox
        android:id="@+id/cbmusic"
        android:layout_width="wrap_content"
        android:layout_height="wrap_content"
        android:layout_marginEnd="14dp"
        android:text="音乐"
        app:layout_constraintBaseline_toBaselineOf="@+id/textView4"
        app:layout_constraintEnd_toStartOf="@+id/cbsports"
        app:layout_constraintStart_toStartOf="@+id/spcitye" />

    <CheckBox
        android:id="@+id/cbsports"
        android:layout_width="wrap_content"
        android:layout_height="wrap_content"
        android:layout_marginEnd="14dp"
        android:text="运动"
        app:layout_constraintBottom_toBottomOf="@+id/cbmusic"
```

```
        app:layout_constraintEnd_toStartOf="@+id/cbbook" />

    <CheckBox
        android:id="@+id/cbbook"
        android:layout_width="wrap_content"
        android:layout_height="wrap_content"
        android:layout_marginEnd="14dp"
        android:text="读书"
        app:layout_constraintBottom_toBottomOf="@+id/cbsports"
        app:layout_constraintEnd_toStartOf="@+id/cbgame" />

    <CheckBox
        android:id="@+id/cbgame"
        android:layout_width="wrap_content"
        android:layout_height="wrap_content"
        android:layout_marginEnd="11dp"
        android:text="游戏"
        app:layout_constraintBottom_toBottomOf="@+id/cbbook"
        app:layout_constraintEnd_toEndOf="parent" />

    <Button
        android:id="@+id/btnRegister"
        android:layout_width="wrap_content"
        android:layout_height="wrap_content"
        android:layout_marginTop="50dp"
        android:onClick="registerClick"
        android:text="注册"
        app:layout_constraintEnd_toEndOf="parent"
        app:layout_constraintStart_toStartOf="parent"
        app:layout_constraintTop_toBottomOf="@+id/cbmusic" />

    <TextView
        android:id="@+id/tvresiter"
        android:layout_width="321dp"
        android:layout_height="285dp"
        android:layout_marginStart="60dp"
        android:layout_marginTop="30dp"
        android:text="用户注册信息："
        app:layout_constraintStart_toStartOf="parent"
        app:layout_constraintTop_toBottomOf="@+id/btnRegister" />

</androidx.constraintlayout.widget.ConstraintLayout>
```

第四步：实现用户信息注册

在 MainActivity 中获取输入用户名，选择性别，所在城市和兴趣爱好，单击命令按钮后添入注册信息文本中，主要的问题就是获取所输入和所选择的值。实现代码如下所示：

```
public class MainActivity extends AppCompatActivity {

    private EditText etname;
    private RadioButton rbboy, rbgirl;
```

```
    private Spinner spcity;
    private CheckBox cbmusic, cbsports, cbbook, cbgame;
    private TextView tvregister;

    @Override
    protected void onCreate(Bundle savedInstanceState) {
        super.onCreate(savedInstanceState);
        setContentView(R.layout.activity_main);
        etname = findViewById(R.id.uname);
        rbboy = findViewById(R.id.rbboy);
        rbgirl = findViewById(R.id.rbgirl);
        spcity = findViewById(R.id.spcitye);
        cbmusic = findViewById(R.id.cbmusic);
        cbsports = findViewById(R.id.cbsports);
        cbbook = findViewById(R.id.cbbook);
        cbgame = findViewById(R.id.cbgame);
        tvregister = findViewById(R.id.tvresiter);
    }

    public void registerClick(View view) {
        String strregister;
        String strname, strsex, strcity, strhobby;
        strname = "\n 姓名：" + etname.getText().toString();
        if (rbboy.isChecked()) {
            strsex = "\n 性别：" + "男";
        } else {
            strsex = "\n 性别：" + "女";
        }
        strcity = "\n 城市：" + spcity.getSelectedItem().toString();
        strhobby = "\n 兴趣爱好：";
        if (cbmusic.isChecked()) {
            strhobby = strhobby + cbmusic.getText().toString() + "  ";
        }
        if (cbsports.isChecked()) {
            strhobby = strhobby + cbsports.getText().toString() + "  ";
        }
        if (cbbook.isChecked()) {
            strhobby = strhobby + cbbook.getText().toString() + "  ";
        }
        if (cbgame.isChecked()) {
            strhobby = strhobby + cbgame.getText().toString() + "  ";
        }
        strregister = tvregister.getText().toString() + strname + strsex + strcity + strhobby;
        tvregister.setText(strregister);
    }
}
```

从上面的代码中可以看到，在 MainActivity 的 onCreate()方法对所要用到的控件进行初始化。在 registerClick()方法中，使用 getText()方法获取用户名输入框中的属性，使用 isChecked()方法判定单选按钮 rbboy(男)是否选中，如果选中则显示性别为男，否则性别为女；使用 getSelectedItem()方法获取 Spinner 控件中所选择城市的名称；使用 isChecked()方法判定 CheckBox 控件对应的兴趣爱好是否被选中，如果选中则添加进 strhobby 字符串中。最后，使用 setText()方法为用户注册信息显示的控件设置显示文本。

第五步：运行程序

运行程序，运行效果如图 3-26 所示，在界面中输入用户名“张三”，选择性别“女”，选择城市为“兰州”，兴趣爱好选择“运动”和“读书”后，点击“注册”命令按钮，效果如图 3-24 所示。

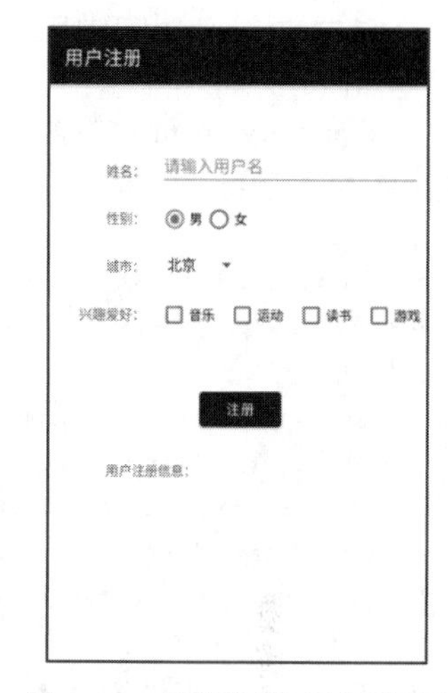

图 3-26　用户注册运行初始图

3.2.10　RecyclerView 控件

RecyclerView 用于在有限的窗口中展示大量数据集，既可以实现数据的纵向排列，也可实现横向排列。为了使用 RecyclerView 控件，我们需要创建一个 Adapter 和一个 LayoutManager。其中 Adapter 继承自 RecyclerView.Adapetr 类，主要用来将数据和布局 item 进行绑定。LayoutManager 是布局管理器，用来设置每一项 view 在 RecyclerView 中的位置布局以及控件 itemview 的显示或者隐藏。当 View 重用或者回收的时候，LayoutManger 都会向 Adapter 请求新的数据来替换原来数据的内容。这种回收重用的机制避免创建很多的 view 或者是频繁调用 findViewById 方法。下面我们通过一个案例来学习 RecyclerView 的使用，效果如图 3-27 所示，可以显示多个数据项，点击某个数据项可以进行相应的操作，实现过程如下。

图 3-27　RecyclerView 的使用效果

分析　在图 3-27 中我们可以看到上面有好多个条目，每个条目有共同的特点，即都包括图片、名字和内容三部分，单击每一个项都有不同反应，我们使用 RecyclerView 控件来实现，操作步骤如下所示。

第一步：新建项目

创建一个新的项目 RecyclerView，包名为 com.example.recyclerview。

第二步：设计用户界面

在 activity_main 中删除自动创建 TextView 控件，从控件组中拖入 RecyclerView 控件，使其充满整个父容器。参考代码如下：

```
<?xml version="1.0" encoding="utf-8"?>
<androidx.constraintlayout.widget.ConstraintLayout  xmlns:android="http://schemas.android.com/apk/
res/android"
    xmlns:app="http://schemas.android.com/apk/res-auto"
    xmlns:tools="http://schemas.android.com/tools"
    android:layout_width="match_parent"
    android:layout_height="match_parent"
    tools:context=".MainActivity">

    <androidx.recyclerview.widget.RecyclerView
        android:id="@+id/recyclerView"
        android:layout_width="match_parent"
        android:layout_height="match_parent" />
</androidx.constraintlayout.widget.ConstraintLayout>
```

注意：在当前 Android Studio 2020.3.1 中可以直接将 RecyclerView 控件拖入界面，但是如果使用 Android Studiod 的版本较早，则需要加入依赖才可以使用。加入依赖的方法如下：

打开“app/build.gradle”文件，在 dependencies 中添加如下内容：

```
dependencies {
    implementation'com.android.support:appcompat-v7:27.1.1'
    implementation'com.android.support:recyclerview-v7:27.1.1'
   ……
}
```

添加完成后，单击“Sync Now”来进行同步。

第三步：准备好界面需要的图片

准备需要显示的图片，复制到 res/drawable 目录中。

第四步：定义实体类

以 RecyclerView 中要显示的内容元素来新建一个实体类，作为 RecyclerView 控件适配器的适配类型。创建一个类先选择当前程序的包，单击右键选择“New”→“Java Class”，弹出如图 3-28 示对话框。

在对话框中输入类名 WeiXin，按回车然后在当前包中就会创建出一个 WeiXin 类，打开文件编写如下代码：

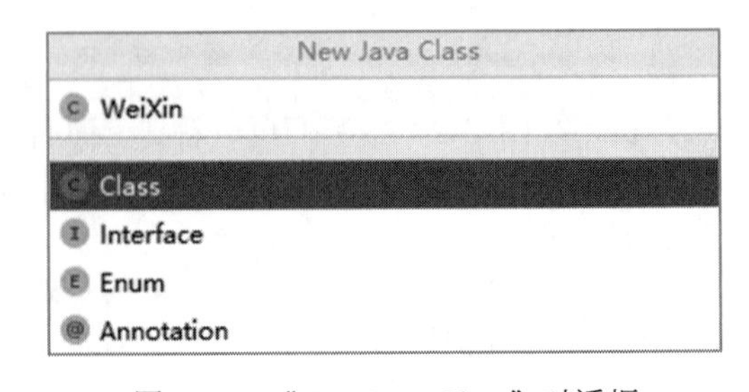

图 3-28 “New Java Class”对话框

```
package com.example.recyclerview;

public class WeiXin {
    int imgId;
    String name;
    String information;

    public WeiXin(int imgId, String name, String information) {
        this.imgId = imgId;
        this.name = name;
        this.information = information;
    }

    public int getImgId() {
        return imgId;
    }

    public String getName() {
        return name;
    }

    public String getInformation() {
        return information;
    }
}
```

在 WeiXin 类中有三个字段，imgId 表示头像对应图片的资源 id，name 表示微信中头像对应的名字，information 表示信息。

第五步：定义 RecyclerView 的子项布局

在 res/layout 目录下新建一个布局 item，作为 RecyclerView 的子项布局。在 item 布局文件中添加一个 ImageView 控件和两个 TextView 控件，具体代码如下所示。

```
<?xml version="1.0" encoding="utf-8"?>
<RelativeLayout xmlns:android="http://schemas.android.com/apk/res/android"
    android:layout_width="match_parent"
    android:layout_height="wrap_content">

    <ImageView
        android:id="@+id/image"
        android:layout_width="80dp"
        android:layout_height="80dp"
        android:layout_margin="10dp" />

    <TextView
        android:id="@+id/tvname"
        android:layout_width="match_parent"
        android:layout_height="wrap_content"
        android:layout_marginTop="30dp"
```

```
        android:layout_toRightOf="@+id/image"
        android:textColor="#000"
        android:textSize="20sp" />

    <TextView
        android:id="@+id/tvinformation"
        android:layout_width="match_parent"
        android:layout_height="wrap_content"
        android:layout_below="@+id/tvname"
        android:layout_alignLeft="@+id/tvname"
        android:textSize="16sp" />
</RelativeLayout>
```

在这个布局中，我们整体使用相对布局 RelativeLayout，ImageView 用来显示头像图片，在左边显示，头像对应的名字 TextView 控件在图片的右边，信息内容放在名字的下面，并与名字左对齐。

第六步：RecyclerView 创建适配器

为 RecyclerView 准备一个适配器，新建 myAdapter 类，让这个适配器继承自 RecyclerView.Adapter，并将泛型指定为 MyAdapter.MyViewHolder。其中 MyViewHolder 是一个内部类。

```
public class MyAdapter extends RecyclerView.Adapter<MyAdapter.MyViewHoder> {
    private Context context;
    private List<WeiXin> weixinList;

    public MyAdapter( Context context,List<WeiXin> weixinList) {
        this.weixinList = weixinList;
        this.context = context;
    }
    public class MyViewHoder extends RecyclerView.ViewHolder {
        TextView tvname;
        TextView tvinformation;
        ImageView imageView;

        public MyViewHoder(@NonNull View itemView) {
            super(itemView);
            tvname = itemView.findViewById(R.id.tvname);
            tvinformation = itemView.findViewById(R.id.tvinformation);
            imageView = itemView.findViewById(R.id.image);
        }
    }

    @NonNull
    @Override
    public MyViewHoder onCreateViewHolder(@NonNull ViewGroup parent, int viewType) {
        View view=View.inflate(context,R.layout.item,null);
        MyViewHoder holder=new MyViewHoder(view);
        return holder;
    }
```

```
    @Override
    public void onBindViewHolder(@NonNull MyViewHoder holder, int position) {
        final WeiXin weixin = weixinList.get(position);
        holder.imageView.setImageResource(weixin.getImgId());
        holder.tvname.setText(weixin.getName());
        holder.tvinformation.setText(weixin.getInformation());
        holder.itemView.setOnClickListener(new View.OnClickListener() {
            //对itemView 设置点击事件
            @Override
            public void onClick(View v) {
                Toast.makeText(context, "您点击了" + weixin.getName(), Toast.LENGTH_LONG).show();
            }
        });
        holder.imageView.setOnClickListener(new View.OnClickListener() {
            @Override
            public void onClick(View v) {
                Toast.makeText(context, "您点击了" + weixin.getName()+"的图片", Toast.LENGTH_LONG).
show();
            }
        });
    }

    @Override
    public int getItemCount() {
        return weixinList.size();
    }
}
```

在上面的代码中可以看到 MyAdapter 类，有两个成员变量 context 和 weixinList，它的构造方法为 MyAdapter(Context context,List<WeiXin> weixinList)，这个方法用于把要展示的数据源传进来，并赋值给一个全局变量 weixinList，我们后续的操作都是在这个数据源的基础上进行的。

在自定义的 MyAdapter 类中定义了一个内部类 MyViewHolder，这个类继承自 RecyclerView.ViewHolder，用于初始化 item 布局中的子控件。需要注意的是，在这个类的构造方法中需要传递 item 布局的 View，这样我们就可以通过 findViewById()方法来获取到布局中的 ImageView 和 TextView 控件了。

由于 MyAdapter 继承自 RecyclerView.Adapter，所以必须重写 onCreateViewHolder()、onBindViewHolder()和 getItemCount()这 3 个方法，其中：

onCreateViewHolder()方法：用于创建 ViewHolder 实例，我们在这个方法中使用 View.*inflate*()方法加载了 item 布局,并创建了 MyViewHolder 实例，把加载的布局传入到构造函数当中，最后将 MyViewHolder 实例返回。

onBindViewHolder()方法：用于对 RecyclerView 子项的数据进行赋值，将在每个子项被滚动到屏幕内的时候执行，这里我们通过 position 参数得到当前项的 WeiXin 实例，然后再将数据设置到 ImageView 和 2 个 TextView 当中。在这个方法中我们通过 holder.itemView.setOnClickListener()对 Recyclerview 子项设置单击事件。

getItemCount()方法：用于设置 RecyclerView 子项个数，直接返回数据源的长度就可以了。

第七步：实现 RecyclerView 数据展示

在 MainActivity 中实现 RecyclerView 数据展示，也就是将要展示的数据与 item 布局绑定，显示在 RecyclerView 控件中，MainActivity 的参考代码如下所示。

```
public class MainActivity extends AppCompatActivity {
    private List<WeiXin> weixinlist = new ArrayList<>();
    private RecyclerView rc;
    private MyAdapter adapter;
    private  int[]  imageId  =  {R.drawable.tx1,  R.drawable.tx2,  R.drawable.tx3,  R.drawable.tx4,
R.drawable.tx5, R.drawable.tx6, R.drawable.tx7, R.drawable.tx8};
    private String[] strname = {"一枝独秀", "黑中红叶", "金秋时节又一年", "秋", "文件传输助手", "一帆风顺
666", "险峰松柏", "红叶片片"};
    private String[] strinformation = {"拜拜……", "明天见", "【图片】", "好玩的事情多分享哦！！！", "【图
片】", "您好！！！", "886", "少壮不努力老大徒伤悲"};

    @Override
    protected void onCreate(Bundle savedInstanceState) {
        super.onCreate(savedInstanceState);
        setContentView(R.layout.activity_main);
        rc = findViewById(R.id.recyclerView);
        initWeixin();
        rc.setLayoutManager(new LinearLayoutManager(this));
        adapter = new MyAdapter(this, weixinlist);
        rc.setAdapter(adapter);
    }

    private void initWeixin() {
        for (int i = 0; i < imageId.length; i++) {
            WeiXin weixin = new WeiXin(imageId[i], strname[i], strinformation[i]);
            weixinlist.add(weixin);
        }
    }
}
```

MainActivity 中定义了 3 个数组 imageId、strname 和 strinformation，分别用来存储微信中头像的图片、名字和信息，在 initWeixin()方法中初始化所有的微信数据。在 onCreate()方法中先获取 RecyclerView 的实例，然后通过 setLayoutManager()方法设置 RecyclerView 实例的布局管理器，这里的 LinearLayoutManager 为线性布局，以垂直滚动列表方式显示项目。最后调用 RecyclerView 的 setAdapter()方法来完成适配器设置，这样 RecyclerView 和数据之间的关联就建立完成了。

第八步：运行程序

运行程序，效果如图 3-27 所示，单击其中某一个则会弹出提示所选项的信息。

补充说明：对于 RecyclerView 的布局管理器设置，除了垂直滚动列表显示外，也可以将布局管理器设置为其他的样式，如水平滚动显示或者是网格显示，效果如图 3-29 所示。

```
rc.setLayoutManager(new LinearLayoutManager(this,RecyclerView.HORIZONTAL,false));
```

使用上面语句替换案例中 rc.setLayoutManager()可以显示为水平滚动效果，如图 3-29(a)所示。

```
rc.setLayoutManager(new GridLayoutManager(this,2));
```

使用上面语句替换案例中 rc.setLayoutManager()可以显示为网格布局，在网格中显示项目，一行显示 2 列内容，如图 3-29(b)所示。

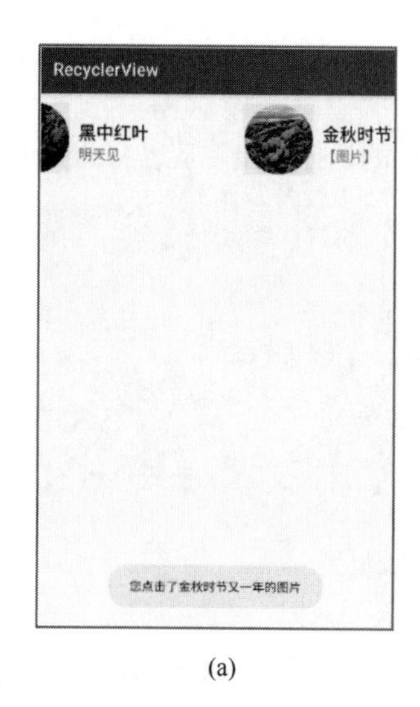

(a)

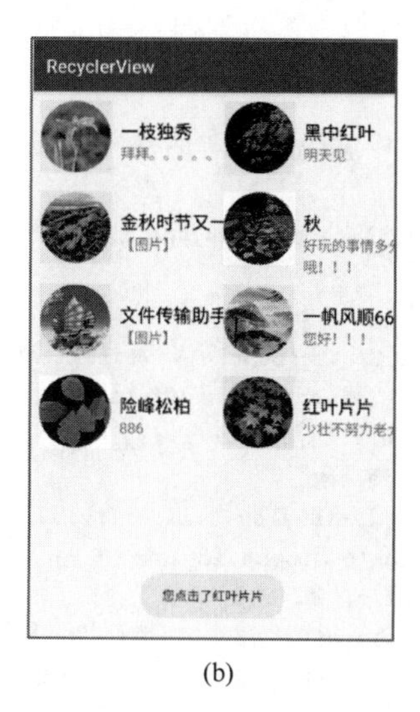

(b)

图 3-29　RecyclerView 的水平滚动效果和网格布局

3.3　Android 的对话框

在图形界面中，对话框是人机交互的一种重要形式，程序可以通过对话框对用户进行一些信息的提示，用户也可以通过对话框和程序进行一些简单的交互操作。 Android 中常用的对话框有普通对话框、单选对话框、多选对话框、自定义对话框等。

3.3.1　普通对话框

普通对话框

普通对话框是出现最多也是最简单的一种对话框，主要用来为用户显示提示信息，然后根据用户的选择，决定后续操作。

以图 3-30 所显示的普通对话框为例，我们来学习如何创建对话框，并对其进行响应。

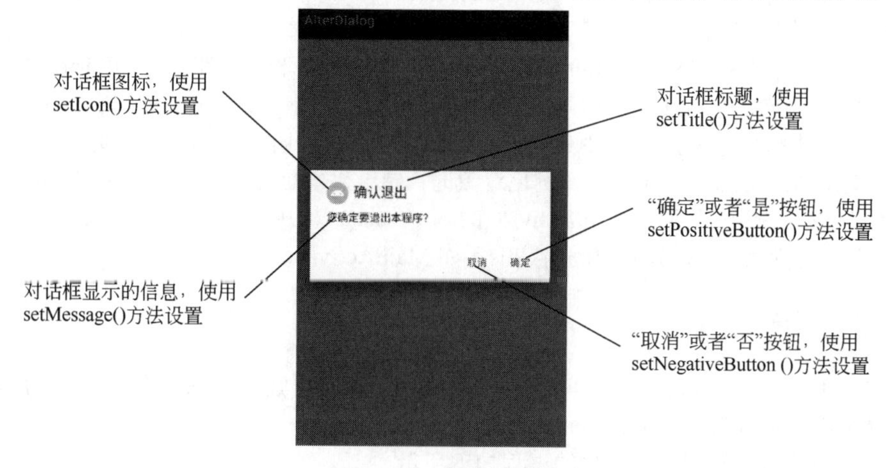

图 3-30　普通对话框

第一步：新建项目

创建一个新的项目 AlterDialog，包名为 com.example.alterdialog。

第二步：显示对话框

在 MainActivity 中编码显示对话框，并根据不同的选择做出响应，MainActivity 中的代码如下。

```
public class MainActivity extends AppCompatActivity {

    @Override
    protected void onCreate(Bundle savedInstanceState) {
        super.onCreate(savedInstanceState);
        setContentView(R.layout.activity_main);
        AlertDialog dialog;
        dialog = new AlertDialog.Builder(this)
                .setIcon(R.mipmap.ic_launcher)
                .setTitle("确认退出")
                .setMessage("您确定要退出本程序？")
                .setPositiveButton("确定", new DialogInterface.OnClickListener() {
                    @Override
                    public void onClick(DialogInterface dialog, int which) {
                        dialog.dismiss();
                        MainActivity.this.finish();
                    }
                }).setNegativeButton("取消", new DialogInterface.OnClickListener() {
                    @Override
                    public void onClick(DialogInterface dialog, int which) {
                        Toast.makeText(MainActivity.this, "继续使用本程序", Toast.LENGTH_LONG).show();
                    }
                }).create();//创建对话框，必不可少
        dialog.show();//显示对话框，必不可少
    }
}
```

在上面的代码中我们可以看到，首先声明了一个 AlertDialog 对象 dialog，创建 AlertDialog.Builde 的对象对其实例化。调用 AlertDialog.Builder 的 setIcon 方法设置对话框图标，使用 setTitle()设置对话框标题，使用 setMessage()方法设置对话框显示的提示内容，使用 setPositiveButton() 和 setNegativeButton() 方法设置对话框按钮。setPositiveButton() 和 setNegativeButton()这两个方法都有两个参数，第一个参数是显示在按钮上的内容，第二个参数用来监听按钮的点击事件，如果不需要对按钮的点击事件进行监听，则第二个参数可以为 null。

注意：使用 AlertDialog.Builder()实例化对象时，如果代码是在某个 Activity 的 onCreate()方法中，则参数为 this，本例就是在 MainActivity 的 onCreate()方法中，所以参数为 this。如果在其他的方法中，则参数为当前 Activity 的名字 this，如 MainActivity.this。

第三步：运行程序

运行程序出现如图 3-30 所示的界面，如果单击“取消”，则对话框消失，显示程序界面；如果单击“确定”，则应用程序关闭。

通过本实例总结实现普通对话框的过程如下所示：

① 声明 AlertDialog 类对象 dialog：AlertDialog dialog;

② 实例化 AlertDialog 类对象：dialog = new AlertDialog.Builder(this);

③ 通过调用 AlertDialog.Builder 的如下方法

- setIcon()方法设置对话框图标；
- setTitle()方法设置对话框标题；
- setMessage()方法设置对话框提示信息；
- setPositiveButton()方法设置对话框的“确定”按钮；
- setNegativeButton()方法设置对话框的“取消”按钮；
- create()方法来创建 AlertDialog 对象；

④ 使用 show()方法显示对话框；

⑤ 使用 dismiss()方法取消对话框。

单选对话框

3.3.2　单选对话框

单选对话框也是对话框的一种，用来显示一组单选列表和按钮，显示单选对话框使用setSingleChoiceItems()方法。以实现如图 3-31 所示单选对话框效果为例，我们来学习单选对话框的使用，操作过程如下。

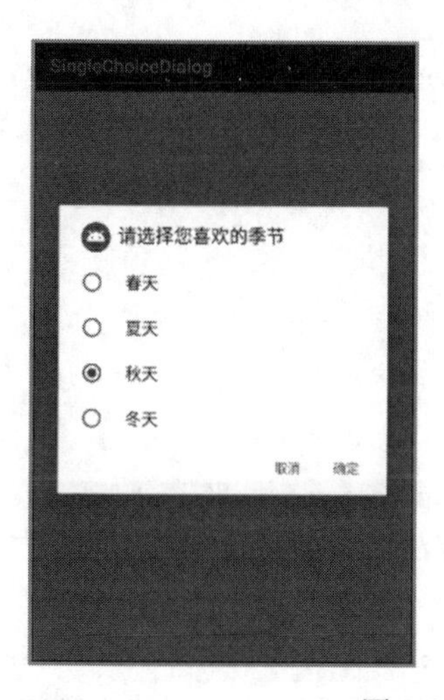

图 3-31　单选对话框

第一步：新建项目

创建项目 SingleChoiceDialog,包名为 com.example.singlechoicedialog。

第二步：界面设计

修改 activity_main 中的 TextView 控件，为它添加 android:id 属性值为 tv1。

第三步：实现界面交互

在 MainActivity 中实现单选对话框的创建与应用，代码如下。

```
public class MainActivity extends AppCompatActivity {
    private TextView tv;
    private String strseason;
    @Override
    protected void onCreate(Bundle savedInstanceState) {
```

```
        super.onCreate(savedInstanceState);
        setContentView(R.layout.activity_main);
        tv = findViewById(R.id.tv1);
        final String arrseason[]= {"春天","夏天","秋天","冬天"};
        strseason = arrseason[0];
           AlertDialog dialog=new AlertDialog.Builder(this)
              .setIcon(R.mipmap.ic_launcher)
              .setTitle("请选择您喜欢的季节")
              .setSingleChoiceItems(arrseason, 0, new DialogInterface.OnClickListener() {
                 @Override
                 public void onClick(DialogInterface dialog, int which) {
                    strseason = arrseason[which];//which 表示所选的列表项的下标。
                 }
              }).setPositiveButton("确定", new DialogInterface.OnClickListener() {
                 @Override
                 public void onClick(DialogInterface dialog, int which) {
                    tv.setText("您喜欢的季节是: "+ strseason);
                 }
              }).setNegativeButton("取消", new DialogInterface.OnClickListener() {
                 @Override
                 public void onClick(DialogInterface dialog, int which) {
                    dialog.dismiss();
                 }
              }).create();//创建对话框
        dialog.show();//显示对话框
    }
}
```

在 MainActivity 中定义了一个字符数组 arrseason 用来存放单选列表项，通过 AlertDialog.Builder 对象调用 setSingleChoiceItem()方法来设置单选对话框，并为该列表设置监听事件。setSingleChoiceItem()方法有三个参数，格式如下所示：

```
setSingleChoiceItems(CharSequence[] items, int checkedItem, final OnClickListener listener)
```

第一个参数 items 表示单选列表项中的所有数据源，这里是定义好的字符数组 arrseason；

第二个参数 chekceditem 表示单选列表中默认的选中项的下标；

第三个参数是 DialogInterface.OnClickListener 接口,实现接口中 OnClick()方法得到被点击的列表项的序号 which，根据 which 的值获取其在数组中对应的字符串，然后通过 setPositiveButton()方法在对话框中添加“确定”按钮，点击确定按钮时，设置界面上 TextView 的内容为所选的内容。最后使用 setNegatvieButton()方法在对话框中添加“取消”按钮，点击取消，使用 dismiss()方法关闭对话框。

第四步：运行程序

运行程序，界面如图 3-31 所示，选择“秋天”选项后点击“确定”按钮，返回程序界面显示“您喜欢的季节是：秋天”。

3.3.3 多选对话框

多选对话框用来提供一组多选列表，在对话框创建中需要调用 setMultiChoiceItems()方法。下面通过实现如图 3-32 所示的运行效果为例，来学习多选对话框的使用。操作步骤如下所示：

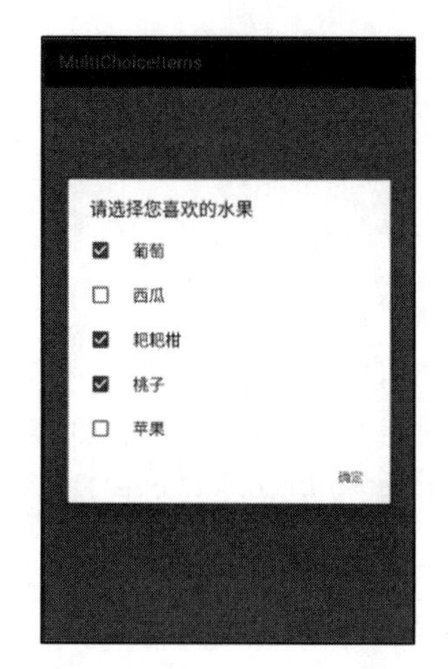

图 3-32　多选对话框

第一步：新建项目

创建新的项目 MultiChoiceItems，包名为 com.example.multichoiceitems.

第二步：设计用户界面

修改默认的 activity_main.xml 文件，为 TextView 控件添加 android:id 属性为 tv,指定 android:text= "您喜欢的水果是：　"。

第三步：显示多选对话框

在 MainActivity 中实现多选对话框的显示，并对选择的选项进行出，代码如下所示。

```
public class MainActivity extends AppCompatActivity {
   private TextView tv;
   @Override
   protected void onCreate(Bundle savedInstanceState) {
      super.onCreate(savedInstanceState);
      setContentView(R.layout.activity_main);
      tv = findViewById(R.id.tv);
      final String arrfruit[]={"葡萄","西瓜","耙耙柑","桃子","苹果"};
      final boolean ischeckeditems[]=new boolean[]{false,true,false,false,true};
      AlertDialog dialog=new AlertDialog.Builder(this)
            .setTitle("请选择您喜欢的水果")
            .setMultiChoiceItems(arrfruit,ischeckeditems, new DialogInterface.OnMultiChoiceClic-
kListener() {
               @Override
               public void onClick(DialogInterface dialog, int which, boolean isChecked) {
                  ischeckeditems[which]=isChecked;
               }
            }).setPositiveButton("确定", new DialogInterface.OnClickListener() {
               @Override
               public void onClick(DialogInterface dialog, int which) {
                  for(int i=0;i<ischeckeditems.length;i++){
                     if(ischeckeditems[i]){
```

```
                    tv.append(arrfruit[i]+" ");
                }
            }
        }
    }).create();
    dialog.show();
  }
}
```

在 MainActivity 中，定义一个字符数组用来存储对话框中要显示的多选列表项，boolean 型数组存储所列的选项是否被选中。在创建的对话框时使用 setMultiChoiceItems()方法设置多选对话框，此方法中有三个参数，格式如下所示：

```
setMultiChoiceItems(CharSequence[] items, boolean[] checkedItems,final OnMultiChoiceClickListener listener)
```

第一个参数是字符数组显示多选列表项，本例中使用了 String 类型数组 arrfruit。

第二个参数是列表项是否选中，当值为 true 时，显示默认选中状态。这里使用 boolean 型数组 ischeckeditems，从其值可以看到第二个选项和最后一个选项为选中状态。

第三个参数是事件监听，其中 onClick()方法的 which 参数为选中项的标号，isChecked 表示是否为选中状态。当点击“确定”按钮时，从第一个选项开始判定是否被选中，如果选中就使用 TextView 对象的 append()方法追加在文本的后面。

第四步：运行程序

运行程序效果如图 3-32 所示，选择图示选项后点击“确定”，返回到程序界面显示选中项的内容。

进度条对话框和消息对话框

3.3.4 自定义对话框

从前面几种对话框可以看到，每种对话框的显示外观相似，如果我们需要设置个性的对话框，如何实现呢？下面以如图 3-33 所示用户登录对话框为例，演示如何自定义对话框。

分析 图 3-33 所示的对话框是一个登录对话框，其中内容由自己设定，这就需要以对话框的形式显示相对应的布局，并能对其进行控制，然后在界面中显示，操作过程如下：

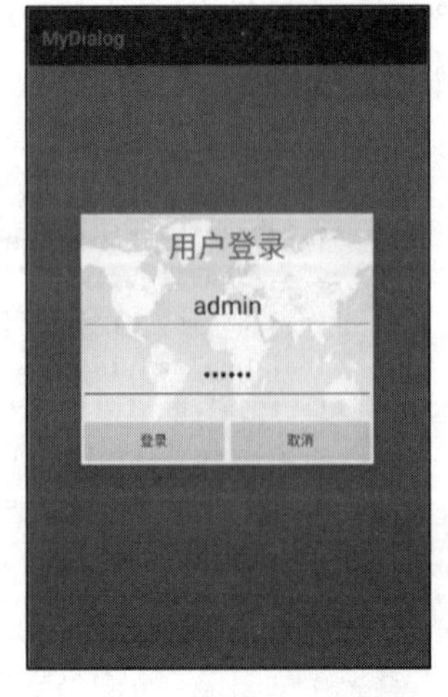

图 3-33 自定义对话框

第一步：创建工程

创建一个项目，名为 MyDialog，包名为 com.example.mydialog.

第二步：创建对话框的布局文件

在 res/layout 目录下新建布局文件 dialog_layout，设计布局代码如下所示。

```
<?xml version="1.0" encoding="utf-8"?>
<androidx.constraintlayout.widget.ConstraintLayout  xmlns:android="http://schemas.android.com/apk/
res/android"
    xmlns:app="http://schemas.android.com/apk/res-auto"
    android:layout_width="match_parent"
    android:layout_height="match_parent"
    android:background="@drawable/bj">

    <TextView
        android:id="@+id/textView"
        android:layout_width="wrap_content"
        android:layout_height="wrap_content"
        android:text="用户登录"
        android:textSize="30dp"
        app:layout_constraintEnd_toEndOf="parent"
        app:layout_constraintStart_toStartOf="parent"
        app:layout_constraintTop_toTopOf="parent" />

    <EditText
        android:id="@+id/userName"
        android:layout_width="409dp"
        android:layout_height="wrap_content"
        android:layout_marginTop="8dp"
        android:ems="10"
        android:gravity="center"
        android:hint="请输入用户名"
        android:inputType="textPersonName"
        android:singleLine="true"
        android:textSize="25dp"
        app:layout_constraintEnd_toEndOf="parent"
        app:layout_constraintStart_toStartOf="parent"
        app:layout_constraintTop_toBottomOf="@+id/textView" />

    <EditText
        android:id="@+id/userPwd"
        android:layout_width="409dp"
        android:layout_height="wrap_content"
        android:layout_marginTop="15dp"
        android:ems="10"
        android:gravity="center"
        android:hint="请输入密码"
        android:inputType="textPassword"
        android:textSize="25dp"
        app:layout_constraintStart_toStartOf="parent"
        app:layout_constraintTop_toBottomOf="@+id/userName" />

    <Button
        android:id="@+id/btnLogin"
        android:layout_width="wrap_content"
        android:layout_height="wrap_content"
        android:layout_marginTop="15dp"
        android:layout_marginEnd="3dp"
        android:text="登录"
        app:layout_constraintEnd_toStartOf="@+id/btnQuit"
```

```
        app:layout_constraintStart_toStartOf="parent"
        app:layout_constraintTop_toBottomOf="@+id/userPwd" />

    <Button
        android:id="@+id/btnQuit"
        android:layout_width="wrap_content"
        android:layout_height="wrap_content"
        android:layout_marginEnd="31dp"
        android:text="取消"
        app:layout_constraintBottom_toBottomOf="@+id/btnLogin"
        app:layout_constraintEnd_toEndOf="parent"
        app:layout_constraintStart_toEndOf="@+id/btnLogin" />
</androidx.constraintlayout.widget.ConstraintLayout>
```

第三步：创建自定义对话框类

在当前包中创建一个 MyDialog 类用来定义自定义对话框，此类继承自 Dialog。MyDialog.java 文件编写代码如下：

```
public class MyDialog extends Dialog {
    private EditText uName, uPwd;
    private Button btnLogin, btnQuit;

    public MyDialog(Context context) {
        super(context);
    }

    @Override
    protected void onCreate(Bundle savedInstanceState) {
        super.onCreate(savedInstanceState);
        requestWindowFeature(Window.FEATURE_NO_TITLE);
        setContentView(R.layout.dialog_layout);
        uName = findViewById(R.id.userName);
        uPwd = findViewById(R.id.userPwd);
        btnLogin = findViewById(R.id.btnLogin);
        btnQuit = findViewById(R.id.btnQuit);
        btnLogin.setOnClickListener(new View.OnClickListener() {
            @Override
            public void onClick(View v) {//点击“登录”是要执行的逻辑
                String strname = uName.getText().toString();
                String strpwd = uPwd.getText().toString();
                if (strname.equals("") || strpwd.equals("")) {//用户名不为空
                    Toast.makeText(getContext(), "用户名或者密码为空，请输入", Toast.LENGTH_LONG).show();
                } else if (strname.equals("admin") && strpwd.equals("admin")) {
                    Toast.makeText(getContext(), "欢迎", Toast.LENGTH_LONG).show();
                } else {
                    Toast.makeText(getContext(), "输入有错", Toast.LENGTH_LONG).show();
                }
            }
        });
        btnQuit.setOnClickListener(new View.OnClickListener() {
            @Override
```

```
            public void onClick(View v) {
                dismiss();
            }
        });
    }
}
```

在自定义的MyDialog类继承自Dialog类，有一个构造方法和一个重写的onCreate()的方法，在onCreate()方法中加载自定义的布局dialog_layout。对布局中的控件的使用方法与MainActivity中的使用方法相同，先初始化，然后设置按钮的单击事件，获取文本的值对其进行判定，根据判定给出不同的提示。对于Toast.makeText()方法的第一个参数，使用getContext()获取运行时的Context。

第四步：显示自定义对话框

在MainActivity中，定义MyDialog对象，并使用show()方法显示自定义的对话框。MainActivity中的代码如下所示。

```
package com.example.mydialog;

import androidx.appcompat.app.AppCompatActivity;
import android.os.Bundle;
public class MainActivity extends AppCompatActivity {
    @Override
    protected void onCreate(Bundle savedInstanceState) {
        super.onCreate(savedInstanceState);
        setContentView(R.layout.activity_main);
        MyDialog myDialog=new MyDialog(this);
        myDialog.show();
    }
}
```

在上面的代码中，首先创建了一个自定义类MyDialog对象，然后使用show()方法使其显示。

第五步：运行程序

运行程序如图3-33所示，根据提示在对话框中输入用户名和密码，单击“登录”按钮对输入的内容进行判断，如果用户名或密码为空，则给出提示，否则如果用户名和密码正确则提示“欢迎”，不正确则提示“输入有错”。

3.4　仿微信底部导航的设计与实现

通过本章的学习我们对用户界面设计有了一定基础，下面以如图3-34所示的仿微信底部导航为例，演示如何使用资源文件与基础控件结合来实现仿微信底部导航。

分析　整个界面主要由三部分组成，顶部为标题，中部为内容区域，底部为导航。当程序开始运行，标题栏为“微信”，底部导航部分“微信”及其对应的图标绿色显示，内容部分显示“微信内容部分”，如果选中底部导航的其他选项，则对应选项文字和图标颜色发生变化，顶部的标题和内容跟随发生变化。具体实现步骤如下：

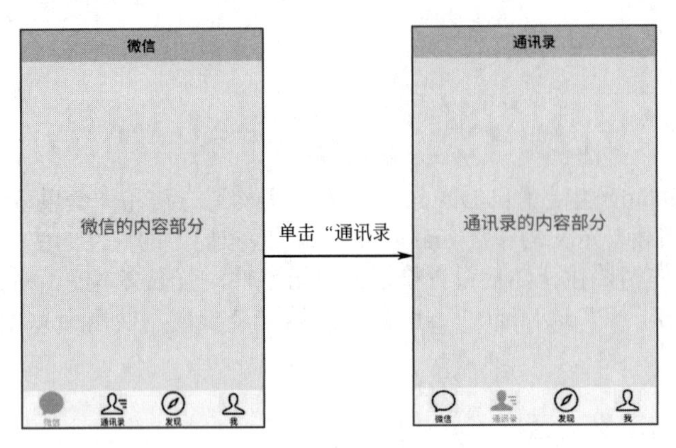

图 3-34　仿微信底部导航运行效果图

第一步：创建项目

创建一个项目名，为 Weixin，项目所在包为 com.example.weixin。

第二步：图片准备

将准备好的图标资源放入 res/drawable 中，注意底部导航每个图标对应两个图片，一个用于未选中状态，一个用于选中状态。

第三步：创建资源文件

创建资源文件用来设置选中状态和未选中状态的文字颜色和图片，由于底部导航的几个选项样式是一样的，所以我们也可以设置一个样式，共同使用。

（1）创建不同状态的文字选择器

选择 res 下的 drawable，单击右键选择"New"→"Drawable Resource File"，在弹出如图 3-35 所示的"New Resource File"对话框中输入 FileName"tab_textcolor"。

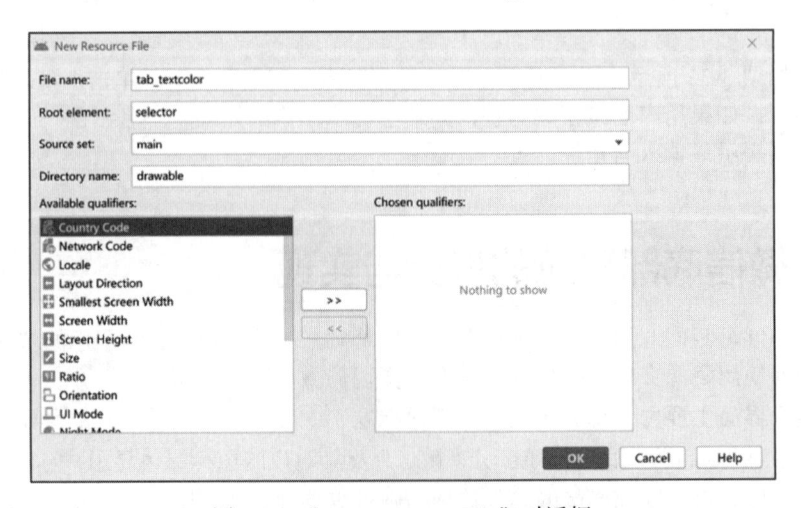

图 3-35　"New Resource File"对话框

编辑新建的文件“tab_textcolor.xml”，代码如下：

```
<?xml version="1.0" encoding="utf-8"?>
<selector xmlns:android="http://schemas.android.com/apk/res/android">
    <item android:color="#000000" android:state_checked="false" />
    <item android:color="#4CAF50" android:state_checked="true" />
</selector>
```

从代码可以看出，如果选中则显示为绿色，否则显示为黑色。

(2) 设置每个选项不同状态的图标

使用同样的方法在drawable下创建一个名为tb_weixin.xml的文件，代码如下：

```
<?xml version="1.0" encoding="utf-8"?>
<selector xmlns:android="http://schemas.android.com/apk/res/android">
    <item android:drawable="@drawable/weixin" android:state_checked="false" />
    <item android:drawable="@drawable/weixin1" android:state_checked="true" />
</selector>
```

当底部导航“微信”选中时，图标显示为weixin1,没有选择中时显示为weixin。依次添加资源文件设置“通讯录”，tb_tongxunlu.xml代码如下：

```
<?xml version="1.0" encoding="utf-8"?>
<selector xmlns:android="http://schemas.android.com/apk/res/android">
    <item android:drawable="@drawable/tongxunlu" android:state_checked="false" />
    <item android:drawable="@drawable/tongxunlu1" android:state_checked="true" />
</selector>
```

“发现”图标选择器资源文件为tb_faxian.xml，代码如下：

```
<?xml version="1.0" encoding="utf-8"?>
<selector xmlns:android="http://schemas.android.com/apk/res/android">
    <item android:drawable="@drawable/faxian" android:state_checked="false" />
    <item android:drawable="@drawable/faxian1" android:state_checked="true" />
</selector>
```

“我”的图标状态对应文件为tb_me.xml，代码如下：

```
<?xml version="1.0" encoding="utf-8"?>
<selector xmlns:android="http://schemas.android.com/apk/res/android">
    <item android:drawable="@drawable/wo" android:state_checked="false" />
    <item android:drawable="@drawable/wo1" android:state_checked="true" />
</selector>
```

(3) 设置通用样式

由于这几个选项按钮的高宽、对齐方式、字体颜色都一样，在布局中我们就不用一个一个设置了，给它们设置一个通用样式大家都可以使用。

打开“/res/values/themes”目录下themes.xml文件，添加如下代码：

```
<style name="mystyle">
      <item name="android:layout_width">0dp</item>
```

```
        <item name="android:layout_height">wrap_content</item>
        <item name="android:button">@null</item>
        <item name="android:layout_weight">1</item>
        <item name="android:gravity">center</item>
        <item name="android:textColor">@drawable/tab_textcolor</item>
    </style>
```

在这个文件中我们自定义了一个样式，定义名字为“mystyle”，指定其中 android:layout_width ，android:layout_height 等的值。其中 android:button 值为@null，用来去掉 RadioButton 前面的圆点。textcolor 属性使用的是我们之前创建好的文字颜色资源文件，也就是根据选中与否显示不同的颜色。

第四步：用户界面设计

在布局文件 activity_main 中加入两个 TextView 分别用来显示标题和内容，添加一个单选按钮组，在此单选按钮组 RadioGroup 控件中添加四个单选按钮 RadioButton。参考代码如下：

```
<?xml version="1.0" encoding="utf-8"?>
<LinearLayout xmlns:android="http://schemas.android.com/apk/res/android"
    xmlns:tools="http://schemas.android.com/tools"
    android:layout_width="match_parent"
    android:layout_height="match_parent"
    android:orientation="vertical"
    tools:context=".MainActivity">

    <TextView
        android:id="@+id/tv_title"
        android:layout_width="match_parent"
        android:layout_height="wrap_content"
        android:background="#CCCACA"
        android:gravity="center"
        android:padding="10dp"
        android:text="微信"
        android:textColor="#000000"
        android:textSize="24sp" />

    <TextView
        android:id="@+id/tv_content"
        android:layout_width="match_parent"
        android:layout_height="0dp"
        android:layout_weight="1"
        android:background="#eeeeee"
        android:textSize="30sp"
        android:gravity="center"
        android:text="微信的内容部分" />

    <RadioGroup
        android:id="@+id/rg_bottom"
        android:layout_width="match_parent"
        android:layout_height="wrap_content"
        android:layout_gravity="bottom"
        android:background="#FAF9F9"
        android:orientation="horizontal">
```

```
        <RadioButton
            android:id="@+id/radio_weixin"
            style="@style/mystyle"
            android:drawableTop="@drawable/tb_weixin"
            android:text="微信" />

        <RadioButton
            android:id="@+id/radio_tongxunlu"
            style="@style/mystyle"
            android:drawableTop="@drawable/tb_tongxunlu"
            android:text="通讯录" />

        <RadioButton
            android:id="@+id/radio_faxian"
            style="@style/mystyle"
            android:drawableTop="@drawable/tb_faxian"
            android:text="发现" />

        <RadioButton
            android:id="@+id/radio_me"
            style="@style/mystyle"
            android:drawableTop="@drawable/tb_me"
            android:text="我" />
    </RadioGroup>
</LinearLayout>
```

在设置界面时有两个 TextView，最上面 tvTitle 用来显示标题，中间部分 tvContent 用来设置选项的详细内容，最下面是一组单选按钮组用来实现底部导航，所用的单选按钮使用同一个的样式，即 mystyle。

第五步：界面交互

我们要求程序运行时根据选项的不同，显示不同的标题和内容，而且要求选中状态和未选中状态选项按钮的显示效果也不同。在 MainActivty 中实现单击导航中选项做出相应的响应，代码如下：

```
public class MainActivity extends AppCompatActivity {
    TextView tv_title, tv_content;
    RadioGroup rgbottom;
    RadioButton rb_weixin, rb_tongxunlu, rb_faxian, rb_wo;

    @Override
    protected void onCreate(Bundle savedInstanceState) {
        super.onCreate(savedInstanceState);
        setContentView(R.layout.activity_main);
        getSupportActionBar().hide();
        tv_title = findViewById(R.id.tv_title);
        tv_content = findViewById(R.id.tv_content);
        rgbottom = findViewById(R.id.rg_bottom);
```

```
        rb_weixin = findViewById(R.id.radio_weixin);
        rb_tongxunlu = findViewById(R.id.radio_tongxunlu);
        rb_faxian = findViewById(R.id.radio_faxian);
        rb_wo = findViewById(R.id.radio_me);
        rgbottom.setOnCheckedChangeListener(new RadioGroup.OnCheckedChangeListener() {
            @Override
            public void onCheckedChanged(RadioGroup group, int checkedId) {
                switch (checkedId) {
                    case R.id.radio_weixin:
                        tv_title.setText("微信");
                        tv_content.setText("微信的内容部分");
                        break;
                    case R.id.radio_tongxunlu:
                        tv_title.setText("通讯录");
                        tv_content.setText("通讯录的内容部分");
                        break;
                    case R.id.radio_faxian:
                        tv_title.setText("发现");
                        tv_content.setText("发现的内容部分");
                        break;
                    case R.id.radio_me:
                        tv_title.setText("我");
                        tv_content.setText("我的内容部分");
                        break;
                }
            }
        });
    }
}
```

代码中使用 getSupportActionBar().hide()隐藏了自带的 ActionBar,然后初始化控件，实现单选按钮组 rgbottom.setOnCheckedChangeListener()方法，根据选中的选项卡进行显示内容的切换。通过本案例我们实现常见的底部导航，学习了 selector 资源文件和通用样式的定义与应用，这种方式在实际开发中应用较多。

第六步：运行程序

运行程序，效果如图 3-34 所示，界面显示“微信”图标处于选中状态，标题为“微信”，内容部分显示“微信的内容部分”，分别点击“通讯录”“发现”和“我”按钮，图标字体颜色、标题和内容跟着变化。在后面我们学习了 Fragment 的使用后，可以将表示的内容的 TextView 用一个 Fragment 对象来替换。

【职业素养拓展3】

职业素养拓展

职业素养拓展内容详见文件“职业素养拓展”。

扫描二维码查看【职业素养拓展 3】

本章小结

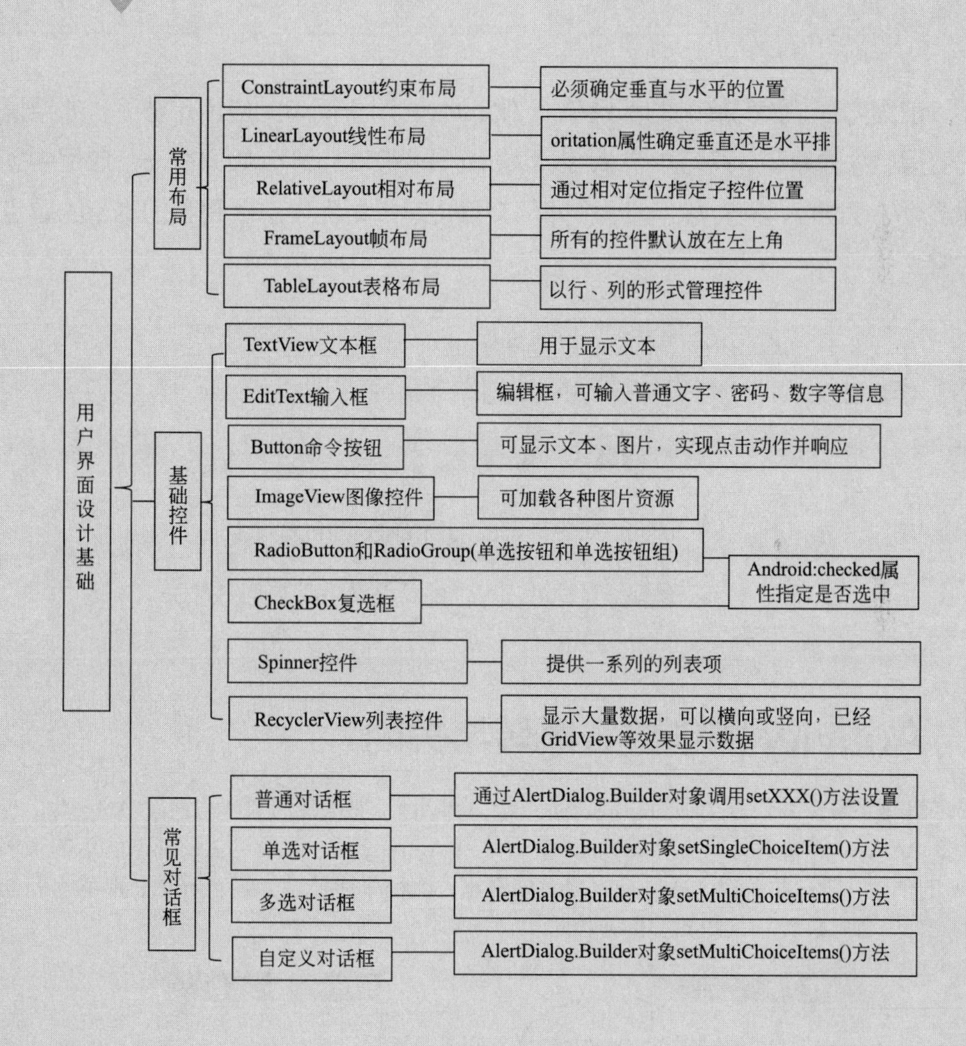

用户界面设计基础
常用布局
ConstraintLayout约束布局
必须确定垂直与水平的位置
LinearLayout线性布局
oritation属性确定垂直还是水平排
RelativeLayout相对布局
通过相对定位指定子控件位置
FrameLayout帧布局
所有的控件默认放在左上角
TableLayout表格布局
以行、列的形式管理控件
基础控件
TextView文本框
用于显示文本
EditText输入框
编辑框，可输入普通文字、密码、数字等信息
Button命令按钮
可显示文本、图片，实现点击动作并响应
ImageView图像控件
可加载各种图片资源
RadioButton和RadioGroup(单选按钮和单选按钮组)
Android:checked属性指定是否选中
CheckBox复选框
Spinner控件
提供一系列的列表项
RecyclerView列表控件
显示大量数据，可以横向或竖向，已经GridView等效果显示数据
常见对话框
普通对话框
通过AlertDialog.Builder对象调用setXXX()方法设置
单选对话框
AlertDialog.Builder对象setSingleChoiceItem()方法
多选对话框
AlertDialog.Builder对象setMultiChoiceItems()方法
自定义对话框
AlertDialog.Builder对象setMultiChoiceItems()方法

第 4 章

Activity

前面内容，我们接触的都是单一 Activity，对于简单的应用足够了。如果情况变得更为复杂了，只有一个 Activity 肯定是不够的，那么如何在一个应用中创建多个 Activity，多个 Activity 中如何交互传递数据呢？这就是我们本章的重点内容。

学习目标：

1. 理解 Activity 的作用；
2. 掌握 Activity 创建、跳转与关闭的方法；
3. 掌握 Activity 之间传递数据的方法；
4. 理解 Activity 的生命周期及其对应的方法。

4.1 Activity 的创建、跳转与关闭

在之前的学习中我们每个项目只涉及了一个 Activity，现在我们要求有两个 Activity，并要实现从一个 Activity 跳转到另一个 Activity，最后关闭 Activity。下面通过如图 4-1 所示的案例学习 Activity 的创建、跳转和关闭。从运行效果可以看出，此程序涉及两个运行界面，那就需要创建新的界面，有界面间跳转和关闭界面的方法。

Activity的创建

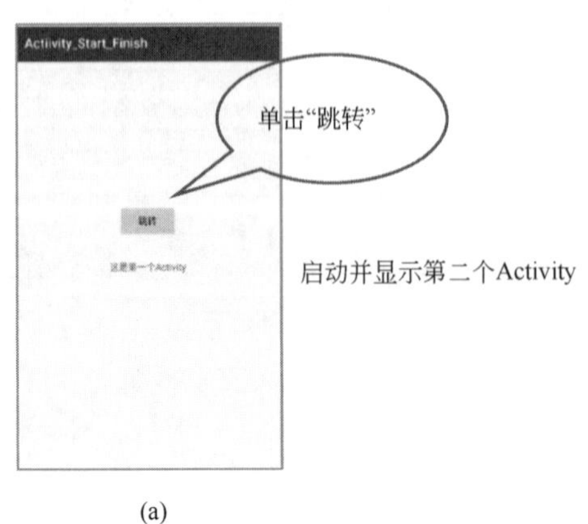

(a)

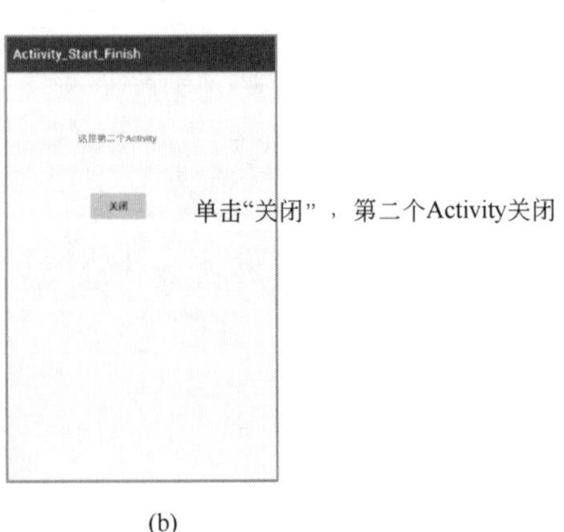

(b)

图 4-1　Activity 的跳转

（1）Activity 间的跳转

首先创建一个意图 Intent 对象，指明要启动的目标 Activity，然后使用 startActivity()方法开启意图。例如：

```
Intent intent=new Intent(MainActivity.this,MainActivity2.class);
startActivity(intent);
```

指定由当前的 MainActivity 跳转到 MainActivity2 对应的界面。

（2）关闭 Activity

关闭当前 Activity 使用 finish()方法。此案例实现步骤如下所示。

第一步：新建项目

新建项目，名称为 Activity_Start_Finish，包名为：com.example.actiivity_start_finish。

第二步：设计用户界面

修改布局文件 activity_main 中 TextView 控件的 android:text 属性为“这是第一个 Activity”，添加一个 Button 按钮更改 android:text 属性为“跳转”，增加 android:onClick 属性设置值为“btnStartActivity”。（提示：简单的界面，可以直接在 Design 模式下，通过拖曳控件到界面，然后选中控件直接单击工具栏上的 图标添加约束）布局参考代码如下：

```
<?xml version="1.0" encoding="utf-8"?>
<androidx.constraintlayout.widget.ConstraintLayout
xmlns:android="http://schemas.android.com/apk/res/android"
    xmlns:app="http://schemas.android.com/apk/res-auto"
    xmlns:tools="http://schemas.android.com/tools"
    android:layout_width="match_parent"
    android:layout_height="match_parent"
    tools:context=".MainActivity">

    <TextView
        android:id="@+id/textView2"
        android:layout_width="wrap_content"
        android:layout_height="wrap_content"
        android:text="这是第一个 Activity"
        app:layout_constraintBottom_toBottomOf="parent"
        app:layout_constraintLeft_toLeftOf="parent"
        app:layout_constraintRight_toRightOf="parent"
        app:layout_constraintTop_toTopOf="parent" />

    <Button
        android:id="@+id/button"
        android:layout_width="wrap_content"
        android:layout_height="wrap_content"
        android:layout_marginBottom="34dp"
        android:text="跳转"
        android:onClick="btnStartActivity"
        app:layout_constraintBottom_toTopOf="@+id/textView2"
        app:layout_constraintEnd_toEndOf="parent"
```

```
        app:layout_constraintStart_toStartOf="parent" />

</androidx.constraintlayout.widget.ConstraintLayout>
```

第三步：创建新的 Activity

创建第二个 Activity，选择当前的包，右击选择“New”→“Activity”→“Empty Activity”选项，如图 4-2 所示。

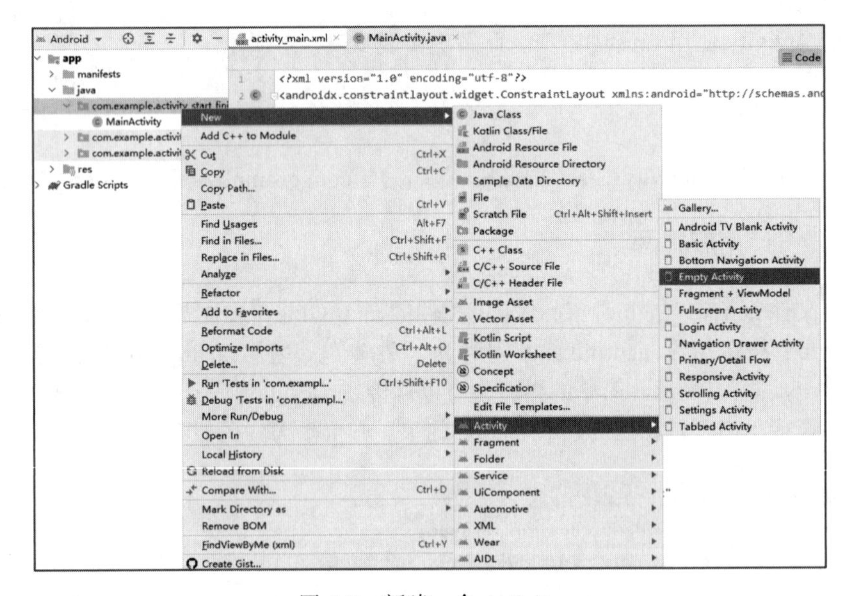

图 4-2　新建一个 Activity

单击“Empty Activity”，弹出如图 4-3 所示对话框，单击“Finish”按钮完成新的 Activity 的创建。

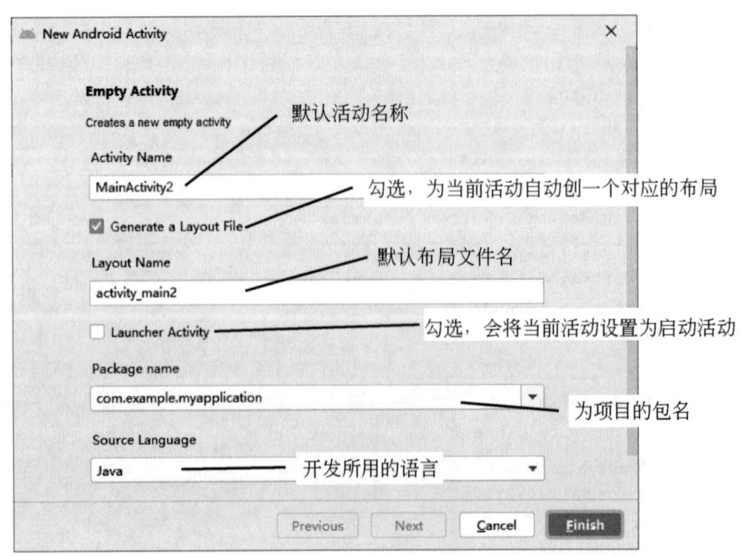

图 4-3　“New Android Activity”对话框

可以看到在当前应用的包下增加了一个新的文件 MainActivity2，在 res/layout 下增加了一个 activity_main2 布局文件，结构如图 4-4 所示。

Activity的加载

图 4-4　增加 MainActivity2 后的程序结构

在AndroidManifest中注册

再来查看 AndroidManifest，这个文件位于 manifests 文件夹中。AndroidManifest 文件包含有关应用的基本信息，如它包含活动、必要的库和其他声明。创建应用时，Android 会自动创建这个文件。添加了 MainActivity2 后代码如下所示。

```
<?xml version="1.0" encoding="utf-8"?>
<manifest xmlns:android="http://schemas.android.com/apk/res/android"
    package="com.example.activity_start_finish">

    <application
        android:allowBackup="true"
        android:icon="@mipmap/ic_launcher"
        android:label="@string/app_name"
        android:roundIcon="@mipmap/ic_launcher_round"
        android:supportsRtl="true"
        android:theme="@style/Theme.Activity_Start_Finish">
        <activity
            android:name=".MainActivity2"
            android:exported="false" />
        <activity
            android:name=".MainActivity"
            android:exported="true">
            <intent-filter>
                <action android:name="android.intent.action.MAIN" />

                <category android:name="android.intent.category.LAUNCHER" />
            </intent-filter>
        </activity>
    </application>

</manifest>
```

从上面的代码可以看出，我们创建的 MainActivity2 已自动在清单文件中声明。在 Android 开发中所有 Activity 都必须在清单文件 AndroidManifest 中声明。如果一个 Activity 没有在这个文件中声明，系统就不会知道它的存在。如果系统不知道一个 Activity 的存在，这个 Activity 就永远不会运行。要在清单文件中声明一个 Activity，需要在<application>元素中包含一个<activity>元素。在<activity>元素中注明 Activity 的名称和其他的属性。

注意：使用 Android Studio 快捷菜单创建的 Activity 在清单文件中自动声明了，如果是手动增加的额外的 Activity，需要自己在清单文件中添加<activity>对其进行声明。

第四步：设置第二个 Activity 的界面

给第二个 Activity 对应的布局文件 activity_main2 添加一个 TextView 文本控件，设置 android:text 属性为“这是第二个 Activity”，添加一个 Button“关闭”命令按钮，给 TextView 控件和 Button 控件添加位置约束，参考代码如下所示。

```
<androidx.constraintlayout.widget.ConstraintLayout
xmlns:android="http://schemas.android.com/apk/res/android"
    xmlns:app="http://schemas.android.com/apk/res-auto"
    xmlns:tools="http://schemas.android.com/tools"
    android:layout_width="match_parent"
    android:layout_height="match_parent"
    tools:context=".MainActivity2">

    <TextView
        android:id="@+id/textView"
        android:layout_width="wrap_content"
        android:layout_height="wrap_content"
        android:layout_marginStart="113dp"
        android:layout_marginLeft="113dp"
        android:layout_marginTop="87dp"
        android:text="这是第二个 Activity"
        app:layout_constraintStart_toStartOf="parent"
        app:layout_constraintTop_toTopOf="parent" />

    <Button
        android:id="@+id/button"
        android:layout_width="wrap_content"
        android:layout_height="wrap_content"
        android:layout_marginTop="66dp"
        android:layout_marginEnd="13dp"
        android:layout_marginRight="13dp"
        android:onClick="btnCloseActivty"
        android:text="关闭"
        app:layout_constraintEnd_toEndOf="@+id/textView"
        app:layout_constraintTop_toBottomOf="@+id/textView" />
</androidx.constraintlayout.widget.ConstraintLayout>
```

在 activity_main2 的命令按钮中添加 android:onClick 属性值为"btnCloseActivty"，我们要实现的是单击这个命令按钮，实现关闭 Activity 的功能。

第五步：从第一个 Activity 跳转到第二个 Activity

实现从 MainActivity 跳转到 MainActivity2，需要在 MainActivity 文件中完成按钮的单击事件 btnStartActivity()，具体代码如下：

```
package com.example.activity_start_finish;

import androidx.appcompat.app.AppCompatActivity;
import android.content.Intent;
import android.os.Bundle;
```

```
import android.view.View;

public class MainActivity extends AppCompatActivity {
    @Override
    protected void onCreate(Bundle savedInstanceState) {
        super.onCreate(savedInstanceState);
        setContentView(R.layout.activity_main);
    }
    public void btnStartActivity(View view){
        Intent intent=new Intent(this,MainActivity2.class);
        startActivity(intent);
    }
}
```

从上面的代码可以看出在 btnStartActivity()方法先创建了一个 Intent 对象，Intent()方法中有两个参数，第一个参数 this 表示当前的 Activity，第二个参数 MainActivity2.class 表示要跳转到的目标 Activity。通过 startActivty()方法将 Intent 对象作为参数，实现由当前 Activity 到 MainActivity2 的跳转。

Intent 是程序中各组件进行交互的一种重要方式，它不仅可以指定当前组件要执行的动作，还可以在不同组件之间进行数据传递。一般用于启动 Activity、Service 以及发送广播等。从第一个 Activity 中启动第二个 Activity，可以向 Android 发送一个意图。Android 会启动第二个 Activity 并传入这个意图。

第六步：关闭 Activity

在当前项目的 MainActivity2 文件中实现命令按钮的单击事件 btnCloseActivity()方法。代码如下所示。

```
package com.example.activity_start_finish;

import androidx.appcompat.app.AppCompatActivity;
import android.os.Bundle;
import android.view.View;

public class MainActivity2 extends AppCompatActivity {
    @Override
    protected void onCreate(Bundle savedInstanceState) {
        super.onCreate(savedInstanceState);
        setContentView(R.layout.activity_main2);
    }
    public void btnCloseActivty(View view){
        finish();
    }
}
```

关闭 Activity 使用 finish()方法，此方法既没有参数，也没有返回值，在需要时直接调用就行。

第七步：运行程序

运行当前程序，出现图 4-1 (a) 所示界面，单击“跳转”命令按钮，跳转到第二个用户界面，

单击“关闭”命令按钮，第二个 Activity 关闭。

4.2 Activity 的数据传递

在上一节我们实现了如何在一个 Activity 中启动另一个 Activity,也就是实现了 Activity 之间的跳转，有时我们需要 Activity 之间跳转的时候携带数据，本节我们将针对数据的传递和回传进行讲解。

4.2.1 Activity 间数据直接传递

在很多应用中我们时常需要将一个界面的数据传递到下一个界面，比如登录界面输入用户名，登录成功在下一个页面中显示登录用户的名字。下面通过一个案例实现由第一个界面跳转到第二个界面，并将第一个界面中的数据显示在第二个界面中。

分析 对于这个案例我们需要在界面跳转的同时携带数据，还有在第二个界面中接收传递过来的数据。我们知道界面跳转是通过意图 Intent 实现，在 Android 开发中 Intent 在跳转时可以携带数据。

（1）在第一个 Activity 中，定义 Intent 对象，并携带数据跳转

```
Intent intent = new Intent(MainActivity.this, MainActivity2.class);//确定跳转目标
intent.setExtra("name","admin");//intent 中携带数据“name”,值为“admin”
startActivity(intent);//跳转
```

（2）在第二个 Activity 中接收数据

```
Intent intent=getIntent(); //获取传递过来的 Intent
String uname=intent.getStringExtra("name");//获取数据
```

将第一个 Activity 中的用户名和电话号码传递到下一个 Activity 并显示出来，运行效果如图 4-5 所示，操作步骤如下所示。

图 4-5 数据传递运行效果

第一步：新建项目

新建一个程序，名为 ActivityDate1,包为 com.example.activitydata1。

第二步：设计用户界面

设计 activity_main.xml 布局文件，添加两个 EdtiText 控件，android:id 属性分别为 etname 和 etphone，设置它们居中对齐。添加两个命令按钮 Button，并为之添加背景颜色。给页面添加一个图片背景，给界面上的控件添加约束，参考布局代码如下所示。

```
    <?xml version="1.0" encoding="utf-8"?>
    <androidx.constraintlayout.widget.ConstraintLayout
 xmlns:android="http://schemas.android.com/apk/res/android"
        xmlns:app="http://schemas.android.com/apk/res-auto"
        xmlns:tools="http://schemas.android.com/tools"
        android:layout_width="match_parent"
        android:layout_height="match_parent"
        android:background="#ddd"
        tools:context=".MainActivity">

        <EditText
            android:id="@+id/etname"
            android:layout_width="409dp"
            android:layout_height="wrap_content"
            android:layout_marginTop="120dp"
            android:background="#fff"
            android:gravity="center"
            android:textSize="20sp"
            android:hint="请输入用户名"
            android:inputType="textPersonName"
            app:layout_constraintEnd_toEndOf="parent"
            app:layout_constraintStart_toStartOf="parent"
            app:layout_constraintTop_toTopOf="parent" />

        <EditText
            android:id="@+id/etphone"
            android:layout_width="409dp"
            android:layout_height="wrap_content"
            android:layout_marginTop="30dp"
            android:background="#fff"
            android:gravity="center"
            android:textSize="20sp"
            android:hint="请输入电话号码"
            android:inputType="phone"
            app:layout_constraintEnd_toEndOf="parent"
            app:layout_constraintStart_toStartOf="parent"
            app:layout_constraintTop_toBottomOf="@+id/etname" />

        <Button
            android:id="@+id/button"
            android:layout_width="wrap_content"
```

```
        android:layout_height="wrap_content"
        android:layout_marginStart="160dp"
        android:layout_marginTop="68dp"
        android:text="传递数据"
        app:layout_constraintStart_toStartOf="parent"
        app:layout_constraintTop_toBottomOf="@+id/etphone" />

</androidx.constraintlayout.widget.ConstraintLayout>
```

第三步：创建第二个界面

选择当前包“New” →“Activity”→“Empty Activity”创建一个新的 Activity 为 MainActivity2。创建结束会在当前包下增加一个 MainActivity2.java 文件，在 res/layout/增加一个 activity_main2.xml 文件。

第四步：设计接收数据的界面

打开布局文件 activity_main2，在界面上添加一个 TextView 控件，为其添加约束，参考代码如下。

```
<?xml version="1.0" encoding="utf-8"?>
<androidx.constraintlayout.widget.ConstraintLayout
xmlns:android="http://schemas.android.com/apk/res/android"
    xmlns:app="http://schemas.android.com/apk/res-auto"
    xmlns:tools="http://schemas.android.com/tools"
    android:layout_width="match_parent"
    android:layout_height="match_parent"
    tools:context=".MainActivity2">

    <TextView
        android:id="@+id/textView"
        android:layout_width="wrap_content"
        android:layout_height="wrap_content"
        android:layout_marginStart="160dp"
        android:layout_marginTop="288dp"
        android:text="TextView"
        app:layout_constraintStart_toStartOf="parent"
        app:layout_constraintTop_toTopOf="parent" />
</androidx.constraintlayout.widget.ConstraintLayout>
```

第五步：实现数据的传递

需要将 MainActivity 中的数据传递到 MainActivity2 中，我们知道由一个 Activity 跳转到第二个 Actvity，使用 startActivity(intent)方法，由 intent 指明跳转的目标，那么数据传递也是需要跳转的，我们能不能同样使用 startActivity()方法来实现呢？在 MainActivity 中编写单击登录命令按钮的代码，代码如下：

```
public class MainActivity extends AppCompatActivity {
    private EditText etname, etphone;

    @Override
    protected void onCreate(Bundle savedInstanceState) {
        super.onCreate(savedInstanceState);
        setContentView(R.layout.activity_main);
        etname = findViewById(R.id.etname);
        etphone = findViewById(R.id.etphone);
        Button btnLogin = findViewById(R.id.button);
        btnLogin.setOnClickListener(new View.OnClickListener() {
            @Override
            public void onClick(View v) {
                String uname = etname.getText().toString();
                String uphone = etphone.getText().toString();
                Intent intent = new Intent(MainActivity.this, MainActivity2.class);
                intent.putExtra("name", uname);
                intent.putExtra("phone", uphone);
                startActivity(intent);
            }
        });
    }
}
```

在登录按钮单击事件中首先获取输入的用户名和电话号码，分别赋值字符串变量 uname 和 uphone，然后创建 Intent 对象，使用 putExtra()方法将传递的数据存储在 Intent 对象 intent 中，最后使用 startActivty(intent)方法实现携带数据跳转到 MainActivity2。

```
putExtra()方法的格式如下：putExtra(key,value);
```

其中第一个参数 key 表示要传递的数据的名称；第二个参数 value 表示要传递的数据的值。

第六步：接收传递来的数据

我们知道在 MainActivity 中使用 startActivity()传递过来了一个携带数据的 Intent 对象，那么我们如何在 MainActivity2 中接收传递过来的数据呢?

我们需要先使用 getIntent()方法来接收传递过来的 Intent 对象，然后再使用 getXxxExtra()方法获取 Intent 对象包含的数据。在 MainActivity2 中接收传递过来的数据，具体代码如下:

```
public class MainActivity2 extends AppCompatActivity {

    @Override
    protected void onCreate(Bundle savedInstanceState) {
        super.onCreate(savedInstanceState);
        setContentView(R.layout.activity_main2);
        TextView tvInformation=findViewById(R.id.textView);
        Intent intent=getIntent();
        String uname=intent.getStringExtra("name");
        String uphone=intent.getStringExtra("phone");
        tvInformation.setText("用户："+uname+"\n 电话："+uphone);
    }
}
```

从上面的代码中可以看到，在 OnCreate()方法中创建一个 Intent 对象，使用 getIntent()方法获取传递过来的 Intent 对象，再使用 getStringExtra()方法获取传递过来的两个字符串数据，分别赋值为字符串变量 uname 和 phone。

第七步：运行程序

运行程序效果如图 4-5 所示，运行之初输入框提示输入信息，根据提示输入用户名和电话号码，单击“传递数据”命令按钮，跳转到第二个 Activity，可以看到刚输入的内容显示在界面上。

注意：可以通过 getXxxExtra()方法来获取传递过来的数据，由于传递过来的数据类型不同，这里 Xxx 不同。例如：传递过来是 int 类型，则使用 getIntExtra()方法，如果是 String 类型，则使用 getStringExtra()方法，如果是 boolean 类型，则需要使用 getBooleanExtra()方法。

还可以使用 Bund 类传递数据。使用 Bundle 实现用户名和电话号码的数据传递，首先在 MainActivity 中登录命令按钮单击事件中编写如下代码，将数据封装在 Bundle 对象中。

```
Intent intent=new Intent(MainActivity.this,MainActivity2.class);
Bundle bundle=new Bundle();//创建 Bundle 对象
bundle.putString("name",uname);//将用户名信息封装到 Bundle 对象中
bundle.putString("phone",uphone);//将电话号码封装到 Bundle 对象中
intent.putExtras(bundle);////将 Bundle 对象封装到 Intent 对象中
startActivity(intent);
```

然后在 MainActivity2 中获取数据，代码如下：

```
Bundle bundle=getIntent().getExtras(); //获取 Bundle 对象
String uname=bundle.getString("name");//获取用户名
String phone= bundle.getString("phone");//获取电话号码
tvInformation.setText("用户："+uname+"\n 电话："+uphone);
```

4.2.2 Activity 间的数据回传

数据回传

有时希望实现类似微信中“所在位置”的选择，效果如图 4-6 所示。即在“发表文字”界面选择“所在位置”，然后在“所在位置”中选择某一地址后，又返回到“发表文字”界面，并在开始的“所在位置”处显示所选择的地址。这就涉及由第一个 Activity 选择数据传递到第二个 Activity，然后再将第二个 Activity 中所选的数据传回第一个 Activity。

数据回传要解决的问题除了从第一个 Activity 跳到第二个 Activity，还需要从第二个 Activity 中返回数据给第一个 Activity 并获取返回的数据。主要涉及以下内容。

（1）startActivityForResult()方法

此方法用来开启一个 Activity，当开启的 Activity 销毁后会回调上一个 Activity 中的 onActivityResult()方法。startActivityForResult()方法语法格式如下：

```
startActivityForResult(Intent intent, int requestCode);
```

此方法中有两个参数，第 1 个参数是 Intent 对象，表示跳转的目标；第 2 个参数是一个 int 类型数据，称作请求码，用于标识请求的来源。

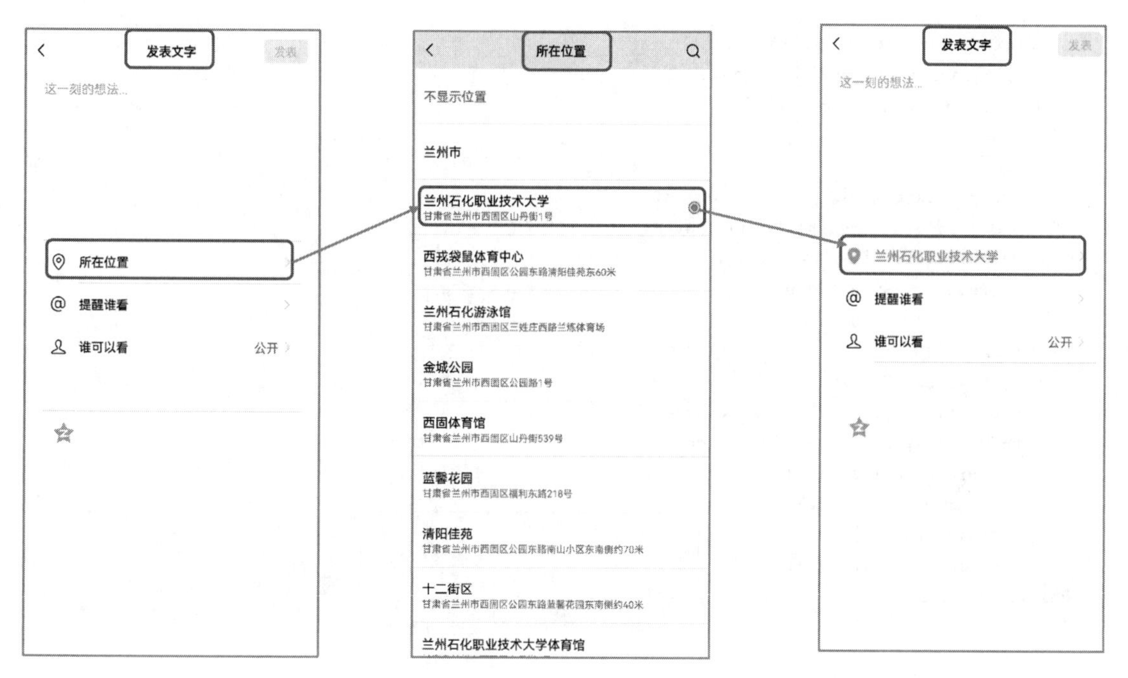

图4-6 微信中“所在位置”选择过程

（2）setResult()方法

此方法用于携带数据进行回传，语法格式如下:

```
setResult(int ResultCode,Intent intent);
```

它有两个参数，第1个参数ResultCode表示返回码；第2参数intent用于携带数据并回传到上一个界面。

（3）onActivityResult()方法

此方法用于接收回传的数据，语法格式如下:

```
onActivityResult(int requestCode, int resultCode, Intent data)
```

此方法中有三个参数：第1个参数是请求码，第2个参数是返回码，第3个参数是返回时携带的数据。当请求码和返回码都正确时，则可以收到数据Intent对象data中携带的数据。

下面通过一个案例学习如何实现数据的回传，实现简化了的位置选择。操作过程如下。

第一步：新建项目

新建一个项目，名为ActivityForResult，项目所在的包为com.example.activityforresult。

第二步：设计第一个界面

在第一个Activity界面中放置两个TextView，其显示的内容分别为“第一个Activity”和“所在位置”。在android:id为tv1的TextView控件中设置了android:onClick属性为“tvClick”。

修改activity_main布局，参考代码如下所示。

```
<?xml version="1.0" encoding="utf-8"?>
<RelativeLayout xmlns:android="http://schemas.android.com/apk/res/android"
    xmlns:app="http://schemas.android.com/apk/res-auto"
    xmlns:tools="http://schemas.android.com/tools"
```

```
    android:layout_width="match_parent"
    android:layout_height="match_parent"
    tools:context=".MainActivity">

    <TextView
        android:layout_width="wrap_content"
        android:layout_height="wrap_content"
        android:text="第一个 Activity"
        android:textSize="40sp" />

    <TextView
        android:id="@+id/tv1"
        android:layout_width="wrap_content"
        android:layout_height="wrap_content"
        android:layout_centerInParent="true"
        android:onClick="tvClick"
        android:text="所在位置"
        android:textSize="40sp" />
</RelativeLayout>
```

第三步：新建第二个 Activity

新建第二个 Activity，名字为 MainActivity2，对应的布局文件为 activity_main2,布局文件的代码如下所示：

```
<?xml version="1.0" encoding="utf-8"?>
<RelativeLayout xmlns:android="http://schemas.android.com/apk/res/android"
    xmlns:app="http://schemas.android.com/apk/res-auto"
    xmlns:tools="http://schemas.android.com/tools"
    android:layout_width="match_parent"
    android:layout_height="match_parent"
    tools:context=".MainActivity2">

    <TextView
        android:layout_width="wrap_content"
        android:layout_height="wrap_content"
        android:text="第二个 Activity"
        android:textSize="40sp" />

    <TextView
        android:id="@+id/tv2"
        android:layout_width="wrap_content"
        android:layout_height="wrap_content"
        android:layout_centerInParent="true"
        android:onClick="tvClick_return"
        android:text="美丽的大海边"
        android:textSize="40sp" />
</RelativeLayout>
```

在代码中看到 android:id 为"tv2"的 TextView 控件设置了 android:onClick 属性为 tvClick_return，需要在 MainActivity2 中实现 tvClick_return()方法。

第四步：从第一个 Activity 中跳转

在 MainActivity 要跳转到下一个 Activity，也就是 MainActivity2，跟之前 Activity 间的跳转的区别是它还要返回数据，所以不能使用 startActivity()方法，而是使用 startActivityForResult()方法。MainActivity 中的代码如下：

```
public class MainActivity extends AppCompatActivity {
    TextView tv1;

    @Override
    protected void onCreate(Bundle savedInstanceState) {
        super.onCreate(savedInstanceState);
        setContentView(R.layout.activity_main);
        tv1 = findViewById(R.id.tv1);
    }

    public void tvClick(View view) {
        Intent intent = new Intent(this, MainActivity2.class);
        startActivityForResult(intent, 1);
    }
}
```

在 onCreate()方法中对 TextView 进行初始化，实现了文本的 android:onClick 属性设置的方法 tvClick(),在这个方法中先定义了一个 Intent 对象，指明 intent 的目标是 MainActivity2.class，然后使用了 startActivityForResult()方法开启一个 Activity，指定请求码为 1，当开启的 Activity 销毁后会回调上一个 Activity 的 onActivityResult()方法。

第五步：返回数据

由第二个 Activity 返回数据到第一个 Activity，我们知道传递数据可以使用 Intent 对象，给这个对象使用 putExtra()方法，让它携带数据就行了，可是如何确定要返回的数据呢？这里使用了 setResult()方法，由 MainActivity2 返回数据到 MainActivity，实现代码如下所示：

```
public class MainActivity2 extends AppCompatActivity {
    TextView tv2;

    @Override
    protected void onCreate(Bundle savedInstanceState) {
        super.onCreate(savedInstanceState);
        setContentView(R.layout.activity_main2);
        tv2=findViewById(R.id.tv2);
    }

    public void tvClick_return(View view) {
        Intent intent=new Intent();
        intent.putExtra("data",tv2.getText().toString());
        setResult(2,intent);
        finish();

    }
}
```

在 tvClick_return()方法中新建了一个 Intent 对象，使用 putExtra()方法使 Intent 对象携带文本显示的内容，然后调用 setResult()方法，指定返回码为 2。

最后使用 finish()方法关闭了 MainActivity2,又回到了 MainActivity 中回调 onActivityResult()方法。

第六步：在 MainActivity 中接收回传的数据

我们在前面提到，使用 startForResult()方法开启的 Activity,在它销毁后会调用开启它的 Activity 的 onActivityResult()方法。接收传回来的数据，需要在MainActivity 中重写 onActivityResult()方法，此方法的代码如下所示：

```
@Override
protected void onActivityResult(int requestCode, int resultCode, @Nullable Intent data) {
    super.onActivityResult(requestCode, resultCode, data);
    if (requestCode == 1 && resultCode == 2) {
        String str = data.getStringExtra("data");
        tv1.setText("您的位置：\n"+str);
    }
}
```

对于当前程序如果请求码为 MainActivity 中 startActivityForResult(intent, 1)方法的第二个参数，也就是 1，返回码为 setResult(2,intent)方法中的第一个参数 2 时，获取传回的数据 data，使用了 getSringExtra()方法，也是由传递回来的数据类型决定的。比如要获取的数据是整型数据，则使用的是 getIntExtra()方法，如果是 boolean 类型的，则使用的是 getBooleanExtra()方法，等等。

第七步：运行程序

运行程序，首先显示“第一个 Activity”，点击“所在位置”跳转到“第二个 Activity”，然后点击“美丽的大海边”，关闭了“第二个 Activity”，返回到“第一个 Activity”，原先的“所在位置”变为了“您的位置：美丽的大海边”，运行效果如图 4-7 所示。

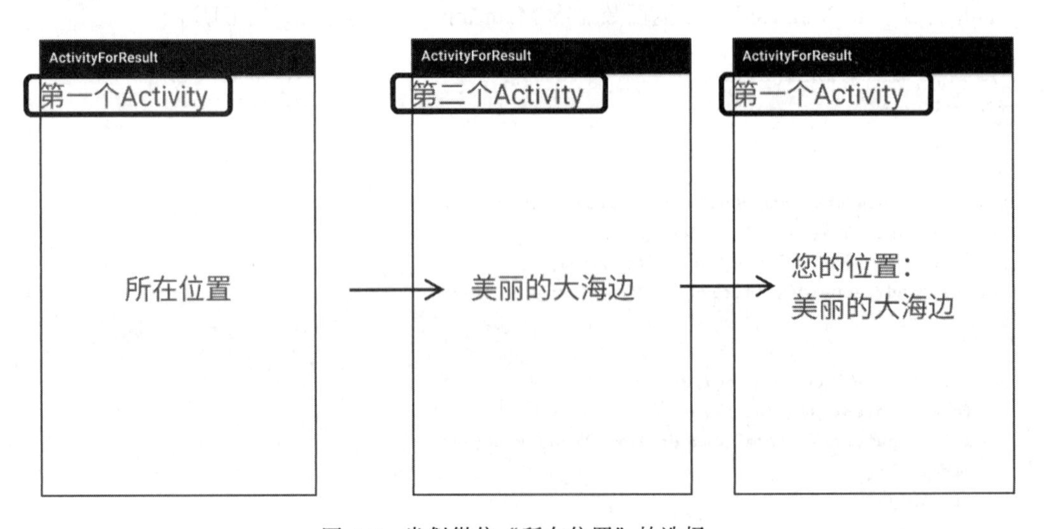

图 4-7　类似微信“所在位置”的选择

4.3 Activity 的生命周期

Activity 是所有 Android 应用的基础。到目前为止，我们已经了解了如何创建 Activity，而且可以使用意图 Intent 让一个 Activity 启动另一个 Activity。不过，底层到底发生了什么?这一节中，我们会更为深入地了解 Activity 生命周期。创建和撤销 Activity 时会发生什么？Activity 出现在前台时要调用哪些方法？Activity 失去焦点隐藏起来时又会调用哪些方法？如何保存和恢复 Activity 的状态?

下面通过一个简单的案例来体会 Activity 的生命周期。

第一步：创建项目

创建一个 Android 项目，名为 ActivityLiefCycle，项目所在的包为 com.example.activitylifecycle。

第二步：重写 Activity 生命周期的方法

打开 MainActivity，在其中重写生命周期的方法，每个方法中使用 Log 来打印信息，通过打印的信息分析这些方法的执行过程。在 MainActivity 编写如下代码：

```
public class MainActivity extends AppCompatActivity {

    @Override
    protected void onStart() {
        super.onStart();
        Log.i("AAA","调用 MainActivity 中的 onStart()方法");
    }

    @Override
    protected void onCreate(Bundle savedInstanceState) {
        super.onCreate(savedInstanceState);
        setContentView(R.layout.activity_main);
        Log.i("AAA","调用 MainActivity 中的 onCreate()方法");
    }

    @Override
    protected void onRestart() {
        super.onRestart();
        Log.i("AAA","调用 MainActivity 中的 onRestart()方法");
    }

    @Override
    protected void onResume() {
        super.onResume();
        Log.i("AAA","调用 MainActivity 中的 onResume()方法");
    }

    @Override
    protected void onPause() {
        super.onPause();
        Log.i("AAA","调用 MainActivity 中的 onPause()方法");
    }
```

```
    @Override
    protected void onStop() {
        super.onStop();
        Log.i("AAA","调用 MainActivity 中的 onStop()方法");
    }

    @Override
    protected void onDestroy() {
        super.onDestroy();
        Log.i("AAA","调用 MainActivity 中的 onDestroy()方法");
    }
}
```

第三步：运行程序

运行程序，当程序启动后在 Logcat 中，过滤日志信息输入“AAA”，可以看到依次执行了 onCreate→onStart→onResume 方法，结果如图 4-8 所示。当调用完 onResume()方法后程序不再向下进行，这时应用程序处于运行状态，等待与用户进行交互。

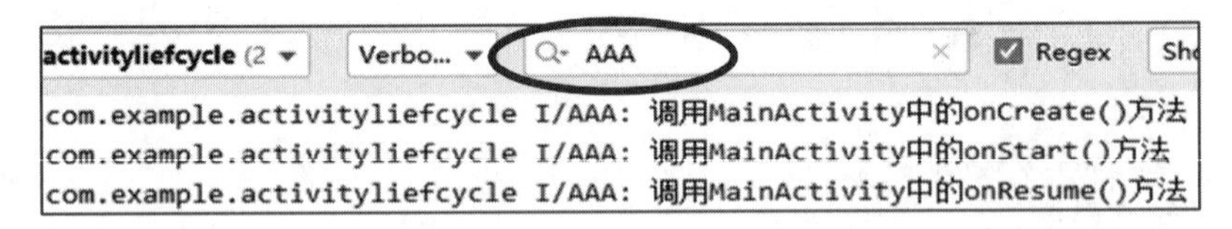

图 4-8　运行项目后 Logcat

单击模拟器“返回”按钮，这时 Activity 会被终止而返回到手机桌面。在 Logcat 中会看到调用过程如图 4-9 所示，执行了 onPause→onStop→onDestory。当调用 onDestory()方法后 Activity 被销毁并清理出内存。

```
com.example.activityliefcycle I/AAA: 调用MainActivity中的onPause()方法
com.example.activityliefcycle I/AAA: 调用MainActivity中的onStop()方法
com.example.activityliefcycle I/AAA: 调用MainActivity中的onDestroy()方法
```

图 4-9　单击“返回”后 Logcat

重新启动这个程序，可以看到这次执行过程与第一次运行的执行过程相同。如图 4-8 所示。

在 Activity 被启动之后，单击 Home 按钮，这时应用会失去焦点而不可见，但是不会被终止，执行了 onPause→onStop，如图 4-10 所示。

```
com.example.activityliefcycle I/AAA: 调用MainActivity中的onPause()方法
com.example.activityliefcycle I/AAA: 调用MainActivity中的onStop()方法
```

图 4-10　单击“Home”按钮后 Logcat

单击 Home 按钮后，又回到桌面，然后再找到这个应用并打开，可以看到依次执行 onRestart→onStart→onResume，执行过程如图 4-11 所示。这次没有执行 onCreate 方法，说明这个 Activity 只是被停止，而并没有被终止。如果需要重新运行这个 Activity，则需要调用 onRestart 方法。

```
com.example.activityliefcycle I/AAA: 调用MainActivity中的onRestart()方法
com.example.activityliefcycle I/AAA: 调用MainActivity中的onStart()方法
com.example.activityliefcycle I/AAA: 调用MainActivity中的onResume()方法
```

图 4-11 重新打开应用后 Logcat

(1) Activity 生命周期的四种状态

Activity 的生命周期指的是 Activity 从创建到销毁的整个过程，这个过程大致可以分为四种状态：运行状态、暂停状态、停止状态和销毁状态。

① 运行状态 一个新 Activity 启动后，它显示在屏幕最前端，此时它处于可见并可和用户交互的激活状态，叫做活动状态或者运行状态。

② 暂停状态 在某些情况下，Activity 对用户仍然可见，但是它无法获取焦点，用户不能对它进行操作，此时的状态叫做暂停状态。例如，当前 Activity 上弹出一个需要反馈的对话框，被覆盖的 Activity 可见，但是如果不响应对话框，就不能操作 Activity，Activity 这时就处于暂停状态。

③ 停止状态 当一个 Activity 完全处于不可见状态时，它处于停止状态。当系统内存需要被用在其他地方的时候，处于停止状态的 Activity 将被强行终止。

④ 销毁状态 当 Activity 结束时，将被清理出内存，需要重新启动才可以显示和使用。

在 Activity 的生命周期状态中，处于运行状态和暂停状态是可见的，处于停止状态和销毁状态是不可见的。

(2) Activity 生命周期的方法

Activity 的生命周期包括创建、可见、失去焦点、不可见、重新可见、销毁等环节，每个环节都有对应方法。Activity 生命周期模型如图 4-12 所示，具体的方法如下：

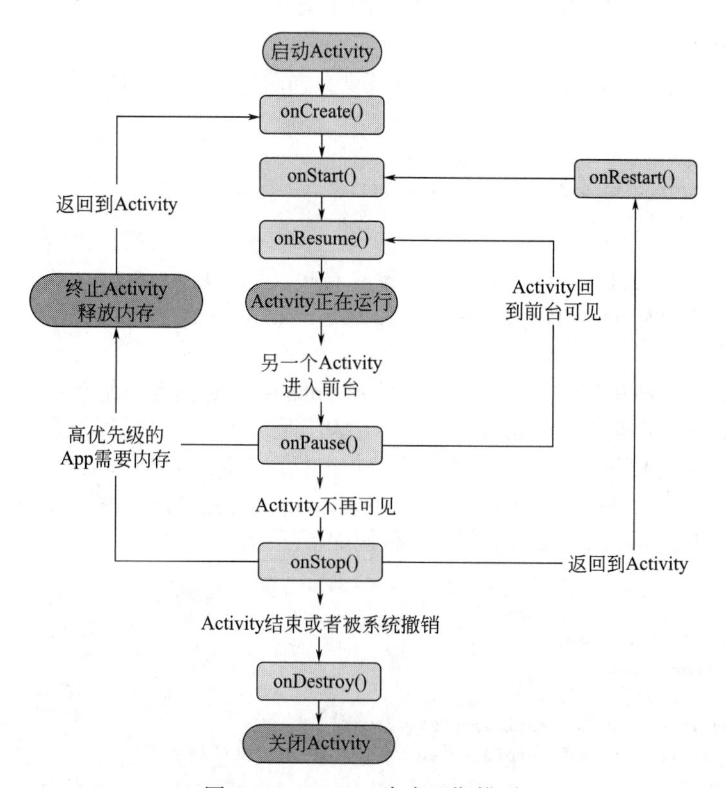

图 4-12 Activity 生命周期模型

① onCreate()方法　创建 Activity 时被调用，我们见得比较多，每个 Activity 中都重写这个方法，一般在此方法中完成 Activity 的初始化操作，如加载布局、绑定事件等。

② onStart()方法　启动 Activity 时，也就是 Activity 由不可见变为可见时被调用。

③ onResume()方法　在 Activity 准备好和用户进行交互的时候调用，此时 Activity 处于运行状态。

④ onPause()方法　当一个 Activity 失去焦点，不能再与用户交互时调用。

⑤ onStop()方法　当 Activity 不再对用户可见时调用，在系统需要内存时终止该 Activity。

⑥ onRestart()方法　Activity 由停止状态变为运行状态之前被调用，也就是 Activity 被重新启动了。

⑦ onDestroy()方法　Activity 销毁时被调用，完成最后的清理工作。

理解了 Activity 生命周期的状态和常用的方法，就可以根据实际需要在 Activity 状态发生变化的时候设置需要的操作。

4.4　Intent 的使用

通过前面的学习我们知道使用 Intent 可以启动 Activity，实现 Activity 的跳转。也可以使用 Intent 对象传递数据和实现数据回传，那么 Intent 是什么，又有哪些用法呢？下面详细介绍一下 Intent 的相关内容。

4.4.1　Intent 简介

显式意图

Intent 又称为意图，可以分为两种：显式意图和隐式意图。

（1）显式意图

显示意图：通过直接指定目标组件的名称实现页面跳转，清单文件中不用添加意图过滤器来开启页面。比如：

```
Intent intent=new Intent(this,MainActivity2.class);
startActivty(intent);
```

在上面的代码中首先创建一个 Intent 对象，直接指明 MainActivity2.class 为目标，在当前 Activity 的基础上打开 MainActivity2.class，然后通过 startActivity()方法来开启这个 Intent。

（2）隐式意图

隐式意图：是指在清单文件中通过意图过滤器<intent-filter>匹配一组动作或数据开启页面。隐式意图不明确指出想要启动哪个活动，而是指定了一系列更为抽象的 action 和 category 等信息，然后交由系统去分析并找出对应的活动去启动。

隐式意图

在清单文件 AndroidManifest 中对目标 MainActivity2 进行设置，添加<intent-filter>标记，并设置 action 和 category 属性，代码如下所示：

```
<activity
    android:name=".MainActivity2"
    android:exported="false" >
    <intent-filter>
        <action android:name="nextactivity"/>
        <category android:name="android.intent.category.DEFAULT"/>
    </intent-filter>
</activity>
```

在 MainActivity 文件中设计如下代码：

```
Intent i=new Intent();
i.setAction("nextactivity");
startActivity(i);
```

注意：清单文件中 action 的内容必须与 java 文件中的 setAction 的内容一致。使用隐式意图开启 Activity 时，系统会默认为该意图添加“android.intent.category.DEFAULT”的 category,因此为了开启的 Activity 能够接收隐式意图,必须在 AndroidManifest 文件的 Activity 标签下的 <intent-filter>中，为被开启的 Activity 指定分类 category 为“android.intent.category.DEFAULT”。

4.4.2 拨打电话

在很多的移动应用中都会用到拨打电话的功能，如何实现呢？这就需要涉及 Intent 的相关知识和 Android 权限处理。

（1）Android 中的权限处理

Android6.0 系统中开始使用动态权限，也就是说，用户不需要在安装软件的时候一次性授权所有申请的权限，而是可以在软件的使用过程中再对某一项权限申请进行授权。比如说一款相机应用在运行时申请了地理位置定位权限，就算你拒绝了这个权限，仍然可以使用这个应用的其他功能。在 Android 中将所有的权限归为两类，一类是普通权限，一类是危险权限。普通权限是指那些不会直接威胁到用户安全和隐私的权限，对于这部分权限申请，系统会自动进行授权，例如网络权限申请<uses-permission android:name="android.permission.INTERNET"/>等。危险权限指那些可能会触及用户隐私或者对设备安全性造成影响的权限，对于这部分权限，必须要用户手动授权才可以，否则程序则无法使用相应的功能。危险权限主要有九类，它们分别是：STORAGE 存储卡上文件读写权限组，MICROPHONE 麦克风录音权限，CAMERA 相机权限组，LOCATION 位置权限组，PHONE 手机权限组，CALENDAR 日历权限组，CONTACTS 联系人权限组，SMS 短信权限组，BODY_SENSORS 传感器权限组。

（2）拨打电话的实现

拨打电话的应用要涉及 PHONE 手机权限组，所以在编写拨打电话逻辑之前先要取得打电话的权限，然后通过 Intent 来使用拨打电话的功能，实现如图 4-13 所示的效果，具体操作步骤如下。

(a) 拨打电话应用界面

(b) 拨打电话运行效果

图 4-13　拨打电话

第一步：新建项目

新建一个项目，名为 CallPhone，包名为 com.example.callphone。

第二步：设计用户界面

在布局文件 activity_main 中从 Palette 窗口的 Text 分组中拖入一个 Phone 控件（EditText 控件的一种，自动设置 Android:inputType 属性为“phone”，只能用来输入电话号码），设置 android:id 属性为 etPhoneNumb，设置输入框的 android:hint 属性为“请输入电话号码”，设置文本对齐方式 android:gravity 属性为"center"；再添加一个 Button 控件，“拨打电话”命令按钮的 onClick 属性为“btnCallPhone”，这就需要在 MainActivity 文件中实现 btnCallPhone()方法。为这两个控件添加位置约束，设计界面效果如图 4-13 所示。

第三步：设置 Android 权限

Android 系统中为了安全的需要，如果需要拨号打电话、发短信或者是访问网络，要先在 AndroidManifest 中配置权限许可，配置方法是在 Application 标记之外添加许可操作标记。打电话权限许可为：

```
<?xml version="1.0" encoding="utf-8"?>
<manifest xmlns:android="http://schemas.android.com/apk/res/android"
    package="com.example.callphone">
    <uses-permission android:name="android.permission.CALL_PHONE"/>
        ......
</manifest>
```

第四步：实现打电话功能

在 MainActivity 先动态申请拨打电话的权限，然后编码实现打电话功能，代码如下：

```
public class MainActivity extends AppCompatActivity {
    EditText etphonenumb;

    @Override
    protected void onCreate(Bundle savedInstanceState) {
        super.onCreate(savedInstanceState);
        setContentView(R.layout.activity_main);
        // setContentView(R.layout.layout);
         etphonenumb = findViewById(R.id.etPhoneNumb);
        String[] permissions={Manifest.permission.CALL_PHONE};
        if (Build.VERSION.SDK_INT > 23) {
            requestPermissions(permissions, 100);
        }
    }

    public void btnCallPhone(View view) {
        String strPhone = etphonenumb.getText().toString();
        Intent intent = new Intent();
        intent.setAction(Intent.ACTION_CALL);
```

```
            intent.setData(Uri.parse("tel:" + strPhone));
            startActivity(intent);
        }
}
```

在 onCreate()方法中对输入框进行初始化，定义了一个 String 类型数组，数组内容为请求授予当前应用程序的权限，如果需要多个请求权限，直接以数组元素的形式加入 permission 就可以了，比如还要使用相机功能，则修改数组定义为 String[] permission= {"android.permission.CALL_PHONE","android.permission.CAMERA"};需要请求几个授权就加入几个权限。在 AndroidManifest 文件中进行了配置，Android 运行权限在 Android6.0 以下，在 AndroidManifest 配置了就默认同意了，但是在 Android6.0 之上就需要手动加入，if(Build.VERSION.*SDK_INT*>=23)进行判断，如果在 23 之上，则调用 requestPermissions()方法申请权限。这个方法有两个参数，第一个参数是要申请的权限，第二个参数是请求码。图 4-14 是 SDK 的版本号。在 btnCallPhone()方法中定义一个意图 Intent 对象，使用 setAction()为意图添加打电话的动作 Intent.ACTION_CALL，使用 setData()方法设置数据为 Uri.*parse*("tel:" + strPhone)，然后开启意图。

我们常见的权限的判定也可以写成 :if(Build.VERSION.*SDK_INT*>Build.VERSION_CODES.*LOLLIPOP_MR1*)，作用与 if(Build.VERSION.*SDK_INT*>=23)相同。

第五步：运行程序

运行程序显示如图 4-15 所示界面，权限请求对话框询问“CallPhone”请求电话权限，是否允许？点击“Allow”显示图 4-13(a) 所示界面，输入电话号码，点击“拨打电话”，进入图 4-13(b) 所示界面即拨打电话中。

e	API Level	Revision
Android Sv2 Preview	Sv2	1
Android API 31	31	1
Android 11.0 (R)	30	3
Android 10.0 (Q)	29	5
Android 9.0 (Pie)	28	6
Android 8.1 (Oreo)	27	3
Android 8.0 (Oreo)	26	2
Android 7.1.1 (Nougat)	25	3
Android 7.0 (Nougat)	24	2
Android 6.0 (Marshmallow)	23	3
Android 5.1 (Lollipop)	22	2
Android 5.0 (Lollipop)	21	2
Android 4.4W (KitKat Wear)	20	2
Android 4.4 (KitKat)	19	4
Android 4.3 (Jelly Bean)	18	3
Android 4.2 (Jelly Bean)	17	3
Android 4.1 (Jelly Bean)	16	5
Android 4.0.3 (IceCreamSandwich)	15	5

图 4-14　SDK 版本号

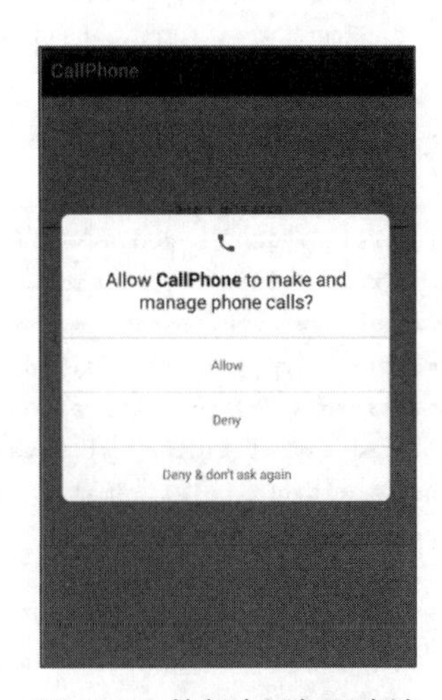

图 4-15　拨打电话权限申请

4.5　用户注册头像的选择

下面通过用户注册头像的选择实现在第一个 Activity 点击“选择头像”进入第二个 Activity，然后在第二个 Activity 所列的头像中单击某个头像，返回到第一个 Activity，并将选择的头像显示在第一个 Activity，运行效果如图 4-16 所示。

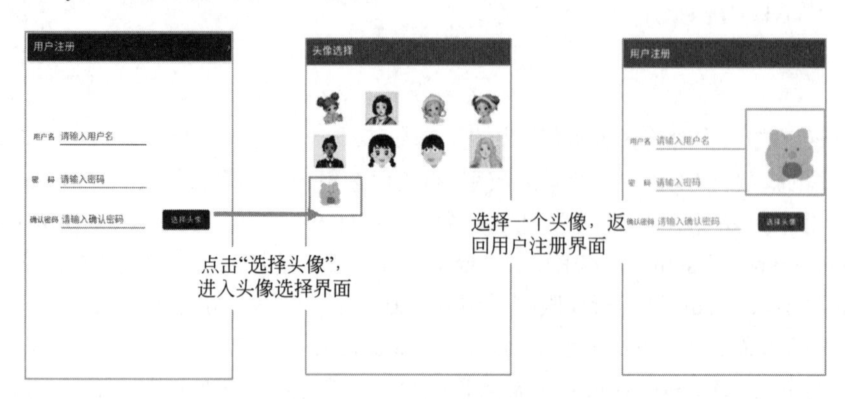

图 4-16 用户注册头像选择运行界面

分析 这个案例涉及两个用户界面，“用户注册”界面和“头像选择”界面，在头像选择界面中使用了 GridView 控件，选择 GridView 中的某一项，将选中项图传回到“用户注册界面”。项目具体步骤如下。

第一步：创建项目

创建一个 Android 项目，名为 RegisterImageChoise，包名为 com.example.registerimagechoise。

第二步：用户界面设计

在布局文件 activity_main 中添加三个 TextView 控件，分别显示“用户名”“密码”和“确认密码”，添加三个 EditText 控件用来输入用户名(android:id 设置为 etname)、密码(android:id 设置为 etpwd)和确认密码(android:id 设置为 etpwd2)。添加一个 ImageView 控件用来显示最终用户选择的头像，添加一个 Button 控件“选择头像”，所有控件添加约束。参考代码如下所示。

```
<?xml version="1.0" encoding="utf-8"?>
<androidx.constraintlayout.widget.ConstraintLayout
xmlns:android="http://schemas.android.com/apk/res/android"
    xmlns:app="http://schemas.android.com/apk/res-auto"
    xmlns:tools="http://schemas.android.com/tools"
    android:layout_width="match_parent"
    android:layout_height="match_parent"
    tools:context=".MainActivity">
    <TextView
        android:id="@+id/textView"
        android:layout_width="wrap_content"
        android:layout_height="wrap_content"
        android:layout_marginEnd="7dp"
        android:layout_marginRight="7dp"
        android:text="用户名"
        app:layout_constraintBaseline_toBaselineOf="@+id/etname"
        app:layout_constraintEnd_toStartOf="@+id/etname" />

    <EditText
        android:id="@+id/etname"
        android:layout_width="0dp"
```

```
        android:layout_height="49dp"
        android:layout_marginTop="20dp"
        android:layout_marginBottom="32dp"
        android:hint="请输入用户名"
        android:inputType="textPersonName"
        app:layout_constraintBottom_toTopOf="@+id/etpwd"
        app:layout_constraintEnd_toEndOf="@+id/etpwd"
        app:layout_constraintStart_toStartOf="@+id/etpwd"
        app:layout_constraintTop_toTopOf="@+id/imageView" />

    <ImageView
        android:id="@+id/imageView"
        android:layout_width="150dp"
        android:layout_height="150dp"
        android:layout_marginTop="162dp"
        android:layout_marginBottom="30dp"
        app:layout_constraintBottom_toTopOf="@+id/button"
        app:layout_constraintEnd_toEndOf="parent"
        app:layout_constraintStart_toEndOf="@+id/etpwd"
        app:layout_constraintTop_toTopOf="parent" />

    <TextView
        android:id="@+id/textView3"
        android:layout_width="wrap_content"
        android:layout_height="wrap_content"
        android:layout_marginStart="9dp"
        android:layout_marginLeft="9dp"
        android:layout_marginEnd="2dp"
        android:layout_marginRight="2dp"
        android:text="确认密码"
        app:layout_constraintBaseline_toBaselineOf="@+id/etpwd2"
        app:layout_constraintEnd_toStartOf="@+id/etpwd2"
        app:layout_constraintStart_toStartOf="parent" />

    <EditText
        android:id="@+id/etpwd"
        android:layout_width="0dp"
        android:layout_height="49dp"
        android:layout_marginEnd="8dp"
        android:layout_marginRight="8dp"
        android:hint="请输入密码"
        android:inputType="textPassword"
        app:layout_constraintBaseline_toBaselineOf="@+id/textView5"
        app:layout_constraintEnd_toStartOf="@+id/imageView"
        app:layout_constraintStart_toEndOf="@+id/textView5" />

    <TextView
        android:id="@+id/textView5"
        android:layout_width="wrap_content"
        android:layout_height="wrap_content"
        android:layout_marginStart="13dp"
        android:layout_marginLeft="13dp"
        android:layout_marginEnd="7dp"
        android:layout_marginRight="7dp"
```

```
        android:layout_marginBottom="41dp"
        android:text="密      码"
        app:layout_constraintBottom_toTopOf="@+id/etpwd2"
        app:layout_constraintEnd_toStartOf="@+id/etpwd"
        app:layout_constraintStart_toStartOf="parent" />

    <EditText
        android:id="@+id/etpwd2"
        android:layout_width="0dp"
        android:layout_height="49dp"
        android:layout_marginEnd="31dp"
        android:layout_marginRight="31dp"
        android:hint="请输入确认密码"
        android:inputType="textPassword"
        app:layout_constraintBottom_toBottomOf="parent"
        app:layout_constraintEnd_toStartOf="@+id/button"
        app:layout_constraintStart_toEndOf="@+id/textView3"
        app:layout_constraintTop_toTopOf="parent" />

    <Button
        android:id="@+id/button"
        android:layout_width="wrap_content"
        android:layout_height="wrap_content"
        android:layout_marginEnd="45dp"
        android:layout_marginRight="45dp"
        android:layout_marginBottom="341dp"
        android:onClick="btnChoise"
        android:text="选择头像"
        app:layout_constraintBottom_toBottomOf="parent"
        app:layout_constraintEnd_toEndOf="parent"
        app:layout_constraintStart_toEndOf="@+id/etpwd2"
        app:layout_constraintTop_toBottomOf="@+id/imageView" />

</androidx.constraintlayout.widget.ConstraintLayout>
```

第三步：新建 Activity

新建一个名为 HeadImageActivity 的新 Activity，设计布局文件 activity_head_image，在布局文件中添加一个 GridView 控件，用于显示可选择的头像列表，设置其属性代码如下：

```
<?xml version="1.0" encoding="utf-8"?>
<androidx.constraintlayout.widget.ConstraintLayout
xmlns:android="http://schemas.android.com/apk/res/android"
    xmlns:app="http://schemas.android.com/apk/res-auto"
    xmlns:tools="http://schemas.android.com/tools"
    android:layout_width="match_parent"
    android:layout_height="match_parent"
    tools:context=".HeadImageActivity">

    <GridView
        android:id="@+id/gridview"
        android:layout_width="match_parent"
```

```
        android:layout_height="wrap_content"
        android:layout_marginTop="50dp"
        android:numColumns="4"
        android:verticalSpacing="15dp"
        app:layout_constraintTop_toTopOf="parent" />

</androidx.constraintlayout.widget.ConstraintLayout>
```

使用 android:numColumns="4"设置 GridView 控件一行分成 4 列，android:verticalSpacing="15dp"设置行间的垂直距离为 15dp。

第四步：可选头像界面设计

在 HeadImageActivity 文件中定义保存头像的数组，然后重写 onCreate()方法，获取 GridView 控件。

```
public class HeadImageActivity extends AppCompatActivity {
    int imgid[] = {R.drawable.toux1, R.drawable.toux2,R.drawable.toux3, R.drawable.toux4, R.drawable.
toux5, R.drawable.toux6, R.drawable.toux7, R.drawable.toux8, R.drawable.toux9};
    private GridView gl;

    @Override
    protected void onCreate(Bundle savedInstanceState) {
        super.onCreate(savedInstanceState);
        setContentView(R.layout.activity_head_image);
        gl = findViewById(R.id.gridview);//初始化 GridView 控件
         MyAdapter myAdapter = new MyAdapter();//创建自定义类MyAdapter 对象
         gl.setAdapter(myAdapter);//GridView 控件关联适配器
         gl.setOnItemClickListener(new AdapterView.OnItemClickListener() {
           @Override//GridView 控件的选择项单击
             public void onItemClick(AdapterView<?> parent, View view, int position, long id) {
              Intent intent = getIntent();
              Bundle bundle = new Bundle();
              bundle.putInt("imageid", imgid[position]);
              intent.putExtras(bundle);
              setResult(2, intent);
              finish();
          }
        });
    }

    class MyAdapter extends BaseAdapter {

        @Override
        public int getCount() {
            return imgid.length;
        }

        @Override
        public Object getItem(int position) {
            return position;
        }
```

```
        @Override
        public long getItemId(int position) {
            return position;
        }

        @Override
        public View getView(int position, View convertView, ViewGroup parent) {
            ImageView imageView;
            if (convertView == null) {
                imageView = new ImageView(HeadImageActivity.this);
                //设置图像的宽度和高度
                imageView.setAdjustViewBounds(true);
                imageView.setMaxHeight(180);
                imageView.setMaxHeight(180);
                //设置 ImageView 的内边距
                imageView.setPadding(5, 5, 5, 5);
            } else {
                imageView = (ImageView) convertView;
            }
            //设置要显示的图片
              imageView.setImageResource(imgid[position]);
            return imageView;
        }
    }
}
```

定义 int 类型数组 imgid 用来保存头像 id，使用 GridView 进行图片集的显示，要添加一个与之关联的适配器，接着再给它添加一个 onItemClickListener 事件监听器，在监听器的 onItemClick()方法中获取 Intent 对象，将选择的头像 ID 保存在一个新建的数据包中，将数据保存到 Intent 中，同时使用 setResult(2,intent)方法设置返回码和返回的 Activity，最后调用 finish()方法关闭当前的 Activity。在本项目的实现中，自定义了 MyAdapter 这个适配器继承自 BaseAdapter。重写 getCount()方法返回保存头像的数据的长度 imgid.length;重写 getItem()方法获取当前选项；重写 getItemId()方法获得当前选项的 ID，重写 getView()方法设置 GridView 中显示的内容。

第五步：头像选择的实现

修改 MainActivity 中的代码，首先实现 Button“选择头像”命令按钮的 onClick 属性 btnChoise()方法，重写 onActivityResult()方法，在该方法中判断请求码和返回码是否与 Activity 跳转和返回方法中设置的值相同，如果相同获取传递回来的数据，设置头像。具体代码如下所示:

```
public class MainActivity extends AppCompatActivity {
    @Override
    protected void onCreate(Bundle savedInstanceState) {
        super.onCreate(savedInstanceState);
        setContentView(R.layout.activity_main);

    }
    public void btnChoise(View view){
        Intent intent=new Intent(this,HeadImageActivity.class);
        startActivityForResult(intent,1);
```

```
    }

    @Override
    protected void onActivityResult(int requestCode, int resultCode, @Nullable Intent data) {
        super.onActivityResult(requestCode, resultCode, data);
        if(requestCode==1||resultCode==2){
            Bundle bundle=data.getExtras();
            int imageid=bundle.getInt("imageid");
            ImageView imageView=findViewById(R.id.imageView);
            imageView.setImageResource(imageid);
        }
    }
}
```

在上面的代码中 btnChoise () 方法是命令按钮的单击事件，使用 startActivityForResult(intent,1) 方法启动 HeadImageActivity,在 HeadImageActivity 使用 setResult(2, intent)方法返回 MainActivity 回调 onActivityResult()方法，在此方法中接收传递过来的数据，设置 imageView 显示所选择的头像。

第六步：修改 Activity 的标签

查看 AndroidManifest，在此文件中只是修改了 application 的 android:label 属性的值和创建头像选择 HeadImageActivity 中 android:label 的属性值。

```
<?xml version="1.0" encoding="utf-8"?>
<manifest xmlns:android="http://schemas.android.com/apk/res/android"
    package="com.example.registerimagechoise">

    <application
        android:allowBackup="true"
        android:icon="@mipmap/ic_launcher"
        android:label="用户注册"
        android:roundIcon="@mipmap/ic_launcher_round"
        android:supportsRtl="true"
        android:theme="@style/Theme.RegisterImageChoise">
        <activity
            android:name=".HeadImageActivity"
            android:exported="false"
            android:label="头像选择"/>
        <activity
            android:name=".MainActivity"
            android:exported="true">
            <intent-filter>
                <action android:name="android.intent.action.MAIN" />
                <category android:name="android.intent.category.LAUNCHER" />
            </intent-filter>
        </activity>
    </application>
</manifest>
```

通过这个案例总结，实现 Activity 间的数据回传需要使用 3 个方法，由第一个 Activity 跳转到第二个 Activity 使用 startActivityForResult()方法，由第二个 Activtiy 回传数据需要使用 setResult()

方法，第一个 Activty 接收第二个 Activity 传递回来的数据需要重写 onActivityResult()方法来接收数据。

【职业素养拓展4】

职业素养拓展

职业素养拓展内容详见文件“职业素养拓展”。

扫描二维码查看【职业素养拓展 4】

本章小结

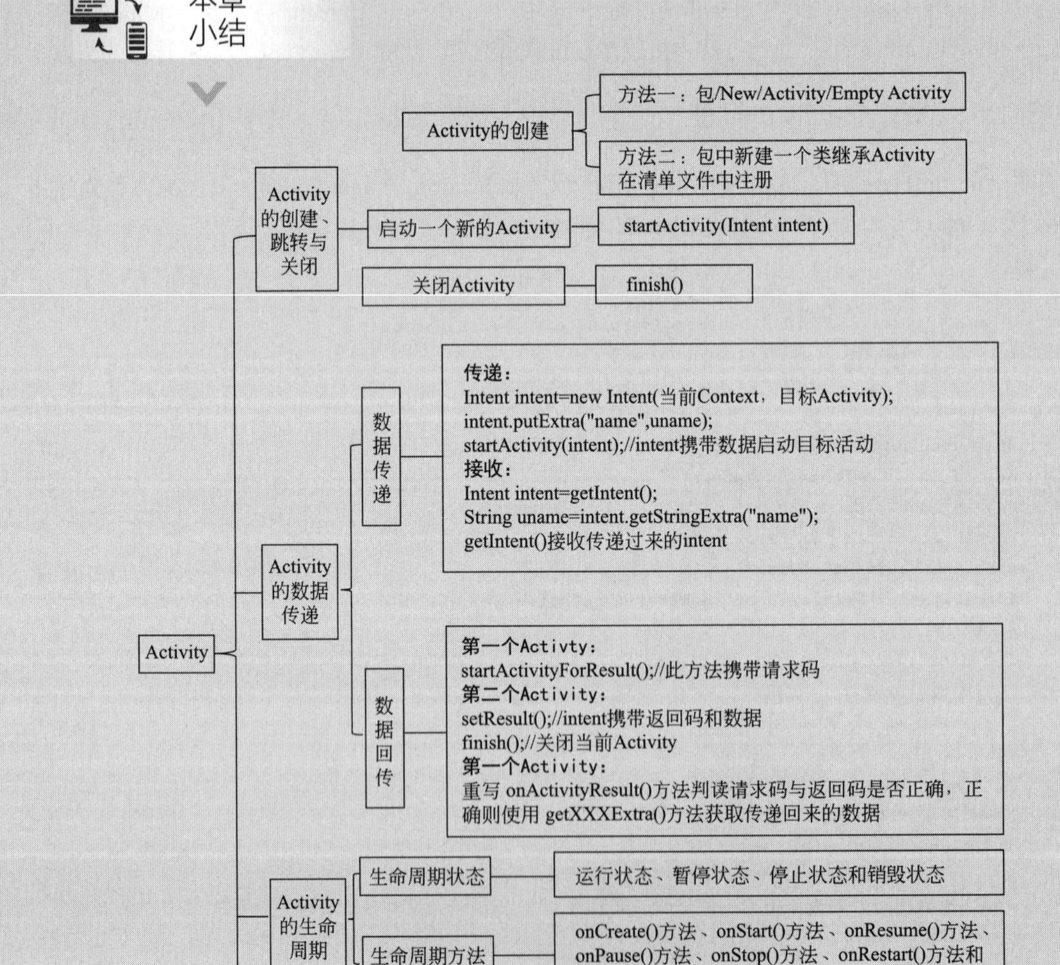

第5章 Fragment

随着移动设备的迅速发展，不仅手机成为人们生活的必需品，就连平板电脑也变得越来越普及。平板电脑和手机的最大差别就在于屏幕的大小，屏幕大小的差别可能会影响同一界面在不同设备上的显示效果，为了同时兼顾到手机开发和平板电脑的开发，Android 提供了 Fragment。本章主要内容为 Fragment 的用法，通过学习要求掌握静态、动态 Fragment 的添加与使用，并能够实现相对复杂的用户界面。

学习目标：

1. 理解 Fragment 的作用；
2. 掌握添加 Fragment 的方法；
3. 能够熟练使用 Fragment 创建常见应用界面。

5.1 Fragment 概述

Fragment 意思为碎片、片段。Fragment 是可以嵌入在 Activity 中当作用户界面的一部分，它可以让程序更加合理和充分地利用屏幕空间。使用 Fragment 可以把屏幕划分成几块，然后进行分组，进行一个模块化的管理。在使用时可以把多个 Fragment 放到一个 Activity 里，也可以把一个 Fragment 放在多个 Activity 里。

Fragment 必须始终托管在 Activity 中，其生命周期直接受宿主 Activity 生命周期的影响。例如，当 Activity 暂停时，Activity 的所有片段也会暂停；当 Activity 被销毁时，所有片段也会被销毁。不过，当 Activity 正在运行（处于已恢复生命周期状态）时，可以独立操纵每个片段，如添加或移除片段。

Android 在 Android3.0（API 级别 11）中引入了片段，主要目的是为大屏幕（如平板电脑）上更加动态和灵活的界面设计提供支持。由于平板电脑的屏幕尺寸远胜于手机屏幕尺寸，因而有更多空间可供组合和交换界面组件。利用片段实现此类设计时，无需管理对视图层次结构做出的复杂更改。

例如，新闻应用可以使用一个片段在左侧显示文章列表，使用另一个片段在右侧显示文章，两个片段并排显示在一个 Activity 中，每个片段都拥有自己的一套生命周期回调方法，并各自处理自己的用户输入事件。因此，用户无需使用一个 Activity 来选择文章，然后使用另一个 Activity

来阅读文章，而是可以在同一个 Activity 内选择文章并进行阅读，如图 5-1 中的平板电脑布局所示。

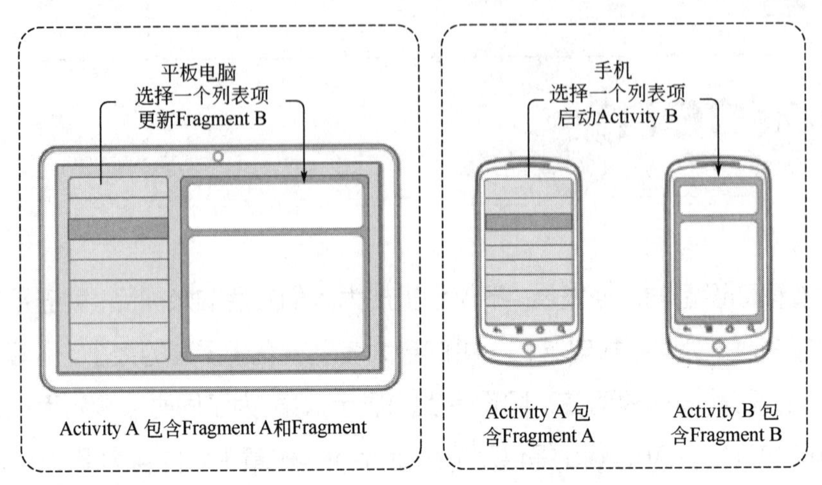

图 5-1　由片段定义的两个界面模块如何适应不同设计的示例

（通过组合成一个 Activity 来适应平板电脑设计，通过单独片段来适应手机设计）

在手机上，如果不能在同一 Activity 内储存多个片段，则可利用单独的片段来实现单窗格界面。仍以新闻应用为例，在平板电脑尺寸的设备上运行时，该应用可以在 Activity A 中嵌入两个片段。不过，手机尺寸的屏幕没有足够的空间来存储两个片段，因此 Activity A 只包含用于显示文章列表的片段，并且当用户选择文章时，它会启动 Activity B，其包含用于阅读文章的第二个片段。因此，应用可通过重复使用不同组合的片段来同时支持平板电脑和手机（如图 5-1 所示）。

具体的实现是在设计时利用不同片段组合适应不同屏幕配置，本章的学习我们只是以 Fragment 的应用为主，没有涉及屏幕适配的问题。

5.2　Fragment 的创建

创建 Fragment 可以通过使用快捷菜单和新建类两种方法实现，下面我们通过一个实例学习创建 Fragment。操作步骤如下所示。

第一步：新建项目

创建一个项目，名为 FragmentTest，项目所在的包为 com.example.fragmenttest。

第二步：新建两个 Fragment

（1）创建第一个 Fragment：使用快捷菜单创建 Fragment

选择当前程序包名，单击鼠标右键选择“New”→“Fragment”→“Fragment(Blank)”会弹出“New Android Fragment”对话框，当输入 Fragment Name 为“Fragment_1”时系统会自动更改 Fragment Layout Name 为“fragment_1”，如图 5-2 所示。此时系统会在当前包下增加一个继承自 Fragment 的类 Fragment1.java，在 res/layout/目录下增加一个布局文件 fragment_1.xml。

打开 Fragment1 看到自动产生的代码，在此我们只关注重写了的 onCreateView()方法，在该方法中加载了布局文件 fragment_1。

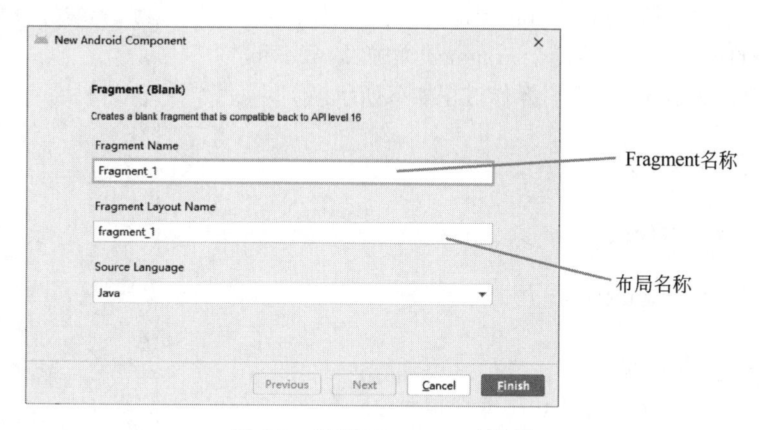

图 5-2　新建 Fragment 对话框

```
@Override
public View onCreateView(LayoutInflater inflater, ViewGroup container,Bundle savedInstanceState) {
    // Inflate the layout for this fragment
    return inflater.inflate(R.layout.fragment_1, container, false);
}
```

（2）创建第二个 Fragment

① 新建一个类 Fragment2 使其继承自 Fragment;

② 创建一个布局文件 fragment_2.xml;

③ Fragment_2 中加载布局 fragment_2 的代码如下所示:

```
public class Fragment_2 extends Fragment {
    @Nullable
    @Override
    public View onCreateView(@NonNull LayoutInflater inflater, @Nullable ViewGroup container,
@Nullable Bundle savedInstanceState) {
        View  view = inflater.inflate(R.layout.fragment_2,container,false);
        return view;
    }
}
```

这里只是重写了 onCreateView()方法，在这个方法中通过 LayoutInflater 的 inflate()方法给 Fragment_2 添加布局 fragment_2，这样 Fragment 就建好了。

5.3　使用 Fragment

在前面我们提到 Fragment 不能独立存在，它必须嵌套在 Activity 中才能使用，那么如何使用 Fragment 呢？有两种方法可以添加 Fragment，一种是使用静态方法添加 Fragment；另一种是用动态方法添加 Fragment。

5.3.1　静态添加 Fragment

静态添加 Fragment 就是在 Activity 的布局文件中引入 Fragment，就当和普通的 View 一样。

下面通过案例演示如何静态添加 Fragment。以上一节内容所建项目 FragmentTest 为基础，将两个 Fragment 添加到 Activity 中，程序运行效果如图 5-3 所示。操作过程如下所示。

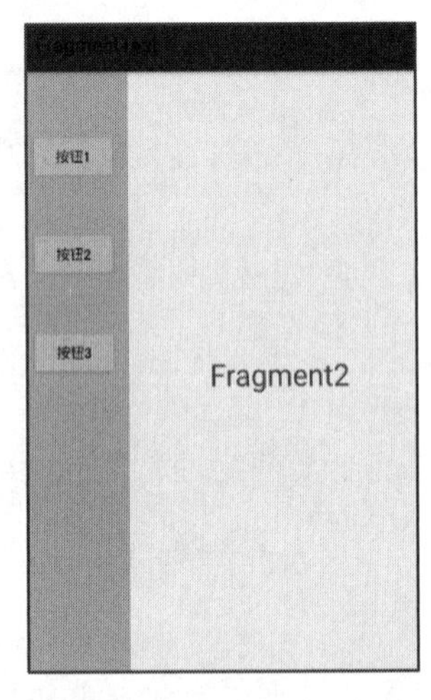

图 5-3 Activity 中加入 Fragment

第一步：打开项目

打开项目 FragmentTest。

第二步：设计 Fragment 界面

（1）fragment_1 布局设计

布局文件 fragment_1 中整体使用线性布局，并设置了背景颜色，添加三个命令按钮,fragment_1 布局文件的代码如下所示。

```
<?xml version="1.0" encoding="utf-8"?>
<LinearLayout xmlns:android="http://schemas.android.com/apk/res/android"
    xmlns:tools="http://schemas.android.com/tools"
    android:layout_width="match_parent"
    android:layout_height="match_parent"
    android:background="#C7C9C8"
    android:orientation="vertical"
    tools:context=".Fragment_1">

    <Button
        android:layout_width="wrap_content"
        android:layout_height="wrap_content"
        android:layout_marginTop="50dp"
        android:text="按钮 1" />

    <Button
        android:layout_width="wrap_content"
        android:layout_height="wrap_content"
        android:layout_marginTop="50dp"
        android:text="按钮 2" />

    <Button
        android:layout_width="wrap_content"
        android:layout_height="wrap_content"
        android:layout_marginTop="50dp"
        android:text="按钮 3" />

</LinearLayout>
```

（2）fragment_2 布局设计

布局文件 fragment_2 中整体使用约束布局，并设置了背景颜色，添加了一个 TextView 控件，标识“Fragment2”,fragment_2 的代码如下所示：

```
<?xml version="1.0" encoding="utf-8"?>
<androidx.constraintlayout.widget.ConstraintLayout
```

```
xmlns:android="http://schemas.android.com/apk/res/android"
    xmlns:app="http://schemas.android.com/apk/res-auto"
    xmlns:tools="http://schemas.android.com/tools"
    android:layout_width="match_parent"
    android:layout_height="match_parent"
    android:background="#F8F3F3">

    <TextView
        android:id="@+id/textView"
        android:layout_width="wrap_content"
        android:layout_height="wrap_content"
        android:text="Fragment2"
        android:textSize="30sp"
        app:layout_constraintBottom_toBottomOf="parent"
        app:layout_constraintEnd_toEndOf="parent"
        app:layout_constraintStart_toStartOf="parent"
        app:layout_constraintTop_toTopOf="parent" />
</androidx.constraintlayout.widget.ConstraintLayout>
```

第三步：在 Activity 中添加 Fragment

在 Activity 中静态添加 Fragment，是通过在 Activity 的布局文件中添加 fragment 标记来实现的。打开 activity_main 布局文件，添加两个<fragment>标记，并设置它们的格式，左边显示 fragment_1，右边显示 fragment_2。activity_main 的布局代码如下所示。

```
<?xml version="1.0" encoding="utf-8"?>
<androidx.constraintlayout.widget.ConstraintLayout
xmlns:android="http://schemas.android.com/apk/res/android"
    xmlns:app="http://schemas.android.com/apk/res-auto"
    xmlns:tools="http://schemas.android.com/tools"
    android:layout_width="match_parent"
    android:layout_height="match_parent"
    tools:context=".MainActivity">

    <fragment
        android:id="@+id/fg1"
        android:name="com.example.fragmenttest.Fragment_1"
        android:layout_width="110dp"
        android:layout_height="0dp"
        app:layout_constraintBottom_toBottomOf="parent"
        app:layout_constraintStart_toStartOf="parent"
        app:layout_constraintTop_toTopOf="parent" />

    <fragment
        android:id="@+id/fg2"
        android:name="com.example.fragmenttest.Fragment_2"
        android:layout_width="309dp"
        android:layout_height="0dp"
        app:layout_constraintBottom_toBottomOf="parent"
        app:layout_constraintStart_toEndOf="@+id/fg1"
        app:layout_constraintTop_toTopOf="parent" />

</androidx.constraintlayout.widget.ConstraintLayout>
```

从代码中可以看到，我们在 activity_main 中添加了两个<fragment/>标记，此标记的属性我们都比较熟悉，只是 android:name 属性的值是 Fragment 的类名，而且必须指明类的包名,还有每个<fragment/>标记中都设置 android:id 属性。

注意：静态添加 Fragment，必须给 fragment 标签设置 id 属性和 name 属性,否则不能加载成功。

第四步：运行程序

从运行效果图 5-3 我们可以看到使用 Fragment 实现了 Activity 界面布局的分割，这种用法比较简单，适合添加固定的 Fragment。

5.3.2 动态添加 Fragment

我们已经学会了在 Activity 的布局文件中通过添加<fragment/>标记来添加 Fragment，但是在实际应用中更多的是在程序运行中根据要求动态添加布局，下面通过实现如图 5-4 所示效果，演示如何动态添加 Fragment。

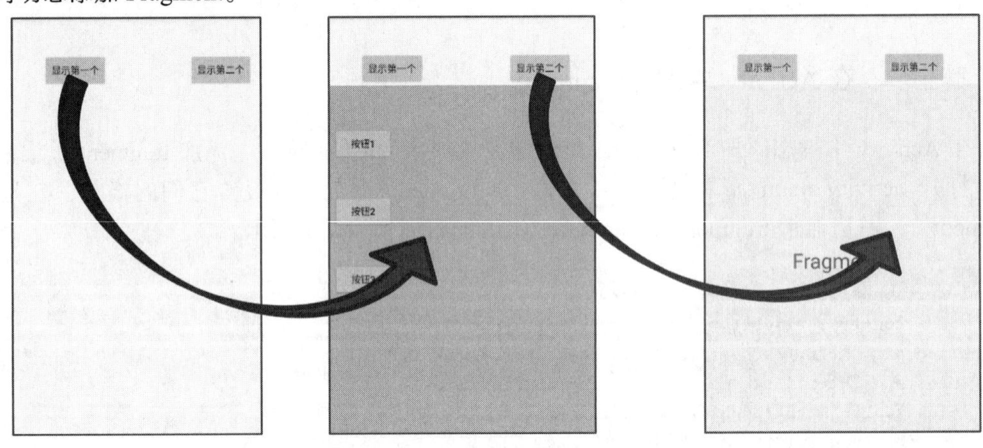

图 5-4　动态 Fragment

分析　程序运行开始，在界面上显示两个命令按钮，如果单击“显示第一个”命令按钮则在下方显示第一个 Fragment 的内容,如果单击“显示第二个”命令按钮则在下方显示第二个 Fragment 内容，根据选择的不同可以加载不同的 Fragment。

动态添加 Fragment 的过程如下所示。

① 创建一个 Fragment 的实例，如：Fragment1 f1=new Fragment1();

② 获取一个 FragmentManager（Fragment 管理器）的实例，通过直接调用 getSupportFragmentManager()方法得到，如：FragmentManager fm=getSupportFragmentManager();

③ 开启 FragmentTransaction（事务）的实例,通过 beginTransaction()方法开启;如：FragmentTransaction ft=fm.beginTransaction();

④ 调用 add()方法或者 repalce()方法向指定容器中添加或者替换 Fragment,此方法中有两个参数，第一个参数为 Activity 的布局容器，第二个参数为要添加或者替换的 Fragment 实例。如：ft.replace(R.id.*fragment_layout*,f1);

⑤ 最后提交事务，调用 commit()方法来完成。如：ft.commit();

注意：动态加载布局一定要使用 commit()方法提交，否则不能完成 Fragment 加载。

实现图 5-4 所示的效果，操作过程如下所示。

第一步：打开项目

打开项目 FragmentTest，使用已有的 Fragment1 和 Fragment2。

第二步：用户界面设计

修改布局文件 activity_main，在布局文件 activity_main 中添加了两个命令按钮 Button 指定 android:id 属性分别为“btn1”和“btn2”，添加一个 FrameLayout 布局用来放置 Fragmemt，设置的 Android:id 属性为“fragment_layout”，作为后面要添加的 Fragment 的容器。修改后 activity_main 的代码如下：

```
<?xml version="1.0" encoding="utf-8"?>
<androidx.constraintlayout.widget.ConstraintLayout
xmlns:android="http://schemas.android.com/apk/res/android"
    xmlns:app="http://schemas.android.com/apk/res-auto"
    xmlns:tools="http://schemas.android.com/tools"
    android:layout_width="match_parent"
    android:layout_height="match_parent"
    tools:context=".MainActivity">
    <Button
        android:id="@+id/btn1"
        android:layout_width="wrap_content"
        android:layout_height="wrap_content"
        android:layout_marginStart="50dp"
        android:layout_marginTop="50dp"
        android:text="显示第一个"
        android:onClick="shwoFragment1"
        app:layout_constraintStart_toStartOf="parent"
        app:layout_constraintTop_toTopOf="parent" />
    <Button
        android:id="@+id/btn2"
        android:layout_width="wrap_content"
        android:layout_height="wrap_content"
        android:layout_marginEnd="50dp"
        android:text="显示第二个"
        android:onClick="showFragment2"
        app:layout_constraintBottom_toBottomOf="@+id/btn1"
        app:layout_constraintEnd_toEndOf="parent" />
    <FrameLayout
        android:id="@+id/framelayout"
        android:layout_width="411dp"
        android:layout_height="0dp"
        app:layout_constraintBottom_toBottomOf="parent"
        app:layout_constraintStart_toStartOf="parent"
        app:layout_constraintTop_toBottomOf="@+id/btn1" />
</androidx.constraintlayout.widget.ConstraintLayout>
```

从代码中可以看出我们为两个 Button 添加了“onClick”属性，分别为“shwoFragment1”和“shwoFragment2”，用来指定命令按钮的单击事件，也就是需要在 MainActivity 文件中通过点击命令按钮加载不同的 Fragment。

第三步：Activity 中动态加载 Fragment

在 MainActivity 中实现 Fragment 的动态加载，MainActivity 的代码如下所示：

```
public class MainActivity extends AppCompatActivity {

    @Override
    protected void onCreate(Bundle savedInstanceState) {
        super.onCreate(savedInstanceState);
        setContentView(R.layout.activity_main);
    }

    public void shwoFragment1(View view) {
        addFragment(new Fragment_1());
    }

    public void showFragment2(View view) {
        addFragment(new Fragment_2());
    }

    private void addFragment(Fragment fragment) {
        FragmentManager fm = getSupportFragmentManager();
        FragmentTransaction ft = fm.beginTransaction();
        ft.replace(R.id.framelayout, fragment);
        ft.commit();
    }
}
```

从上面代码可以看出，定义了一个 addFragment(Fragment fragment)方法，此方法的参数是一个 Fragment 对象，在方法体中调用 getSupportFragmentManager()方法得到 FragmentManager 的实例，然后使用 beginTransaction()方法开启 FragmentTransaction 的实例，再调用 replace()方法替换 framelayout 中的 Fragment 对象，最后调用 commit()方法完成事务提交。两个命令按钮 Button 的单击事件都调用了 addFragment()这个方法，而这个方法是我们自定义的，主要作用是添加指定参数（即 Fragment）到布局文件 Activity_main 中 id 为“fragment_layout”的布局中去。

第四步：运行程序

运行程序，分别点击界面上的命令按钮，则会显示不同的 Fragment 实例，效果如图 5-4 所示。

5.4 Fragment 的生命周期

Fragment 必须依存于 Activity 而存在，Activity 有生命周期，同样 Fragment 也有生命周期，它和 Activity 的生命周期很像，只有一些细微的区别。下面我们通过一个案例来学习 Fragment 的生命周期。

在之前项目打开项目 FragmentTest 的基础上，修改 Fragment_1 中的代码如下所示：

```
public class Fragment_1 extends Fragment {

    @Override
    public void onAttach(Context context) {
        super.onAttach(context);
        Log.i("Fragment1:", "onAttach()方法");
    }

    @Override
    public void onCreate(Bundle savedInstanceState) {
        super.onCreate(savedInstanceState);
        Log.i("Fragment1:", "onCreate()方法");
    }

    @Override
    public View onCreateView(LayoutInflater inflater,ViewGroup container,Bundle savedInstanceState) {
        Log.i("Fragment1:", "onCreateView()方法");
        return inflater.inflate(R.layout.fragment_1, container, false);
    }

    @Override
    public void onActivityCreated(Bundle savedInstanceState) {
        super.onActivityCreated(savedInstanceState);
        Log.i("Fragment1:", "onActivityCreated()方法");
    }

    @Override
    public void onStart() {
        super.onStart();
        Log.i("Fragment1:", "onStart()方法");
    }

    @Override
    public void onResume() {
        super.onResume();
        Log.i("Fragment1:", "onResume()方法");
    }

    @Override
    public void onPause() {
        super.onPause();
        Log.i("Fragment1:", "onPause()方法");
    }

    @Override
    public void onStop() {
        super.onStop();
        Log.i("Fragment1:", "onStop()方法");
    }

    @Override
    public void onDestroyView() {
        super.onDestroyView();
```

```
        Log.i("Fragment1:", "onDestroyView()方法");
    }

    @Override
    public void onDestroy() {
        super.onDestroy();
        Log.i("Fragment1:", "onDestroy()方法");
    }

    @Override
    public void onDetach() {
        super.onDetach();
        Log.i("Fragment1:", "onDetach()方法");
    }
}
```

我们在每个方法中都使用 Log.i()打印日志，运行程序单击“显示第一个”，观察 Logcat 中的打印信息，如图 5-5 所示：

```
com.example.framgment I/Fragment1:: onAttach()方法
com.example.framgment I/Fragment1:: onCreate()方法
com.example.framgment I/Fragment1:: onCreateView()方法
com.example.framgment I/Fragment1:: onActivityCreated()方法
com.example.framgment I/Fragment1:: onStart()方法
com.example.framgment I/Fragment1:: onResume()方法
```

图 5-5　启动程序时的打印日志

可以看到当 Fragment1 第一次被加载到屏幕上时，会依次执行 onAttach()、onCreate()、onCreateView()、onActivityCreated()、onStart()和 onResume()方法。然后点击“显示第二个”命令按钮，此时打印信息如图 5-6 所示。

```
com.example.framgment I/Fragment1:: onPause()方法
com.example.framgment I/Fragment1:: onStop()方法
com.example.framgment I/Fragment1:: onDestroyView()方法
com.example.framgment I/Fragment1:: onDestroy()方法
com.example.framgment I/Fragment1:: onDetach()方法
```

图 5-6　替换为另一个 Fragment 时的打印日志

由于 Fragmeng2 替换了 Fragmeng1，依次调用了 Fragmeng1 的 onPause()、onStop()、onDestroyView()、onDestroy()和 onDetach()方法，此时 Fragment1 就进入了销毁状态。

如果我们对 Fragment 的动态添加的方法进行修改，修改后的代码如下所示：

```
private void addFragment(Fragment fragment) {
    FragmentManager fm = getSupportFragmentManager();
    FragmentTransaction ft = fm.beginTransaction();
    ft.replace(R.id.framelayout, fragment);
    ft.addToBackStack(null);
    ft.commit();
}
```

再次运行程序，单击“显示第一个”命令按钮，观察 Logcat 中的打印信息，如图 5-5 所示，然后点击“显示第二个”命令按钮，可以看到，运行结果如图 5-7 所示。

```
com.example.framgment I/Fragment1:: onPause()方法
com.example.framgment I/Fragment1:: onStop()方法
com.example.framgment I/Fragment1:: onDestroyView()方法
```

图 5-7　替换为另一个 Fragment 时的打印日志

从运行结果可以看出，添加了 addToBackStack()方法使得 Fragment2 替换 Fragment1 时，没有调用 onDestroy()和 onDetach()方法。

addToBackStack()方法的作用：当移除或替换一个片段并向返回栈添加事务时，系统会停止（而非销毁）移除的片段。如果用户执行回退操作进行片段恢复，该片段将重新启动。如果不向返回栈添加事务，则系统会在移除或替换片段时将其销毁。

接着按 Back 键，Fragment1 重新回到了屏幕，打印信息如图 5-8 所示。

```
com.example.framgment I/Fragment1:: onCreateView()方法
com.example.framgment I/Fragment1:: onActivityCreated()方法
com.example.framgment I/Fragment1:: onStart()方法
com.example.framgment I/Fragment1:: onResume()方法
```

图 5-8　返回 Fragment1 时的打印日志

由于 Fragment1 又重新回到运行状态，因此 onCreateView()、onActivityCreated()、onStart()和 onResume()方法会得到执行。注意此时 onCreate()方法并没有执行，因为我们借助 addToBackStack()方法使得 Fragment1 没有被销毁。再次按下“Back”键，则打印信息如图 5-6 所示，Fragment1 将被销毁。

通过案例我们对 Fragment 的生命周期和常用的方法有了一定的了解，下面再来看看 Fragment 的生命周期图，如图 5-9 所示。

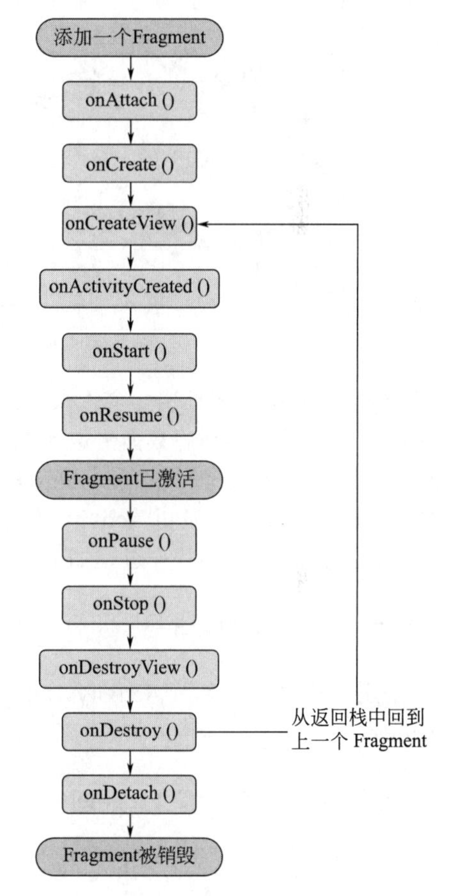

图 5-9　Fragment 生命周期模型

从生命周期图中可以看到 Fragment 和 Activity 有 onCreate()、onStop()方法等多个相同的方法，除了相同的方法还多了几个额外的生命周期回调方法，在此我们只对多的这几个方法进行介绍。

onAttach()方法：当 Fragment 与 Activity 发生关联时调用。

onCreateView()：加载该 Fragment 的布局时调用。

onActivityCreated()：当 Activity 的 onCreate 方法返回时调用。

onDestoryView()：与 onCreateView 相对应，当该 Fragment 的视图被移除时调用。

onDetach()：与 onAttach 相对应，当 Fragment 与 Activity 关联被取消时调用。

5.5　Fragment 与 Activity 间的数据互传

虽然 Fragment 作为 Activity 的一部分存在，但是很多时候我们需要在 Activity 和 Fragment 之间传输数据。比如微信中用户登录界面可以使用 Activity，登录成功需要将用户信息传到“我”对应的 Fragment 中，再要查看个人详细信息，又可以将 Fragment 中的数据传送到“个人信息”

Activity 中。

下面我们以如图 5-10 所示的效果图为例，学习如何实现 Activity 和 Fragment 之间相互传递数据。

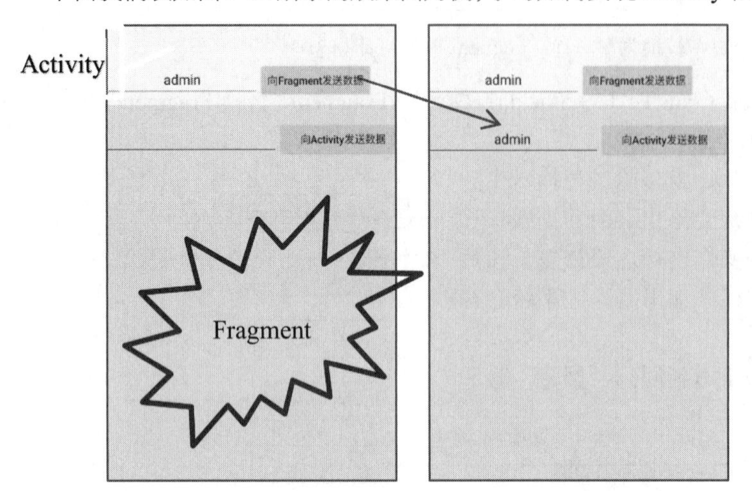

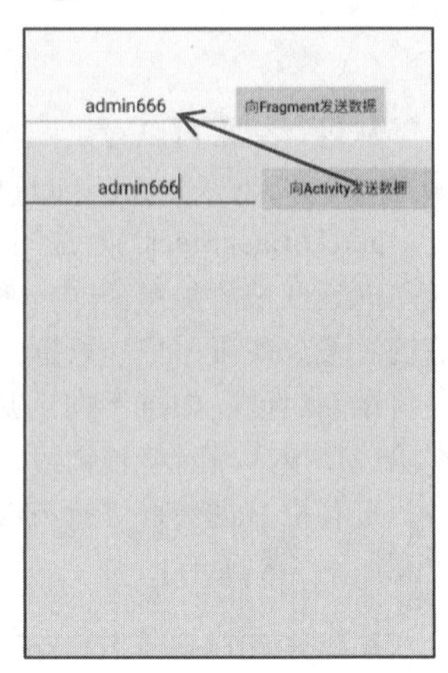

图 5-10 Activity 与 Fragment 数据传递

分析 Activity 界面由一个输入框、一个命令按钮和一个 FrameLayout 布局组成，在输入框输入内容，点击命令按钮将数据传送到 Fragment 的输入框中，在 Fragment 中修改传入的数据，再返回到 Activity 的输入框中。我们在之前学习的数据传递是通过携带数据的 Intent 对象在 Activity 之间传数据的，这里涉及由 Activity 传递数据到 Fragment 和由 Fragment 传递数据到 Activity，又是如何来实现的呢？具体操作过程如下。

第一步：创建项目

创建一个项目，名为 Activity_Data_Fragment，项目所在的包名为 com.example.activity_data_fragment。

第二步：创建一个 Fragment

新建一个 Fragment 名为 Fragment1，对应的布局文件为 fragment_1，布局文件的代码如下：

```
<?xml version="1.0" encoding="utf-8"?>
<LinearLayout xmlns:android="http://schemas.android.com/apk/res/android"
    xmlns:tools="http://schemas.android.com/tools"
    android:layout_width="match_parent"
    android:layout_height="match_parent"
    android:background="#E3E9EC"
    tools:context=".Fragment1">

    <EditText
        android:id="@+id/et_fragment"
        android:layout_width="200dp"
        android:layout_height="wrap_content"
        android:layout_marginTop="20dp"
        android:layout_weight="1"
        android:gravity="center"
        tools:ignore="SpeakableTextPresentCheck,TouchTargetSizeCheck" />
```

```
    <Button
        android:id="@+id/btn_fragment"
        android:layout_width="wrap_content"
        android:layout_height="wrap_content"
        android:layout_marginTop="20dp"
        android:layout_weight="1"
        android:text="向 Activity 发送数据"
        android:textAllCaps="false" />
</LinearLayout>
```

其中EditText用来显示Activity传来的数据，也作为向Activity发送数据的数据源，Button按钮用来发送数据。

第三步：设计用户界面

在布局文件 activty_main 界面上添加一个输入框 EditText，一个命令按钮 Button，一个FragmentContainerView控件。EditText控件的android:id属性值为et_Activity，android:text的属性值为“admin”。命令按钮 Button 控件的 android:id 属性值为“btn_Activity”，设置单击事件android:onClick属性值为“btn_activity_click”。FragmentContainerView控件的android:id属性值为fcv,android:name属性值为com.example.activity_data_fragment.Fragment1。修改后activity_main代码如下：

```
<?xml version="1.0" encoding="utf-8"?>
<androidx.constraintlayout.widget.ConstraintLayout
xmlns:android="http://schemas.android.com/apk/res/android"
    xmlns:app="http://schemas.android.com/apk/res-auto"
    xmlns:tools="http://schemas.android.com/tools"
    android:layout_width="match_parent"
    android:layout_height="match_parent"
    tools:context=".MainActivity">

    <EditText
        android:id="@+id/et_Activity"
        android:layout_width="wrap_content"
        android:layout_height="wrap_content"

        android:layout_marginTop="50dp"
        android:ems="10"
        android:gravity="center|center_horizontal"
        android:inputType="textPersonName"
        android:minHeight="48dp"
        android:text="admin"
        app:layout_constraintStart_toStartOf="parent"
        app:layout_constraintTop_toTopOf="parent" />

    <Button
        android:id="@+id/btn_Activity"
        android:layout_width="wrap_content"
        android:layout_height="wrap_content"
        android:text="向 Fragment 发送数据"
        android:onClick="btn_activity_click"
        android:textAllCaps="false"
```

```
        app:layout_constraintBaseline_toBaselineOf="@+id/tv_Activity"
        app:layout_constraintStart_toEndOf="@+id/tv_Activity" />

    <androidx.fragment.app.FragmentContainerView
        android:id="@+id/fcv"
        android:name="com.example.activity_data_fragment.Fragment1"
        android:layout_width="413dp"
        android:layout_height="625dp"
        android:layout_marginTop="10dp"
        app:layout_constraintEnd_toEndOf="parent"
        app:layout_constraintTop_toBottomOf="@+id/tv_Activity" />
</androidx.constraintlayout.widget.ConstraintLayout>
```

从代码中可以看到这个布局有两个部分：一个是 TextView 控件，用来显示从 Activity 传送到 Fragment 的数据和从 Fragment 传递过来接收到数据；另一个是 FragmentContainerView 控件，用作 Fragment 的容器，如果使用控件报错，在 build.gradle 文件中添加如下依赖。

```
    implementation 'androidx.fragment:fragment:1.3.2'
```

第四步：Activity 向 Fragmemt 发送数据

MainActivity 向 Fragment 发送数据，主要是在“向 Fragment 发送数据”命令按钮中实现。

```
public void btn_activity_click(View view) {
    Fragment1 f1=new Fragment1();
    Bundle bundle=new Bundle();
    bundle.putString("name",etactivity.getText().toString());
    f1.setArguments(bundle);
    FragmentManager fm=getSupportFragmentManager();
    FragmentTransaction ft=fm.beginTransaction();
    ft.replace(R.id.fcv,f1);
    ft.commit();
}
```

在以上代码中可以看到，创建 Bundle 对象，把要传递的数据存入 bundle 对象，然后通过 Fragment 的 setArguments（bundle）传到 Fragment 对象 f1 中。

第五步：Fragment 中接收数据

在 Fragment 中接收数据，Activity 中使用 setArguments（bundle）传到 Fragment，那么在 Fragment 中接收数据需要使用 getArguments()方法，获取传递过来的 bundle。重写 Fragment 中的 onCreateView(),实现代码如下：

```
@Override
    public View onCreateView(LayoutInflater inflater, ViewGroup container,
                             Bundle savedInstanceState) {
        View view=inflater.inflate(R.layout.fragment_1, container, false);
       EditText etfragment=view.findViewById(R.id.et_fragment);
       Button btnfragment=view.findViewById(R.id.btn_fragment);
       Bundle bundle=getArguments();
       if(bundle!=null){
           etfragment.setText(bundle.getString("name"));
       }
```

```
        return view;
    }
}
```

在代码中创建 Bundle 对象，使用 getArguments()方法接收 Activity 传递过来的 Bundle 对象，当有数据传过来时，设置 EditText 对象显示内容为传递过来的数据。

第六步：由 Fragment 向 Activity 传递数据

由 Fragment 向 Activity 传递数据。在 MainActivity 中定义了一个方法 sendString()，在此方法中设置输入框的内容为 sendString()中的参数;在 Fragment1 的命令按钮单击事件中获取 Fragment 中文本的内容传递给 MainActivity 中 sendString()方法，实现从 Fragment 到 Activity 中的数据传递。

（1）在 MainActivity 中添加一个方法 sendString(),代码如下：

```
public void sendString(String s){
    etactivity.setText(s);
}
```

完整的 MainActivity 代码如下所示：

```
public class MainActivity extends AppCompatActivity {
    private EditText etactivity;

    @Override
    protected void onCreate(Bundle savedInstanceState) {
        super.onCreate(savedInstanceState);
        setContentView(R.layout.activity_main);
        etactivity = findViewById(R.id.et_activity);
    }

    public void btn_activity_click(View view) {
        Fragment1 f1 = new Fragment1();
        Bundle bundle = new Bundle();
        bundle.putString("name", etactivity.getText().toString());
        f1.setArguments(bundle);
        FragmentManager fm = getSupportFragmentManager();
        FragmentTransaction ft = fm.beginTransaction();
        ft.replace(R.id.fcv, f1);
        ft.commit();
    }

    public void sendString(String s) {
        etactivity.setText(s);
    }
}
```

（2）在 Fragment1 中调用 MainActivity 中的方法，代码如下：

```
String s=etfragment.getText().toString();
MainActivity mainActivity=(MainActivity)getActivity();
mainActivity.sendString(s);
```

Fragment1 的代码如下所示：

```
public class Fragment1 extends Fragment {

    @Override
    public View onCreateView(LayoutInflater inflater, ViewGroup container,
                             Bundle savedInstanceState) {
        View view = inflater.inflate(R.layout.fragment_1, container, false);
        EditText etfragment = view.findViewById(R.id.et_fragment);
        Button btnfragment = view.findViewById(R.id.btn_fragment);
        Bundle bundle = getArguments();
        if (bundle != null) {
            etfragment.setText(bundle.getString("name"));
        }
        btnfragment.setOnClickListener(new View.OnClickListener() {
            @Override
            public void onClick(View v) {
                String s = etfragment.getText().toString();
                MainActivity mainActivity = (MainActivity) getActivity();
                mainActivity.sendString(s);
            }
        });
        return view;
    }
}
```

第七步：运行程序

运行效果如图 5-10 所示，程序开始运行，在 Activity 界面中加载了 Fragment，显示的是布局文件中设置的内容，点击 Activity 中“向 Fragment 发送数据”则将 Activity 中文本的内容通过 Bundle 对象传给了 Fragment，然后修改 Fragment 中接收到的数据，单击“向 Activity 发送数据”，可以看到将 Fragment 中的数据传递给了 Activity 中的文本。

对 Fragment 和 Activity 之间的数据传递总结如下：

① 由 Activity 向 Fragment 传递数据：在 Activity 中将要传递的数据封装在 Bundle 中，然后在 Activity 中使用 Fragment 的实例通过 setArgument(Bundel bundel) 方法绑定传递，在要传递到的 Fragment 中使用 getArgment()方法得到传递的 Bundle，从而获取传递。

② 由 Fragment 向 Activity 传递数据:在 MainActivity 中定义了一个方法 sendString(String s)，在此方法中设置文本的内容;在 Fragment 中调用 MainActivity 中的 sendString()方法将 Fragment 中的值传到了 Activity。

5.6 仿微信应用中 Fragment 的使用

通过本章内容的学习，我们可以在一个 Activity 中加入 Fragment。在第 3 章我们实现了仿微信底部导航的应用和 RecyclerView 控件的使用，现在我们在仿微信底部导航和 RecyclerView 基础上，加入 Fragment，通过底部导航实现仿微信应用界面间的切换，效果如图 5-11 所示，实现步骤如下所示。

分析 对于本案例需要根据选择的“微信”“通讯录”或“发现”和“我”切换显示内容，

这里不同的界面使用不同 Fragment 来实现。

图 5-11　仿微信主界面

第一步：打开项目

打开第 3 章我们做好的“仿微信底部导航的项目 Weixin”。

第二步：创建 Fragment

新建一个 Fragment,名为 Weixin_Fragment,并添加一个布局文件 fragment_weixin。

第三步：修改主界面布局

修改布局 activity_main，将之前界面中部的 TextView 控件换成布局 FragmentContainerView 控件，其他部分不变，代码如下所示。

```
<?xml version="1.0" encoding="utf-8"?>
<LinearLayout
xmlns:android="http://schemas.android.com/apk/res/android"
    xmlns:tools="http://schemas.android.com/tools"
    android:layout_width="match_parent"
    android:layout_height="match_parent"
    android:orientation="vertical"
    tools:context=".MainActivity">

    <TextView
        android:id="@+id/tv_title"
        android:layout_width="match_parent"
        android:layout_height="wrap_content"
        android:background="#AAAAAA"
        android:gravity="center"
        android:padding="10dp"
        android:text="微信"
        android:textColor="@color/white"
        android:textSize="24sp" />

    <androidx.fragment.app.FragmentContainerView
        android:id="@+id/fcv"
        android:name="com.example.weixin.Weixin_Fragment"
        android:layout_width="match_parent"
        android:layout_height="0dp"
        android:layout_weight="1"/>

    <RadioGroup
        android:id="@+id/rg_bottom"
        android:layout_width="match_parent"
        android:layout_height="wrap_content"
        android:layout_gravity="bottom"
        android:background="#F4F2F2"
        android:orientation="horizontal">

        <RadioButton
            android:id="@+id/radio_weixin"
            style="@style/mystyle"
```

```
            android:drawableTop="@drawable/tb_weixin"
            android:checked="true"
            android:text="微信" />

        <RadioButton
            android:id="@+id/radio_tongxunlu"
            style="@style/mystyle"
            android:drawableTop="@drawable/tb_tongxunlu"
            android:text="通讯录" />

        <RadioButton
            android:id="@+id/radio_faxian"
            style="@style/mystyle"
            android:drawableTop="@drawable/tb_faxian"
            android:text="发现" />

        <RadioButton
            android:id="@+id/radio_me"
            style="@style/mystyle"
            android:drawableTop="@drawable/tb_me"
            android:text="我" />
    </RadioGroup>
</LinearLayout>
```

第四步：设计“微信”布局

在 Weixin_Fragment 对应的布局文件 fragment_weixin 布局中添加一个 RecyclerView 控件，其代码如下：

```
<?xml version="1.0" encoding="utf-8"?>
<androidx.constraintlayout.widget.ConstraintLayout
xmlns:android="http://schemas.android.com/apk/res/android"
    xmlns:app="http://schemas.android.com/apk/res-auto"
    xmlns:tools="http://schemas.android.com/tools"
    android:layout_width="match_parent"
    android:layout_height="match_parent">

    <androidx.recyclerview.widget.RecyclerView
        android:id="@+id/recycler"
        android:layout_width="0dp"
        android:layout_height="0dp"
        app:layout_constraintBottom_toBottomOf="parent"
        app:layout_constraintEnd_toEndOf="parent"
        app:layout_constraintStart_toStartOf="parent"
        app:layout_constraintTop_toTopOf="parent" />
</androidx.constraintlayout.widget.ConstraintLayout>
```

布局文件中只有一个 RecyclerView 控件，用于显示当前选项要显示的数据。

第五步：设计“微信”记录实体类、item 布局和适配器

打开第 3 章的 RecyclerView 项目，复制其中实体类 Weixin 和适配器 MyAdapter 到当前包中，复制 RecyclerView 的 item 布局文件 item.xml 到 res/layout 目录下。

第六步：在 Weixin_Fragment 中显示内容

在前面我们所作的 RecyclerView 是在 Activity 中，本案例是在 Fragment 中，Weixin_Fragment 中的逻辑如下所示。

```
public class Weixin_Fragment extends Fragment {
    private int[] imageId;
    private String[] strname;
    private String[] strinformation;
    private MyAdapter adapter;
    private List<WeiXin> list;
    private RecyclerView rc;

    @Nullable
    @Override
    public View onCreateView(@NonNull LayoutInflater inflater, @Nullable ViewGroup container,
@Nullable Bundle savedInstanceState) {
        View view = inflater.inflate(R.layout.fragment_weixin, null);
        rc = view.findViewById(R.id.recycler);
        initData();
        return view;
    }

    private void initData() {
        imageId = new int[]{R.drawable.tx1, R.drawable.tx2, R.drawable.tx3, R.drawable.tx4,
R.drawable.tx5, R.drawable.tx6, R.drawable.tx7, R.drawable.tx8};
        strname = new String[]{"一枝独秀", "黑中红叶", "金秋时节又一年", "秋", "文件传输助手", "一帆风顺 666",
"险峰松柏", "红叶片片"};
        strinformation = new String[]{"拜拜……", "明天见", "【图片】", "好玩的事情多分享哦！！！", "【图片】
", "您好！！！", "886", "少壮不努力老大徒伤悲"};
        list = new ArrayList<WeiXin>();
        initWeixin();
        rc.setLayoutManager(new LinearLayoutManager(getContext()));
        adapter = new MyAdapter(getContext(), list);
        rc.setAdapter(adapter);
    }

    private void initWeixin() {
        for (int i = 0; i < imageId.length; i++) {
            WeiXin weixin = new WeiXin(imageId[i], strname[i], strinformation[i]);
            list.add(weixin);
        }
    }
}
```

将案例 RecyclerView 中 MainActivity 对数据的初始化和控件的使用放在了 WeixinFragment 的 onCreate()方法中 ，其他的和之前的做法一样。

第七步：将 Fragment 嵌入 Activity

使用与 Weixin_Fragment 同样的方法添加“发现”Faxian_Fragment、“通讯录”Tongxunlu_Fragment、“我”Me_Fragment 的 Fragment 和对应的布局，修改 MainActivity 代码如下所示：

```
public class MainActivity extends AppCompatActivity {
    TextView tv_title, tv_content;
    RadioGroup rgbottom;
    RadioButton rb_weixin, rb_tongxunlu, rb_faxian, rb_wo;

    @Override
    protected void onCreate(Bundle savedInstanceState) {
        super.onCreate(savedInstanceState);
        setContentView(R.layout.activity_main);
        getSupportActionBar().hide();
        tv_title = findViewById(R.id.tv_title);
    //  tv_content = findViewById(R.id.tv_content);
        rgbottom = findViewById(R.id.rg_bottom);
        rb_weixin = findViewById(R.id.radio_weixin);
        rb_tongxunlu = findViewById(R.id.radio_tongxunlu);
        rb_faxian = findViewById(R.id.radio_faxian);
        rb_wo = findViewById(R.id.radio_me);
        rgbottom.setOnCheckedChangeListener(new RadioGroup.OnCheckedChangeListener() {
            @Override
            public void onCheckedChanged(RadioGroup group, int checkedId) {
                switch (checkedId) {
                    case R.id.radio_weixin:
                        tv_title.setText("微信");
                        addFragment( new Weixin_Fragment());
                        break;
                    case R.id.radio_tongxunlu:
                        tv_title.setText("通讯录");
                        addFragment(new Tongxunlu_Fragment());
                        break;
                    case R.id.radio_faxian:
                        tv_title.setText("发现");
                        addFragment( new Faxian_Fragment());
                        break;
                    case R.id.radio_me:
                        tv_title.setText("我");
                        addFragment( new Me_Fragment());
                        break;
                }
            }
        });
    }
    private void addFragment(Fragment fragment) {
        FragmentManager fm = getSupportFragmentManager();
        FragmentTransaction ft = fm.beginTransaction();
        ft.replace(R.id.fcv, fragment);
        ft.commit();
    }
}
```

第八步：运行程序

运行程序，效果如图 5-11 所示。从这个案例我们可以看出，将之前在 Activity 中的操作放在了 Fragment 中,Fragment 中的使用与 Activity 类似，我们只需要在 Activity 中添加或者替换 Fragment 就可以实现案例所示的效果了。

职业素养拓展

职业素养拓展内容详见文件“职业素养拓展”。

扫描二维码查看【职业素养拓展 5】

本章小结

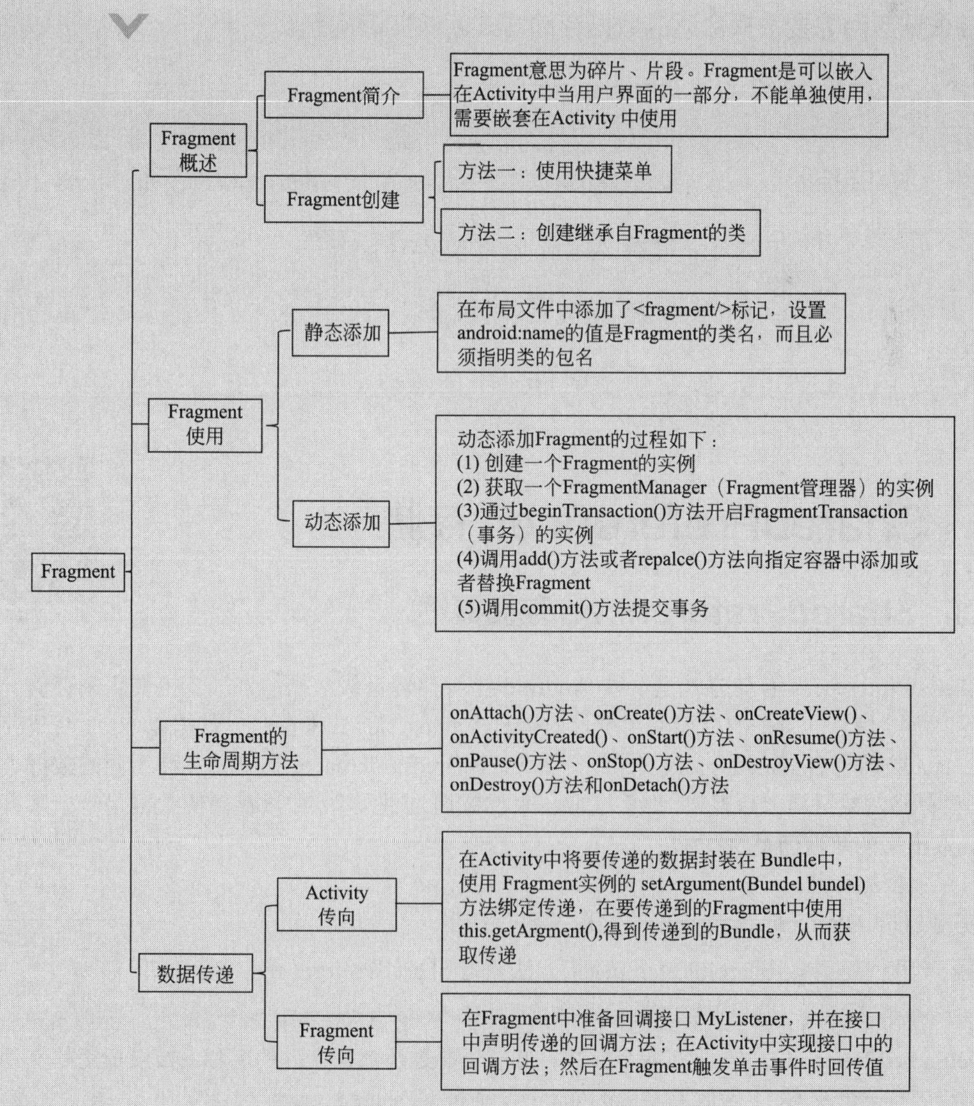

第6章

数据存储

大部分的应用程序都会涉及数据存储，Android 应用也不例外，本章主要介绍 Android 中的几种数据存储方式：SharedPreferences 存储、文件存储和 SQLite 数据库存储以及 Jetpack 中的 ORM 组件 Room 的使用。通过学习能够根据应用需要选择合适的数据存储方式解决实际问题。

学习目标：

1. 掌握 Android 中 SharedPreferences 存储方法；
2. 掌握 Android 中文件存储方法；
3. 掌握 SQLite 数据库的操作方法；
4. 了解使用 Room 的基本过程。

6.1 SharedPreferences 存储

SharedPreferences 简介

6.1.1 SharedPreferences 存储概述

SharedPreferences 存储方式是一种 Android 上存储轻量级数据的方式，主要用来存储一些简单的程序配置信息，数据的存储使用键值对<key,Value>的方式来存储，数据以 xml 格式存放在“data/data/当前包名/shared_prefs/目录下。另外，SharedPreferences 还支持多种不同的数据类型存储，如果存储的数据类型是整型，那么读取出来的数据也是整型；如果存储的数据是一个字符串，那么读取出来的数据仍是字符串。

（1）数据的存储

使用 SharedPreferences 存数据的过程主要有几步。

第一步：使用 getSharedPreferenecs()方法获取 SharedPreferences 对象，如：

```
SharedPreferences sp=getSharedPreferences("mydata",MODE_PRIVATE);
```

getSharedPreferenecs()方法有两个参数，第一个参数是数据文件的名称，如果指定的文件不存在则会创建一个文件，存放在“data/data/当前包名/shared_prefs/”目录下；第二个参数是 MODE_PRIVATE，表示只有当前应用程序才能使用这个 SharedPreferences 文件。

第二步：调用 SharedPreferences 类的 edit()方法，获取 SharedPreferences.Editor 对象，如：

```
SharedPreferences.Editor editor=sp.edit();
```

第三步：使用一个 SharedPreferences.Editor 对象的 putXxx()的方法存入数据,如:

```
editor.putString("name","张三");
```

putString()方法用于添加字符串类型的 value,如要添加其他类型的 value,则需要替换 String。例如：如要添加 int 类型的 value，则需要使用 putInt(key,value)方法。

第四步：使用 commit()方法提交数据

```
editor.commit();
```

操作完数据，一定要调用 commit()方法进行数据的提交，否则所做操作不生效。

数据存好了，如果要使用，就必须取出来。

（2）数据的读取

SharedPreferences 读取数据的过程主要有两步:

第一步：获取 ShaedPreferences 对象

```
SharedPreferences sp=getSharedPreferences("mydata",MODE_PRIVATE);
```

第二步：使用 getXxx()方法获取数据

```
String name=sp.getString("name","");
```

getXxx()方法的第二个参数为缺省值，如果 sp 中不存在该 key，将返回缺省值，如 getString(“name”,“”);如果 name 不存在，则 key 就返回空字符串。获取数据的 key 值与存入数据的 key 值数据类型要一致，否则查找不到指定数据。

6.1.2　记住账号和密码

下面通过常见的记住用户密码的实例，学习使用 SharedPreferences 方式存储数据的应用。具体要求：如果在登录的时候选择了“记住密码”复选框，则再次登录的时候用户名和密码直接显示在输入框中，否则需要输入数据，运行效果如图 6-1 所示。

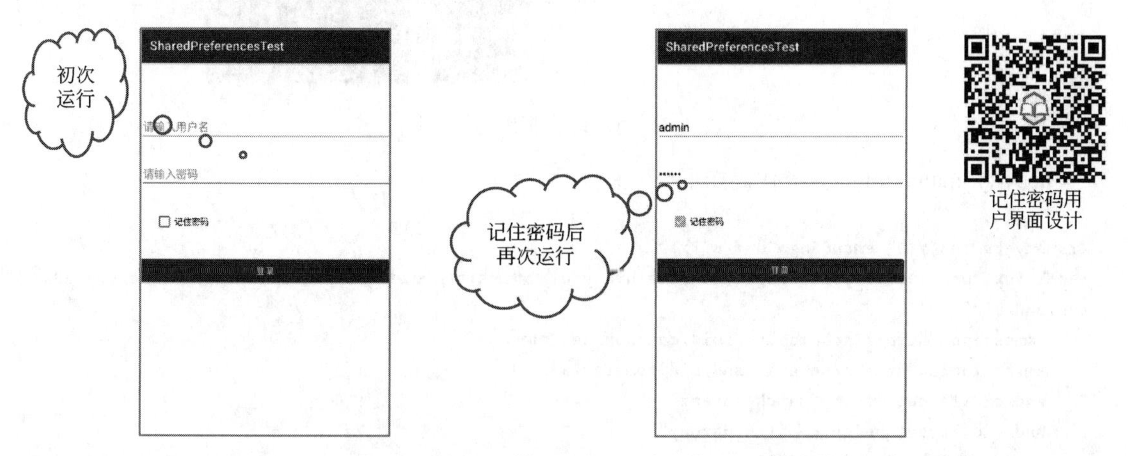

图 6-1　用户记住密码运行效果

分析　程序要求当用户输入用户名(admin)和密码(123456)并且选择记住密码复选框，登录成功一次之后，如果重新启动应用程序，在登录界面显示已经记住的用户名和密码。使用

SharedPreferences 将用户名和密码保存到文件中，再次登录则从文件取出数据显示在对应的控件中，实现步骤如下所示。

第一步：新建项目

新建一个项目，名称为 SharedPreferencesTest，所在包为 com.example.sharedpreferencestest。

第二步：设计用户界面

在 activity_main 界面上添加两个 EditText 控件，分别用来输入用户名和密码。输入用户名的控件 android:id 为 etuser，设置 android:hint 属性为“请输入用户名”；输入密码的 EditText 的 android:id 为 etpwd，设置 android:hint 属性为“请输入密码”；添加一个复选框 CheckBox,id 为“checkbox”,android:text 为“记住密码”；添加一个“登录”命令按钮 Button，设置 android:text 属性为“登录”。选中所有控件添加约束，界面效果如图 6-2 所示。

记住密码登录按钮的实现

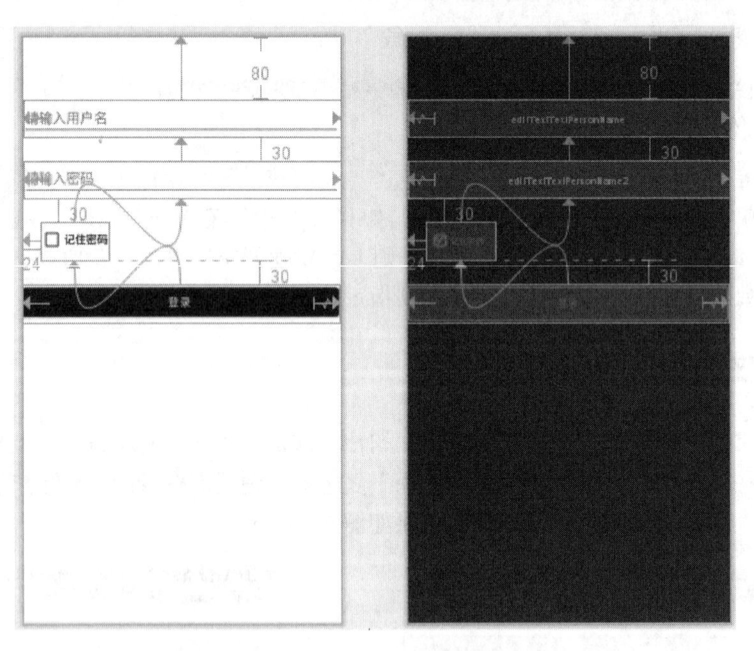

图 6-2　用户记住密码界面设计

activity_main.xml 的参考代码如下：

```
<?xml version="1.0" encoding="utf-8"?>
<androidx.constraintlayout.widget.ConstraintLayout  xmlns:android="http://schemas.android.com/apk/
res/android"
    xmlns:app="http://schemas.android.com/apk/res-auto"
    xmlns:tools="http://schemas.android.com/tools"
    android:layout_width="match_parent"
    android:layout_height="match_parent"
    tools:context=".MainActivity">

    <EditText
        android:id="@+id/etuser"
        android:layout_width="409dp"
```

```
        android:layout_height="wrap_content"
        android:layout_marginTop="80dp"
        android:ems="10"
        android:hint="请输入用户名"
        android:inputType="textPersonName"
        app:layout_constraintEnd_toEndOf="parent"
        app:layout_constraintStart_toStartOf="parent"
        app:layout_constraintTop_toTopOf="parent" />

    <EditText
        android:id="@+id/etpwd"
        android:layout_width="409dp"
        android:layout_height="wrap_content"
        android:layout_marginTop="30dp"
        android:ems="10"
        android:hint="请输入密码"
        android:inputType="textPassword"
        app:layout_constraintEnd_toEndOf="parent"
        app:layout_constraintStart_toStartOf="parent"
        app:layout_constraintTop_toBottomOf="@+id/etuser" />

    <CheckBox
        android:id="@+id/checkBox"
        android:layout_width="wrap_content"
        android:layout_height="wrap_content"
        android:layout_marginTop="30dp"
        android:text="记住密码"
        app:layout_constraintStart_toStartOf="parent"
        app:layout_constraintTop_toBottomOf="@+id/etpwd" />

    <Button
        android:id="@+id/button"
        android:layout_width="409dp"
        android:layout_height="wrap_content"
        android:layout_marginTop="30dp"
        android:text="登录"
        app:layout_constraintEnd_toEndOf="parent"
        app:layout_constraintStart_toStartOf="parent"
        app:layout_constraintTop_toBottomOf="@+id/checkBox" />
</androidx.constraintlayout.widget.ConstraintLayout>
```

第三步：实现记住用户密码功能

实现记住用户密码功能

实现记住用户密码就是将数据存储进文件中，需要的时候再来读取。在MainActivity中使用SharedPreferences实现记住用户密码功能。具体代码如下所示：

```
public class MainActivity extends AppCompatActivity {
    private SharedPreferences preferences;
    private EditText user, password;
    private Button login;
    private CheckBox cbremember;
    private boolean isRemember;
```

```
    private SharedPreferences.Editor editor;

    @Override
    protected void onCreate(Bundle savedInstanceState) {
        super.onCreate(savedInstanceState);
        setContentView(R.layout.activity_main);
        user = findViewById(R.id.etuser);
        password = findViewById(R.id.etpwd);
        login = findViewById(R.id.button);
        cbremember = findViewById(R.id.checkBox);
        preferences = getSharedPreferences("data", MODE_PRIVATE);
        isRemember = preferences.getBoolean("rempass", false);
        //实现登录功能
        login.setOnClickListener(new View.OnClickListener() {
            @Override
            public void onClick(View v) {
                String username = user.getText().toString();
                String pw = password.getText().toString();
                if (username.equals("admin") && pw.equals("123456")) {
                    editor = preferences.edit();
                    if (cbremember.isChecked()) {//检查复选框是否被选中
                        editor.putBoolean("rempass", true);
                        editor.putString("username", username);
                        editor.putString("password", pw);
                    } else {
                        editor.clear();
                    }
                    editor.commit();
                } else {
                    Toast.makeText(MainActivity.this, "用户名或密码不正确，请重新输入！", Toast.LENGTH
_LONG).show();
                }
            }
        });
        //记住密码后，重新登录，显示用户名和密码
        if (isRemember) {
            String username = preferences.getString("username", "");
            String pw = preferences.getString("password", "");
            user.setText(username);
            password.setText(pw);
            cbremember.setChecked(true);
        }
    }
}
```

上面的代码先对界面中的控件进行初始化，然后使用 getSharedPreferences()方法获取 SharedPreferences 对象，如果登录成功并且记住密码复选框选中，则使用 edit()方法获取编辑器，接着使用 putString()方法存入选中状态 rempass 为 true，存入用户名 username 和密码 password 的值分别为输入框中获取的值，并使用 commit()方法提交数据。如果复选框没有选中，则使用 editor.clear()方法删除所有数据。当再次登录时先获取文件中选中状态 rempass 的值，如果为 true，则使用 getString()方法，取出用户名和密码的值填入其对应的输入框。

第四步：运行程序

运行程序，运行效果如图 6-1 所示。在用户名和密码输入框中分别输入 admin 和 123456，选中记住密码复选框，单击“登录”命令按钮。退出程序，重新运行程序用户名和密码已经填充，复选框选中。单击复选框取消选中状态，再次单击登录，清除了数据。再一次运行程序，用户名和密码如初始状态，只显示输入提示信息。

6.2 文件存储

6.2.1 文件存储概述

文件存储

在 Android 中使用文件存储读取/写入文件是通过 I/O 流的形式实现的，与 Java 中的文件存储类似。在文件中存取数据首先要创建文件，然后使用写入和读取的方法实现对数据的存取。这里用到了 openFileOutput()方法和 openFileInput()方法。

① openFileOutput()方法用于将数据存储到指定的文件中，此方法格式如下所示：

```
FileOutStream fos=openFileOutput(String name, int mode);
```

这个方法设有两个参数，第一个参数用于指定文件名，如果文件不存在，Android 会自动创建它，创建的文件存储在/data/data/<package name>/files 目录中；第二个参数是文件的操作模式，常用的模式有四种：

文件存储实例分析与界面设计

- MODE_PRIVATE 是默认的操作模式，表示该文件是私有的，只能被应用本身访问，在该模式下写入的内容将会覆盖原文件中的内容；
- MODE_APPEND 也是私有的，如果要创建的文件存在，则新写入的数据不会覆盖以前的数据；
- MODE_WORLD_READABLE 模式可以被其他应用程序读取；
- MODE_WORLD_WRITEABLE 模式可以被其他应用程序写入。

② openFileInput()方法用于打开应用程序对应的输入流，读取指定文件中的数据，此方法的使用格式如下：

```
FielInputStream fis=openFileInput(String name);
```

openFileInput()方法只接收一个参数，即要读取的文件名。然后系统会自动到默认目录下去加载这个文件，并返回一个 FileInputStream 对象，得到了这个对象之后再通过 java 流的方式就可以将数据读取出来了。

6.2.2 保存账号和密码

在本节课我们使用文件存储的方式实现如图 6-1 所示的案例，具体实现过程如下：

第一步：新建项目

新建一个项目，名为 FileSaveInformattion，所在包为 com.example.filesaveinformattion。

第二步：设计用户界面

用户界面设计如图 6-2 所示，与项目 SharedPreferencesTest 中的界面一样。

第三步：实现记住账号和密码工程

在 MainActivity 中实现保存数据，读取数据的功能，具体代码如下所示：

实例编码实现1

实例编码实现2

```
public class MainActivity extends AppCompatActivity {
    private EditText user, password;
    private Button login;
    private CheckBox cbRemeber;
    private FileOutputStream out;
    private String content;
    private File file;

    @Override
    protected void onCreate(Bundle savedInstanceState) {
        super.onCreate(savedInstanceState);
        setContentView(R.layout.activity_main);
        user = findViewById(R.id.etuser);
        password = findViewById(R.id.etpwd);
        cbRemeber = findViewById(R.id.checkBox);
        login = findViewById(R.id.button);
        try {
            FileInputStream in = openFileInput("mydata.txt");
            BufferedReader br = new BufferedReader(new InputStreamReader(in));
            content = br.readLine();
            String[] infor = content.split(":");
            if (infor[2].equals("true")) {
                cbRemeber.setChecked(true);
                user.setText(infor[0]);
                password.setText(infor[1]);
                in.close();
            }
        } catch (IOException e) {
            e.printStackTrace();
        }
        //登录命令按钮的单击事件
        login.setOnClickListener(new View.OnClickListener() {
            @Override
            public void onClick(View v) {
                String username = user.getText().toString();
                String pw = password.getText().toString();
                if (username.equals("admin") && pw.equals("123456")) {
                    if (cbRemeber.isChecked()) {//检查复选框是否被选中
                        try {
                            out = openFileOutput("mydata.txt", MODE_PRIVATE);
                            OutputStreamWriter outwrite = new OutputStreamWriter(out);
                            BufferedWriter bw = new BufferedWriter(outwrite);
                            PrintWriter printWriter = new PrintWriter(bw);
                            printWriter.println(username + ":" + pw + ":" + "true");
                            printWriter.flush();
                        } catch (IOException e) {
                            e.printStackTrace();
                        }
                    } else {
```

```
                    file = getFileStreamPath("mydata.txt");
                    Log.i("aa", file.toString());
                    if (file != null) {
                        file.delete();
                    }
                }
            } else {
                Toast.makeText(MainActivity.this, "用户名或密码不正确，请重新输入！", Toast.LENGTH_
LONG).show();
            }
        }
    });
  }
}
```

上面的代码中，先通过 openFileInput()方法获取到输入流对象，创建 BufferedReader 对象从输入流对象读取数据，然后使用 readLine()方法将读取的数据存入字符串 content，使用字符串的 split()方法以“:”为分隔符将 content 分为 String 数组 infor，infor[0]为用户名，infor[1]为密码，infor[2]为复选框的状态。在命令按钮的单击事件中，如果用户名和密码都正确，复选框选中时使用 openFileOutput()方法获取到输出流对象，使用 PrintWriter 对象的 println()方法将用户名、密码和记住密码的状态使用“：”分隔写入文件。如果没有选中记住密码复选框，则使用 File 的 delete()方法删除文件。

第四步：运行程序

程序的运行效果与使用 SharedPreferenceTest 案例中的运行效果完全一致。

注意：在这只是让同学们理解文件存储的方法，实际开发中还需要根据需求细化完善应用程序的功能。

6.3 SQLite 数据库存储

SQLite基本操作

前面我们学习了文件存储和 SharedPreferences 存储方式，但这两种方式都适用于保存一些简单的数据和键值对数据，当我们需要存储大量复杂数据的时候，就可以使用 Android 中的数据库技术 SQLite。

SQLite 数据库是 Android 系统内嵌的数据库，是一款轻量级的关系型数据库，运算速度非常快，占用资源少，适合在移动设备上使用。SQLite 支持标准的 SQL 语法，支持 NULL、INTEGER、REAL、TEXT 和 BLOB 等数据类型，如果已经学过关系型数据库的操作方法，那么对 SQLite 数据库的操作将会变得很简单。对于 Android 平台来说，系统内置了丰富的 API 来供开发人员操作 SQLite，使用这些开发工具可以轻松完成数据的存取，接下来我们学习 Android 中 SQLite 数据库的操作方法。

6.3.1 SQLite 数据库创建

SQLite数据库创建

创建数据库我们需要自定义一个类继承自 SQLiteOpenHelper 类，由于 SQLiteOpenHelper 是一个抽象类，所以还必须实现它的两个抽象方法：OnCreate()方法和 onUpgrade()方法。除了实现抽象的方法我们自定义的类中还需要创建一个构造方法，指定创建的数据库的名称等内容。

创建数据库的步骤如下所示。

第一步：创建项目

新建一个项目，名为 MyContacts，包名为 com.example.mycontacts。

第二步：新建 MyHelper 类

选择当前包 com.example.mycontacts，单击右键“New”→“java Class”，弹出“New Java Class”对话框，创建的类名为“MyHelper”，然后按回车键。使得 MyHelper 类继承自 SQLiteOpenHelper 类，重写 SQLiteOpenHelper 中的抽象方法，并添加一个构造方法。MyHelper.java 详细代码如下所示。

```
public class MyHelper extends SQLiteOpenHelper {

    public MyHelper(@Nullable Context context) {
        super(context, "mycontacts.db", null, 1);
    }

    @Override
    public void onCreate(SQLiteDatabase db) {
        db.execSQL("create table contancts(_id integer primary key autoincrement,name varchar(20),
phone varchar(20))");
    }

    @Override
    public void onUpgrade(SQLiteDatabase db, int oldVersion, int newVersion) {
    }
}
```

在上面的代码中，首先是一个构造方法 MyHelper，调用了父类的构造方法，方法的格式如下：

```
SQLiteOpenHelper(Context context,String name, CursorFactory factory, int version)
```

此方法有四个参数，第一个参数是 Context，第二个参数是数据库的名字，第三个参数一般是 null，第四个参数是当前数据库的版本号。

onCreate()方法是在数据库第一次建立时被调用，一般用于类创建数据库中的表，并做适当的初始化工作。使用时需要一个 SQLiteDatabase 对象作为参数，根据需要对这个对象填充表和初始化数据。在此我们通过调用 SQLiteDatabase 对象的 execSQL()方法，执行创建表的 SQL 命令。

onUpgrade()方法在数据库版本号增加时使用。

创建了 MyHelper 类来实现数据库的创建，要对数据库操作还需获取对数据进行可读写操作的对象，使用 getReadableDatabase()方法获取可读 SQLiteDatabase 对象，使用 getWritableDatabase()方法获取可读可写的 SQLiteDatabase 对象。数据库打开了再对其中的表进行增删改查的操作。

6.3.2 仿手机通讯录

在我们日常使用的手机应用中，很多都有通讯录或者是联系人、用户管理等内容，使用它们我们可以对相关的信息进行添加、修改、删除、查找等操作，下面我们以简化的手机通讯录为例，学习 SQLite 中数据操作的方法，实现如图 6-3 所示的效果。

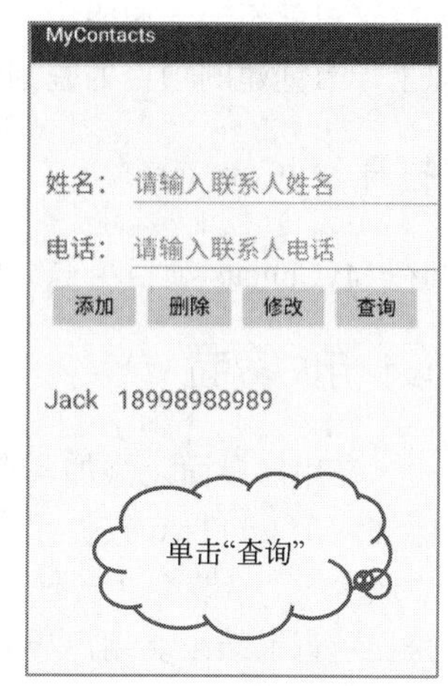

通讯录数据库
创建

图 6-3　通讯录运行界面

分析　根据运行效果，我们要实现对输入的联系人信息进行添加，删除指定的联系人，修改联系人的电话，并查询联系人的信息。要实现操作，必须知道 SQLite 中对数据操作所要调用的增删改查的方法。

（1）添加数据

添加使用方法 insert()，此方法的语法格式如下：

```
insert(String table, String nullColumnHack, ContentValues values)
```

insert()方法有三个参数，第一个参数为表名，第二个参数表示如果插入的数据每一列都为空，需要指定此行中某一列的名称，系统将此列设置为 NULL，一般此参数直接传入 null 即可，第三个参数为一个 ContentValues 对象，数据通过 put()方法封装在对象中。

（2）删除数据

```
delete(String table,String whereClause,String whereArgs);
```

delete()方法有三个参数，第一个参数为表名，第二个参数为删除条件，第三个参数为占位符的实际参数值。

（3）更新数据

```
update(String table,ContentValues values,String whereClause,String whereArgs);
```

update()方法有四个参数，第一个参数为表名，第二个参数为更新数据，第三个参数为更新条件，此条件的设置可使用占位符（？），第四个参数为占位符的实际参数值。

（4）查询数据

```
    query(String table, String[] columns, String selection, String[] selectionArgs, String groupBy,
String having, String orderBy, String limit)
```

query()方法有八个参数，第一个参数 table：查询的表名；第二个参数 columns：查询的列名；第三个参数 selection：where 条件；第四个参数 selectionArgs：为 where 条件中的占位符提供具体的值；第五个参数 groupBy：指定分组的列名；第六个参数 having：指定分组的条件；第七个参数 orderBy：指定排序的列名；第八个参数 limit：指定偏移量和获取的记录数。

说明：当执行了查询操作后，此方法返回一个 Cursor 对象，查询到的所有数据都将从这个对象中取出，使用 Cursor 中的 moveToNext()和 getXxx()等方法，可以获取到查询到的记录值，Cursor 中的方法很多，在这里就不一一说明了，如果大家想学习请参看 Android API 文档。

下面我们以上一节创建的项目为基础来实现联系人应用中数据的操作，步骤如下所示。

第一步：打开 MyContacts

上一节我们在 MyContats 项目中已经创建了 MyHelper 这个类，在此我们省略数据库的创建。

第二步：设计用户界面

通讯录页面布局

在用户界面上添加两个 TextView 控件用来显示提示信息“姓名”和“电话”，添加两个 EditText 控件用来输入“联系人姓名”和“联系人电话”，设置“添加”“删除”“修改”“查询”四个命令按钮用来操作数据库，最后还有一个文本框用来显示查询结果。用户界面设计参考如图 6-4 所示。

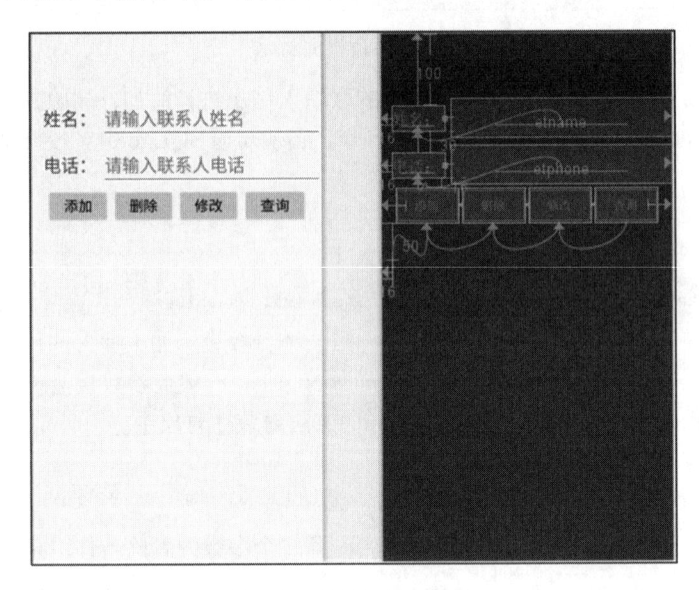

图 6-4　通讯录界面设计

activity_main.xml 的参考代码如下所示：

```
<?xml version="1.0" encoding="utf-8"?>
<androidx.constraintlayout.widget.ConstraintLayout
xmlns:android="http://schemas.android.com/apk/res/android"
    xmlns:app="http://schemas.android.com/apk/res-auto"
    xmlns:tools="http://schemas.android.com/tools"
    android:layout_width="match_parent"
    android:layout_height="match_parent"
    tools:context=".MainActivity">


    <TextView
        android:id="@+id/tvname"
        android:layout_width="wrap_content"
        android:layout_height="wrap_content"
```

```
        android:layout_marginStart="16dp"
        android:layout_marginLeft="16dp"
        android:layout_marginTop="100dp"
        android:text="姓名："
        android:textSize="25sp"
        app:layout_constraintStart_toStartOf="parent"
        app:layout_constraintTop_toTopOf="parent" />

    <EditText
        android:id="@+id/etname"
        android:layout_width="0dp"
        android:layout_height="wrap_content"
        android:layout_marginStart="8dp"
        android:layout_marginLeft="8dp"
        android:hint="请输入联系人姓名"
        android:inputType="textPersonName"
        android:textSize="25sp"
        app:layout_constraintBaseline_toBaselineOf="@+id/tvname"
        app:layout_constraintEnd_toEndOf="parent"
        app:layout_constraintStart_toEndOf="@+id/tvname" />

    <TextView
        android:id="@+id/tvphone"
        android:layout_width="wrap_content"
        android:layout_height="wrap_content"
        android:layout_marginStart="16dp"
        android:layout_marginLeft="16dp"
        android:layout_marginTop="30dp"
        android:text="电话："
        android:textSize="25sp"
        app:layout_constraintStart_toStartOf="parent"
        app:layout_constraintTop_toBottomOf="@+id/tvname" />

    <EditText
        android:id="@+id/etphone"
        android:layout_width="0dp"
        android:layout_height="wrap_content"
        android:layout_marginStart="8dp"
        android:layout_marginLeft="8dp"
        android:hint="请输入联系人电话"
        android:inputType="phone"
        android:textSize="25sp"
        app:layout_constraintBaseline_toBaselineOf="@+id/tvphone"
        app:layout_constraintEnd_toEndOf="parent"
        app:layout_constraintStart_toEndOf="@+id/tvphone" />

    <Button
        android:id="@+id/btn_add"
        android:layout_width="wrap_content"
        android:layout_height="wrap_content"
        android:layout_marginStart="16dp"
        android:layout_marginLeft="16dp"
        android:layout_marginTop="16dp"
        android:text="添加"
```

```
        android:onClick="btn_insert"
        android:textSize="20sp"
        app:layout_constraintEnd_toStartOf="@+id/btn_del"
        app:layout_constraintStart_toStartOf="parent"
        app:layout_constraintTop_toBottomOf="@+id/tvphone" />

    <Button
        android:id="@+id/btn_del"
        android:layout_width="wrap_content"
        android:layout_height="wrap_content"
        android:text="删除"
        android:onClick="btn_delete"
        android:textSize="20sp"
        app:layout_constraintBottom_toBottomOf="@+id/btn_add"
        app:layout_constraintEnd_toStartOf="@+id/btn_upd"
        app:layout_constraintStart_toEndOf="@+id/btn_add" />

    <Button
        android:id="@+id/btn_upd"
        android:layout_width="wrap_content"
        android:layout_height="wrap_content"
        android:text="修改"
        android:onClick="btn_update"
        android:textSize="20sp"
        app:layout_constraintBottom_toBottomOf="@+id/btn_del"
        app:layout_constraintEnd_toStartOf="@+id/btn_select"
        app:layout_constraintStart_toEndOf="@+id/btn_del" />

    <Button
        android:id="@+id/btn_select"
        android:layout_width="wrap_content"
        android:layout_height="wrap_content"
        android:layout_marginEnd="16dp"
        android:layout_marginRight="16dp"
        android:text="查询"
        android:onClick="btn_select"
        android:textSize="20sp"
        app:layout_constraintBottom_toBottomOf="@+id/btn_upd"
        app:layout_constraintEnd_toEndOf="parent"
        app:layout_constraintStart_toEndOf="@+id/btn_upd" />

    <TextView
        android:id="@+id/myContacts"
        android:layout_width="wrap_content"
        android:layout_height="wrap_content"
        android:layout_marginStart="16dp"
        android:layout_marginLeft="16dp"
        android:layout_marginTop="50dp"
        android:textSize="25sp"
        app:layout_constraintStart_toStartOf="parent"
        app:layout_constraintTop_toBottomOf="@+id/btn_add" />
</androidx.constraintlayout.widget.ConstraintLayout>
```

在界面设计中为每个命令按钮控件设置了 android:onClick 属性，“添加”命令按钮为“btn_insert”，“删除”命令按钮为“btn_delete”，“修改”命令按钮为“btn_update”，“查询”命令按钮为“btn_select”。

第三步：实现数据操作

在 MainActivity 中使用数据库操作的 insert()方法、delete()方法、update()方法和 query()方法实现对“mycontacts.db”数据库中的数据表 contancts 进行操作。MainActivty 代码如下所示。

```
public class MainActivity extends AppCompatActivity {
    private EditText etname, etphone;
    private TextView myContacts;
    private MyHelper myHelper;
    String name, phone;
    SQLiteDatabase db;
    ContentValues values = new ContentValues();

    @Override
    protected void onCreate(Bundle savedInstanceState) {
        super.onCreate(savedInstanceState);
        setContentView(R.layout.activity_main);
        myHelper = new MyHelper(this);
        etname = (EditText) findViewById(R.id.etname);
        etphone = (EditText) findViewById(R.id.etphone);
        myContacts = (TextView) findViewById(R.id.myContacts);
    }

    public void btn_insert(View view) {
        name = etname.getText().toString();
        phone = etphone.getText().toString();
        db = myHelper.getWritableDatabase();
        values.put("name", name);
        values.put("phone", phone);
        db.insert("contancts", null, values);
        Toast.makeText(this, "联系人已添加", Toast.LENGTH_LONG).show();
        etname.setText("");
        etphone.setText("");
        db.close();
    }

    public void btn_delete(View view) {
        db = myHelper.getWritableDatabase();
        db.delete("contancts", "name=?", new 
String[]{etname.getText().toString()});
        Toast.makeText(this, "联系人已删除", Toast.LENGTH_LONG).show();
        etname.setText("");
        db.close();
    }

    public void btn_update(View view) {
        db = myHelper.getWritableDatabase();
        values.put("phone", etphone.getText().toString());
        db.update("contancts", values, "name=?", new 
String[]{etname.getText().toString()});
```

```
        Toast.makeText(this, "联系人信息已修改", Toast.LENGTH_LONG).show();
        etname.setText("");
        etphone.setText("");
        db.close();
    }

    public void btn_select(View view) {
        db = myHelper.getReadableDatabase();
        Cursor cursor = db.query("contancts", null, null, null, null, null, null, null);
        myContacts.setText("");
        if (cursor.getCount() == 0) {
            Toast.makeText(this, "没有联系人信息", Toast.LENGTH_LONG).show();
        } else {
            while (cursor.moveToNext()) {
                myContacts.append(cursor.getString(1) + "    " +
cursor.getString(2) + "\n");
            }
            cursor.close();
            db.close();
        }
    }
}
```

通讯录的功能实现

从 MainActivity 的实现代码中可以看出，在 onCreate()方法中创建 MyHelper 对象，并对界面上的控件进行初始化。在命令按钮的单击事件中，进行增删改查操作。

“添加”命令按钮逻辑中使用 getWritableDatabase()方法获取可读写的 SQLiteDatabase 对象，将要添加的数据使用 put()方法放入 ContentValues 对象 values 中，调用 insert()方法，将 values 中的数据添加到表 contacts 中。接着设置用户名和手机号的输入框为空，使用 db.close()方法关闭数据库。

“删除”命令按钮逻辑中使用 getWritableDatabase()方法获取可读写的 SQLiteDatabase 对象，调用 delete()方法，删除 contacts 表中“name”为 etname（用户名输入框）中输入的名字，清除输入框中的内容，使用 db.close()方法关闭数据库。

“修改”命令按钮逻辑中使用 getWritableDatabase()方法获取可读写的 SQLiteDatabase 对象，将要修改的电话号码使用 put()方法放入 ContentValues 对象 values 中，调用 update()方法，修改 contacts 表中用户名输入框中用户的电话号码为 values 中的值。

“查询”命令按钮逻辑中使用 getReadableDatabase()方法获取可读的 SQLiteDatabase 对象，使用 query()方法查询表 contacts 中的数据，在 query()方法中除了表名其他所有的参数设置为 null，表示查询表中的所有数据，相当于“select * from contacts”的作用，将查询的结果赋值给 Cursor 对象。这里 Cursor 用于数据库的查询读取，Cursor 是每行的集合。getCount()方法返回 couser 中的行数，如果 cursor.getCount()返回的值为 0，则数据表里没有数据，否则依次追加到显示结果的 myContacts 控件中。

第四步：运行程序

运行程序，运行效果如图 6-4 所示。输入用户名和电话号码，单击“添加”命令按钮，提示“联系人已添加”；选择“查询”可以显示已经添加的用户，然后输入已经添加的“联系人姓名”，输入新电话号码，选择“修改”，再“查询”发现用户的电话已经修改了，在姓名输入框输入已经添加了的姓名，选择“删除”，再次“查询”删除的信息已不存在了。

扩展 本项目只是简单使用 insert()方法、delete()方法、update()方法和 query()方法。程序还有很多细节没有处理。

比如：查询查的是所有联系人的信息，如果需要查询指定联系人的电话，则将 cursor 改为：

```
Cursor cursor = db.query("contancts", new String[]{"phone"},"name=?" , new String[]{etname.
getText().toString()}, null, null, null, null);
```

query()方法的第二个参数是查询的列，如果要查表中的多列，则 String 数据中的元素用“，”号分隔即可；第三个参数是要满足的添加的列，第四个参数是查询的条件。

还有每次添加、删除、修改完成后，下面显示的内容没有立即改变，等等，这都需要再细化完善项目的，这里我们掌握数据操作的方法，大家可以自行完善程序。

补充 针对会 SQL 语言的开发者来说，在 SQLiteDatabase 中还提供了直接操作 SQL 语言的方法，对于前面几种操作方法，对应 SQL 语言的操作如下。

添加数据方法：

```
db.execSQL( “insert into contacts(name,phone) values(?,?)” ,new String[]{” weihua” , ” 189966969” );
```

更新数据方法：

```
db.execSQL( “update contacts set phone=? where name=?” ,new String[]{” weihua” ,” 139966969” );
```

删除数据方法：

```
db.execSQL( “delete from contacts where name=?” ,new String[]{ “weihua” });
```

查询数据方法：

```
db.rawQuery( “select * from contacts” ,null);
```

这几种方法中，除了查询时调用 SQLiteDatabase 的 rawQuery()方法，其他操作都是调用的 execSQL()方法。

6.4 使用 Room 实现通讯录

Android 采用 SQLite 作为数据库存储，SQLite 代码写起来繁琐且容易出错，所以开源社区里逐渐出现了各种 ORM（Object Relational Mapping）库。这些开源 ORM 库都是为了方便 SQLite 的使用，包括数据库的创建、升级、增删改查等。常见的 ORM 有 ORMLite、GreenDAO 等。

Room 是 Jetpack 中的 ORM 组件。Room 可以简化 SQLite 数据库操作。Room 由 Entity、Dao 和 Database 三部分组成。

Entity：封装实际数据的实体类，每个实体类都会在数据中对应一张表，并且表中的列是根据实体类中的字段自动生成的。

Dao：是数据访问对象的意思，通常会在这里对数据库的各项操作进行封装，在实际编程时，逻辑层就不需要和底层数据库打交道了，直接和 Dao 层进行交互即可。

Database：用于定义数据库中的关键信息，包括数据库的版本号，包含哪些实体类以及提供 Dao 层的访问实例。Database 是 Entity 和 Dao 的集合，代表一个 SQLite 数据库，是我们访问 Dao

和 Entity 的入口。

下面我们使用 Room 来实现与 MyContacts 项目运行效果一样的功能，通过比较更好地理解 Room 的使用。实现过程如下所示：

第一步：创建一个项目

新建一个项目，名为 RoomContacts，包名为 com.example.roomcontacts。

第二步：添加依赖

在 build.gradle 文件中添加依赖：

```
implementation 'androidx.room:room-runtime:2.2.2'
annotationProcessor 'androidx.room:room-compiler:2.2.2'
```

第三步：定义用户界面

直接复制项目 MyContacts 中的 activity_main 界面设计代码即可。

第四步：定义 Entity

选择 com.example.roomcontacts 包，新建一个类，名为 Contacts，在类中编写如下代码：

```
package com.example.roomcontacts;

import androidx.room.ColumnInfo;
import androidx.room.Entity;
import androidx.room.Ignore;
import androidx.room.PrimaryKey;

@Entity(tableName = "Contacts")
public class Contacts {
    @PrimaryKey(autoGenerate=true)
    @ColumnInfo(name="_id")
    private int id;
    @ColumnInfo(name="name")
    private String name;
    @ColumnInfo(name="phone")
    private String phone;
    public Contacts(){}
    @Ignore
    public Contacts(String name,String phone){
        this.name=name;
        this.phone=phone;
    }

    public int getId() { return id;    }

    public void setId(int id) {  this.id = id;    }
```

```
    public String getName() {  return name;    }

    public void setName(String name) {  this.name = name;    }

    public String getPhone() {  return phone;    }

    public void setPhone(String phone) {  this.phone = phone;    }

    @Override
    public String toString(){
        return "Contacts{id=" + id+", name='" + name +", price=" +phone +'}';
    }
}
```

创建一个关于联系人的 Contacts，即创建一张联系人信息表。在类文件的最上方需要加上@Entity 标签，通过该标签将该类与 Room 中表关联起来。

tableName 属性为该表设置名字。

@PrimaryKey 标签用于指定该字段作为表的主键。

@ColumnInfo 标签可用于设置该字段存储在数据库表中的名字。

@Ignore 标签用来告诉系统忽略该字段或者方法。由于 Room 只能识别和使用一个构造器，如果希望定义多个构造器，可以使用 Ignore 标签，让 Room 忽略这个构造器。同样，@Ignore 标签还可用于字段，使用@Ignore 标签标记过的字段，Room 不会持久化该字段的数据。

第五步：定义 Dao

针对以上联系人 Entity，需要定义一个 Dao 接口文件，以完成对 Entity 的访问。注意:在文件的上方，需要加入@Dao 标签。

在当前包 com.example.roomcontacts 中定义个接口 interface 名为 ContactsDao，代码如下所示：

```
package com.example.roomcontacts;

import androidx.room.Dao;
import androidx.room.Delete;
import androidx.room.Insert;
import androidx.room.OnConflictStrategy;
import androidx.room.Query;
import androidx.room.Update;
import java.util.List;

@Dao
public interface ContactsDao {
    //如果冲突--替换
    @Insert(onConflict = OnConflictStrategy.REPLACE)
    void insertContacts(Contacts contacts);

    //查询所有的信息
    @Query("SELECT * FROM contacts")
    List<Contacts> getAll();
```

```
    //查询指定名字的电话
    @Query("SELECT * FROM contacts WHERE name ==:name")
    List<Contacts> getPhone(String name);

    @Query("DELETE FROM contacts WHERE name ==:name")
    void deleteContacts(String name);

    @Delete
    int deleteContacts(Contacts contacts);

    //更新 返回更新的个数
    @Update(onConflict = OnConflictStrategy.REPLACE)
    int updateContacts(Contacts contacts);

    @Query("UPDATE contacts set phone=:phone Where name==:name")
    void updateContacts(String name, String phone);
}
```

第六步：定义 Database

定义好 Entity 和 Dao 后，接下去就是创建数据库了。在包 com.example.roomcontacts 创建一个继承自 RoomDatabase 的抽象类 MyDatabase，创建的数据表是实体 Contacts。在这个抽象类中定义了一个抽象方法 contactsDao()。

```
package com.example.roomcontacts;

import androidx.room.Database;
import androidx.room.RoomDatabase;

@Database(entities = Contacts.class,version = 1,exportSchema = false)
public abstract  class MyDatabase  extends RoomDatabase {
    public abstract ContactsDao contactsDao();
}
```

在上面的代码中@Database 标签用于告诉系统这是 Room 数据库对象，entities 属性用于指定该数据库有哪些表，若需建立多张表，以逗号相隔开。version 属性用于指定数据库版本号，后续数据库的升级正是依据版本号来判断的。该类需要继承自 RoomDatabase，在类中我们创建的 Dao 对象，在这里以抽象方法的形式返回。使用 Room 框架其中一个好处是，如果创建过程中有问题，在编译期间编辑器就会提示你，而不用等到程序运行时。

至此，数据库和表的创建工作完成了。接下去，我们来看看 Room 框架下的数据库增删改查。

第七步：定义接口 IContactsMode

在包 com.example.roomcontacts 中定义一个接口设计数据库操作的方法。代码如下所示：

```
package com.example.roomcontacts;

import java.util.List;
```

```
public interface IContactsMode {
    void add(Contacts contacts);

    int delete(Contacts contacts);

    void delete(String name);

    int update(Contacts contacts);

    List<Contacts> findAll();

    List<Contacts> findPhone(String name);

    void update(String name, String phone);
}
```

在此接口中设计添加数据的方法 add()、删除数据的方法 delete()方法、修改数据的 update()方法、查询数据的 findAll()方法和 findPhone()方法。

第八步：实现接口

实现第七步的接口 IContactsMode，新建一个类 ContactsMode 实现接口 IContactsMode 中的抽象方法。代码如下所示：

```
package com.example.roomcontacts;

import android.content.Context;
import androidx.room.Room;
import java.util.List;

public class ContactsMode implements IContactsMode {
    private static volatile ContactsMode contactsMode;
    private Context context;

    private final MyDatabase myDatabase;
    private ContactsDao contactsDao;

    public ContactsMode(Context context) {
        myDatabase = Room.databaseBuilder(context, MyDatabase.class, "Contacts.db")
                .allowMainThreadQueries().build();
        contactsDao = myDatabase.contactsDao();
    }

    public static ContactsMode getInstance(Context context) {

        if (contactsMode == null) {
            synchronized (ContactsMode.class) {
                if (contactsMode == null) {
                    contactsMode = new ContactsMode(context);
                }
            }
```

```
        }
        return contactsMode;
    }

    @Override
    public void add(Contacts contacts) {
        contactsDao.insertContacts(contacts);
    }

    @Override
    public int delete(Contacts contacts) {
        return contactsDao.deleteContacts(contacts);
    }

    @Override
    public void delete(String name) {
        contactsDao.deleteContacts(name);
    }

    @Override
    public int update(Contacts contacts) {
        return contactsDao.updateContacts(contacts);
    }

    @Override
    public List<Contacts> findAll() {
        return contactsDao.getAll();
    }

    @Override
    public List<Contacts> findPhone(String name) {
        return contactsDao.getPhone(name);
    }

    @Override
    public void update(String name, String phone) {
        contactsDao.updateContacts(name, phone);
    }
}
```

第九步：实现数据库操作

在 MainActivity 中调用 ContactsMode 中的方法实现数据的增删改查。MainActivity 的代码如下所示。

```
public class MainActivity extends AppCompatActivity {
    private EditText etname, etphone;
    private TextView myContacts;
    String name, phone;
    ContactsMode cm;
    Contacts contacts;

    @Override
```

```
    protected void onCreate(Bundle savedInstanceState) {
        super.onCreate(savedInstanceState);
        setContentView(R.layout.activity_main);
        etname = (EditText) findViewById(R.id.etname);
        etphone = (EditText) findViewById(R.id.etphone);
        myContacts = findViewById(R.id.myContacts);
        cm = new ContactsMode(this);
    }

    public void btn_insert(View view) {
        name = etname.getText().toString();
        phone = etphone.getText().toString();
        contacts = new Contacts(name, phone);
        cm.add(contacts);
    }

    public void btn_delete(View view) {
        name = etname.getText().toString();
        cm.delete(name);
    }

    public void btn_update(View view) {
        name = etname.getText().toString();
        phone = etphone.getText().toString();
        cm.update(name, phone);
    }

    public void btn_select(View view) {
        myContacts.setText("");
        cm.findAll();
        for (Contacts c : cm.findAll()) {
            myContacts.append(c.getName() + "  " + c.getPhone() + "\n");
        }
    }
}
```

第十步：运行程序

程序运行效果与 MyContacts 的运行效果完全一样。通过这个案例我们学会了如何在 Android 项目中利用 Room 创建数据库，以及对数据库进行增删改查等基本操作，Room 使得我们在 Android 中使用 SQLite 变得非常容易。

【职业素养拓展6】

职业素养拓展

职业素养拓展内容详见文件“职业素养拓展”。
扫描二维码查看【职业素养拓展 6】

本章小结

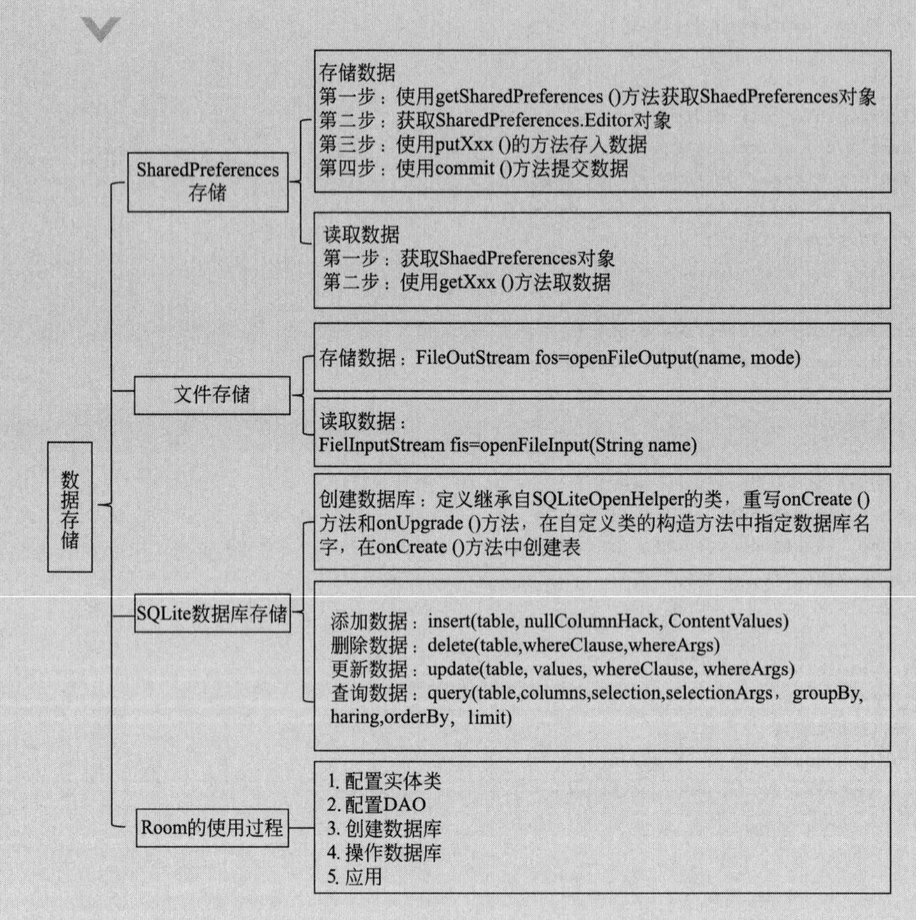
数据存储
SharedPreferences存储
存储数据
第一步：使用getSharedPreferences ()方法获取ShaedPreferences对象
第二步：获取SharedPreferences.Editor对象
第三步：使用putXxx ()的方法存入数据
第四步：使用commit ()方法提交数据
读取数据
第一步：获取ShaedPreferences对象
第二步：使用getXxx ()方法取数据
文件存储
存储数据：FileOutStream fos=openFileOutput(name, mode)
读取数据：
FielInputStream fis=openFileInput(String name)
SQLite数据库存储
创建数据库：定义继承自SQLiteOpenHelper的类，重写onCreate ()方法和onUpgrade ()方法，在自定义类的构造方法中指定数据库名字，在onCreate ()方法中创建表
添加数据：insert(table, nullColumnHack, ContentValues)
删除数据：delete(table,whereClause,whereArgs)
更新数据：update(table, values, whereClause, whereArgs)
查询数据：query(table,columns,selection,selectionArgs， groupBy, haring,orderBy， limit)
Room的使用过程
1. 配置实体类
2. 配置DAO
3. 创建数据库
4. 操作数据库
5. 应用

第7章

内容提供者

在前面我们学习了数据库的相关知识，可以在应用程序中对数据进行存储并处理，但是那些数据都是同一个应用中的数据，虽然这样安全性高，却不能满足不同应用之间数据的共享。Android 提供了内容提供者 ContentProvider，可以实现不同应用之间的数据共享。本章主要介绍内容提供者的使用，学习访问系统数据和自定义应用之间的数据共享。

学习目标：

1. 理解跨应用数据共享的工作原理；
2. 掌握内容提供者的创建和定义；
3. 能够实现不同应用程序之间的数据共享。

7.1 内容提供者概述

内容提供者简介

在 Android 中，每一个应用程序的数据都是采用私有的形式进行操作的，不管这些数据是用文件还是数据库保存，都不能被外部应用程序访问。在很多情况下，用户需要在不同应用程序之间进行数据的共享，为了解决这类问题，Android 专门提供了 ContentProvider，以实现不同应用程序之间的数据共享。

内容提供者是不同应用程序之间数据共享的一种接口机制，使用 ContentProvider 类将不同的应用程序的数据操作标准统一起来，这些标准在类中明确通过方法进行定义。其他应用程序在不知数据来源、路径的情况下，可以对共享数据进行增加、删除、修改和查询等操作。

通常，会在以下两种场景中使用内容提供者：一种是通过代码访问其他应用中的现有内容提供者，例如 Android 系统的通讯录、通话记录等内置数据通过 ContentProvider 提供给了用户，用户可以在自己的应用中访问这些数据；另一种是在应用中创建新的内容提供者，从而与其他应用共享数据，我们可以在自己的应用中创建内容提供者供其他应用共享。

7.2 访问系统通讯录

在 Android 中系统的许多应用都提供了内容提供者，其他应用程序可以访问其中的数据。下面我们通过一个实例来访问系统的通讯录。

分析 对于系统通讯录的访问，由于通讯录属于危险权限，在应用时需要动态申请权限，实现步骤如下所示。

第一步：新建程序

创建应用程序 Read_System_CONTACTS,包名为 com.example.read_system_contacts。

第二步：在清单文件中申请授权

打开清单文件 AndroidManifest，在清单文件中申请授权，代码如下所示。

```
<?xml version="1.0" encoding="utf-8"?>
<manifest xmlns:android="http://schemas.android.com/apk/res/android"
    package="com.example.read_contacts">
    <uses-permission android:name="android.permission.READ_CONTACTS"/>
        ……
   </manifest>
```

在清单文件中申请授权读取系统通讯录，所以我们要添加一个<uses-permission/>,内容为"android.permission.READ_CONTACTS"。

第三步：设计用户界面

打开 activity_main 文件，在布局中添加一个 TextView 控件，用来显示访问到的内容。

```
<?xml version="1.0" encoding="utf-8"?>
<androidx.constraintlayout.widget.ConstraintLayout  xmlns:android="http://schemas.android.com/apk/
res/android"
    xmlns:app="http://schemas.android.com/apk/res-auto"
    xmlns:tools="http://schemas.android.com/tools"
    android:id="@+id/linearLayout"
    android:layout_width="match_parent"
    android:layout_height="match_parent"
    tools:context=".MainActivity">

    <TextView
        android:id="@+id/tv_contacts"
        android:layout_width="0dp"
        android:layout_height="wrap_content"
        android:text=""
        android:textSize="30sp"
        app:layout_constraintEnd_toEndOf="parent"
        app:layout_constraintStart_toStartOf="parent"
        app:layout_constraintTop_toTopOf="parent" />

</androidx.constraintlayout.widget.ConstraintLayout>
```

第四步：在 MainActivity 中访问系统通讯录

访问内容提供者

在前面我们提到 Android 系统为通讯录提供内容提供者，那么如何去使用才能访问系统通讯录中的内容呢?

打开 MainActivity，修改代码如下所示。

```
public class MainActivity extends AppCompatActivity {
      private TextView tvcontacts;

      @Override
      protected void onCreate(Bundle savedInstanceState) {
          super.onCreate(savedInstanceState);
          setContentView(R.layout.activity_main);
          tvcontacts = findViewById(R.id.tv_contacts);
          ActivityCompat.requestPermissions(this, new String[]{Manifest.permission.READ_CONTACTS},
101);
          Cursor cursor = getContentResolver().query(ContactsContract.CommonDataKinds.Phone.CONTENT_
URI, null, null, null, null);
          if (cursor != null) {
              while (cursor.moveToNext()) {
                  String name = cursor.getString((cursor.getColumnIndex(ContactsContract.Common-
DataKinds.Phone.DISPLAY_NAME)));
                  String phonenum = cursor.getString((cursor.getColumnIndex(ContactsContract.Common-
DataKinds. Phone.NUMBER)));
                  tvcontacts.append(name + " : " + phonenum + "\n");
              }
          }
      }
  }
```

在上面代码中主要可分为两个部分，第一部分是动态申请权限部分，第二部分为访问系统通讯录部分。

（1）动态申请权限

这里动态申请权限使用了 requestPermissions()方法,此方法的原型如下所示:

```
requestPermissions(final @NonNull Activity activity, final @NonNull String[] permissions, final
@IntRange(from = 0) int requestCode)
```

方法有三个参数：第一个参数为 Activity；第二个参数为 String[]，表示需要申请的权限，如果有多个权限，数组中添加数据元素就可以了；第三个参数为 int 类型，表示请求码。

（2）访问系统通讯录

对于每一个应用程序来说，如果想要访问内容提供者中共享的数据，就一定要借助 ContentResolver 类。ContentResolver 可以与任意内容提供者进行会话，与其合作来对所有相关交互通讯进行管理。当外部应用需要对 ContentProvider 中的数据进行添加、删除、修改和查询操作时，通过 getContentResolver()方法获取到 ContentResolver 类的实例。ContentResolver 中提供了一系列的方法用于对数据进行增删改查操作。在上面的代码中主要有以下几点。

① getContentResolver()方法：获取 ContentResolver 类的实例。

② query()方法：用来查询数据，它的用法与我们在 SQLite 中的用法一样，它的第一个参数是要查询的数据表。

③ ContactsContract.CommonDataKinds.Phone.CONTENT_URI。其中 ContactsContract. Common-DataKinds. Phone 是系统封装好的，CONTENT_URI 是一个常量，查看它的定义为：

public static final Uri CONTENT_URI = Uri.withAppendedPath(Data.CONTENT_URI, "phones");

可以看到 CONTENT_URI 就是 Uri 对象，我们在自定义内容提供者时会提到设置 Uri，在访问共享数据时通过 Uri 来找，而这里直接使用系统定义好的。

④ ContactsContract.CommonDataKinds.Phone.DISPLAY_NAME 和 ContactsContract.Common-

DataKinds. Phone.NUMBER：分别为系统通讯录中的用户名和号码。

第五步：运行程序

我们需要访问系统通讯录，所以先打开通讯录，添加如图 7-1 所示联系人。

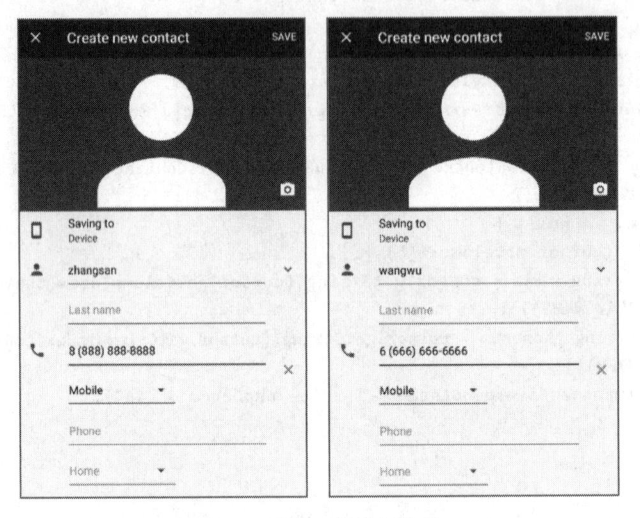

图 7-1　添加联系人信息

给通讯录添加了联系人，然后运行程序，界面启动弹出如图 7-2 所示的“Allow Read_System_CONTACTS to access your contacts?”对话框，选择“Allow”允许访问系统通讯录，出现如图 7-3 所示的运行效果。

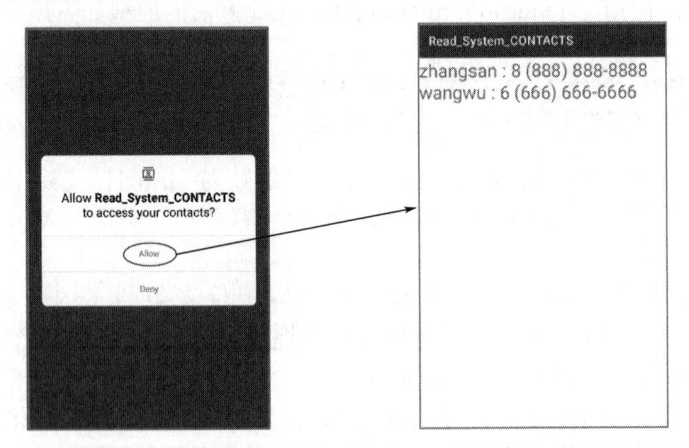

图 7-2　请求“Read_System-CONTACTS”权限　图 7-3　访问系统通讯录效果图

7.3　程序间数据共享

创建一个内容提供者

7.3.1　创建内容提供者

内容提供者是 Android 的四大组件之一，它的创建与 Activity 的类似。下面我们演示如何创建内容提供者，操作步骤如下所示。

第一步：新建程序

新建一个程序 Create_ContentProvider，所在的包名 com.example.create_contentprovider。

第二步：创建内容提供者

选择当前应用程序的包：com.example.create_contentprovider;单击右键“New”→“Other”→“Content Provider”，弹出如图 7-4 所示对话框。

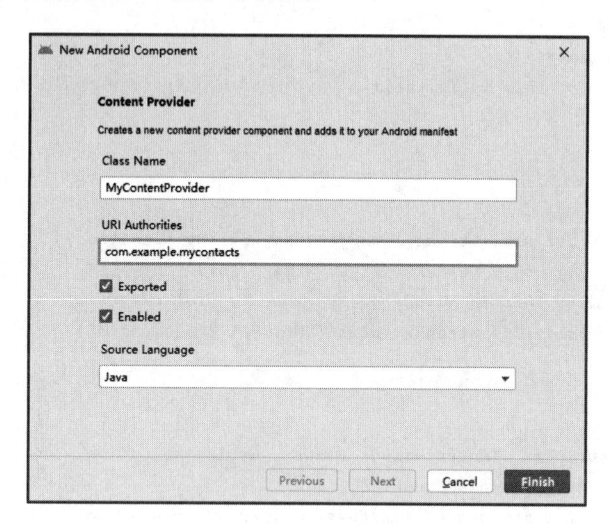

图 7-4　“Content Provide”对话框

在上面的对话框中，“Class Name”使用默认的值“MyContentProvider”，指定“URI Authorites”的值为“com.example.mycontacts”，点击“Finish”完成内容提供者的创建。我们观察应用程序的包，发现多了一个“MyContentProvider.java”文件，打开查看代码，如下所示。

```
package com.example.mycontentprovider;

import android.content.ContentProvider;
import android.content.ContentValues;
import android.database.Cursor;
import android.net.Uri;

public class MyContentProvider extends ContentProvider {
    public MyContentProvider() {
    }

    @Override
    public int delete(Uri uri, String selection, String[] selectionArgs) {
        // Implement this to handle requests to delete one or more rows.
        throw new UnsupportedOperationException("Not yet implemented");
    }

    @Override
    public String getType(Uri uri) {
        // TODO: Implement this to handle requests for the MIME type of the data
        // at the given URI.
```

```
        throw new UnsupportedOperationException("Not yet implemented");
    }

    @Override
    public Uri insert(Uri uri, ContentValues values) {
        // TODO: Implement this to handle requests to insert a new row.
        throw new UnsupportedOperationException("Not yet implemented");
    }

    @Override
    public boolean onCreate() {
        // TODO: Implement this to initialize your content provider on startup.
        return false;
    }

    @Override
    public Cursor query(Uri uri, String[] projection, String selection,
                        String[] selectionArgs, String sortOrder) {
        // TODO: Implement this to handle query requests from clients.
        throw new UnsupportedOperationException("Not yet implemented");
    }

    @Override
    public int update(Uri uri, ContentValues values, String selection,
                      String[] selectionArgs) {
        //TODO: Implement this to handle requests to update one or more rows.
        throw new UnsupportedOperationException("Not yet implemented");
    }
}
```

Uri简介

从上面的代码中可以看到我们创建的类继承自 ContentProvider 类，重写了 insert()、delete()、update()和 query()等的方法，这些方法参数都有一个是 Uri 对象，那么什么是 Uri，在这里的作用是什么呢?

Uri 是统一资源定位符，在内容提供者中 Uri 代表了要操作的数据，每个内容提供者中都会对外提供一个公共的 Uri 对象，当应用程序数据需要共享时，其他的应用程序就可以通过 ContentProvider 传入这个 Uri 来对数据进行操作。Uri 的格式如下所示：

content://authority//path，由三个部分组成。

第一部分：“content://”，是系统规定的固定写法。

第二部分：authority，是一个标志，是内容提供者的字符串，也就是该内容提供者的标识符，相当于我们每个人都有的姓名。

第三部分：是内容提供者提供数据的某个子集，因为内容提供者有时会提供多个数据集，我们具体是要访问哪个子集，需要把具体路径标出来。如：

```
content://my_content_provider/contancts
```

表示可以访问自定义的 authority 为“my_content_provider”的内容提供者中的数据表“contancts”，这也是我们在访问内容提供者的共享数据时要用到的值。

对于其中涉及的方法，含义如下：

onCreate()方法：在初始化内容提供者时调用。通常在这里完成数据库的创建和升级等操作。

insert()方法：向内容提供者中添加一条数据。使用 uri 参数确定要添加的表，待添加的数据保存在 values 参数中，添加完成后，返回一个用于表示这条新纪录的 Uri。

delete()方法：从内容提供者中删除数据。使用 uri 参数来确定删除哪一张表中的数据，selection 和 selectionArgs 是删除的约束条件。返回值为整数，表示删除的记录的个数。

update()方法：更新内容提供者中已有的数据，使用 uri 参数确定要修改的数据表，values 为要更新的数据，返回修改的记录的个数。

query()方法：此方法是从内容提供者中查询数据。使用 uri 确定要查询的表，projection 参数用于确定查询哪些列,selection 和 selectionArgs 是查询的条件,sortOrder 参数表示是否对查询结果排序。

第三步：清单文件中声明内容提供者

打开清单文件 AndroidManifest，显示如下代码：

```
<?xml version="1.0" encoding="utf-8"?>
<manifest xmlns:android="http://schemas.android.com/apk/res/android"
    package=" com.example.mycontentprovider">
……
        <provider
            android:name=".MyContentProvider"
            android:authorities="com.example.mycontacts"
            android:enabled="true"
            android:exported="true">
        </provider>
</manifest>
```

在上面的代码中，有一对标记<provider></provider>是对创建的内容提供者的声明。使用这种方式创建的内容提供者，在清单文件中会自动声明了。如果是通过新建类创建的内容提供者，需要手动在清单文件中进行声明。

provider 标记的各属性含义如下：

name:指明内容提供者的名称。

authorities：用于标识内容提供者提供的数据，也是在其他应用程序中访问共享数据时，设置 Uri 对象需要使用的值，这个属性通常设置为包名。

enabled：标识内容提供者能否被系统实例化，默认值为 true 标识，可以被系统实例化。当值为 false 时，不能被系统实例化。

exported：表示内容提供者能否被其他程序使用，如果属性值为 true，表示任何应用程序都可以通过 URI 访问内容提供者，也是默认值。

这样我们的内容提供者 MyContentProvider 类就创建好了，如果要给其他的应用提供共享数据，还需要实现此类当中数据增删改查的相关方法。

7.3.2　访问自定义通讯录

读取联系人

以我们在 SQLite 中实现的通讯录程序 MyContacts 为基础，新建一个项目共享 MyContatcs 中的数据，运行程序效果如图 7-5 所示，在这个项目中我们可以对数据进行增删改查。下面我们再新建一个程序来操作 MyContacts 中的数据，实现不同应用程序间的数据共享。

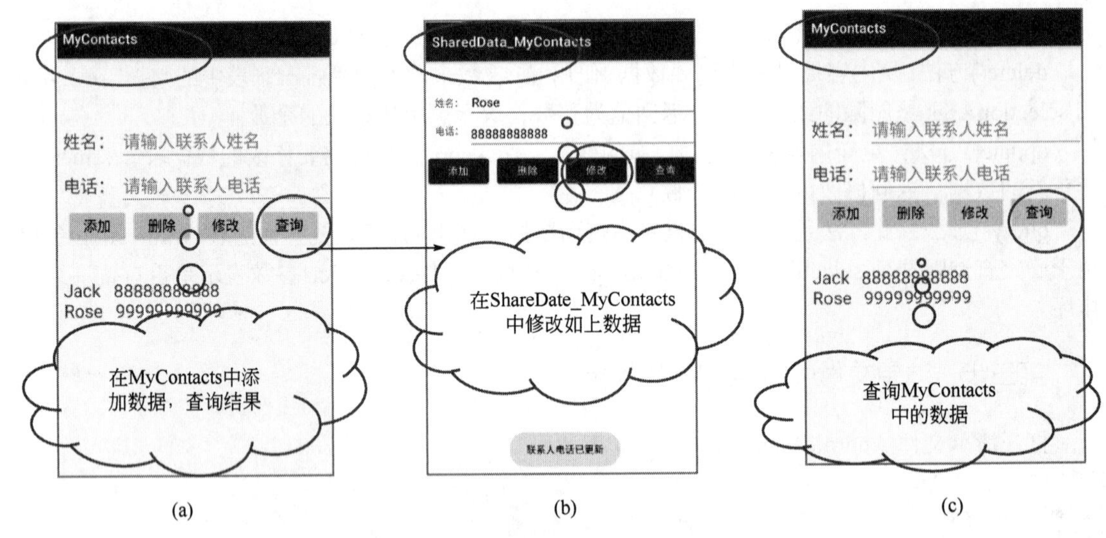

(a) (b) (c)

图 7-5 程序间数据共享运行效果

分析 实现不同程序间的数据共享，要解决两个问题，第一个是如何将数据提供给其他程序，第二个问题是如何访问并操作其他程序提供的数据。对于第一个问题，需要在应用程序中添加内容提供者并实现数据操作的方法，以供其他程序访问；对于第二个问题先使用 Uri.parse()方法将一个内容提供者解析成 Uri 对象，再使用 getContentResolver()获取 ContentResolver 实例，然后使用 ContentResolver 实例的各种操作数据的方法。

不同应用程序之间数据共享的实现过程如下所示。

第一步：打开程序 MyContacts

打开第 6 章的项目 MyContacts。

第二步：创建内容提供者并实现数据操作方法

在 MyContants 包 com.example.mycontacts 中“New”→“Other”→“Content Provide”创建内容提供者，Class Name 为“MyContentProvider”，URI Authorities 为“com.example.mycontacts1”，其余为默认选项，点击“Finish”创建“MyContentProvider”类，再实现数据操作的方法，代码如下所示。

```
public class MyContentProvider extends ContentProvider {
    private MyHelper myHelper;//创建 MyHelper 对象
    public MyContentProvider() {
    }

    @Override
    public int delete(Uri uri, String selection, String[] selectionArgs) {
        SQLiteDatabase db=myHelper.getWritableDatabase();
        int num=db.delete("contancts",selection,selectionArgs); //num 返回删除的记录数
        return num;
    }

    @Override
```

```
    public String getType(Uri uri) {
        // TODO: Implement this to handle requests for the MIME type of the data
        // at the given URI.
        throw new UnsupportedOperationException("Not yet implemented");
    }

    @Override
    public Uri insert(Uri uri, ContentValues values) {
        SQLiteDatabase db=myHelper.getWritableDatabase();
        db.insert("contancts",null,values);
        return null;
    }

    @Override
    public boolean onCreate() {
      myHelper=new MyHelper(getContext());
        return true;
    }

    @Override
    public Cursor query(Uri uri, String[] projection, String selection,
                        String[] selectionArgs, String sortOrder) {
        SQLiteDatabase db=myHelper.getReadableDatabase();
        Cursor cursor=db.query("contancts",projection,selection,selectionArgs,null,null,sortOrder);
        return cursor;
    }

    @Override
    public int update(Uri uri, ContentValues values, String selection,
                      String[] selectionArgs) {
      SQLiteDatabase db=myHelper.getWritableDatabase();
      int num=db.update("contancts",values,selection,selectionArgs);//num 返回修改的记录数
      return num;
    }
}
```

在 MyContentProvider 中定义了成员变量 myHelper，在 onCreate()方法中对 myHelper 进行初始化，getContext()方法返回上下文 context;在 insert()方法、delete()方法和 update()方法中使用 getWritableDatabase()方法分别获取可写 SQLiteDatabase 对象 db。然后分别对数据表“contancts”进行增加、删除和修改的操作，在 query()方法中查询“contancts”表数据。也就是说这个内容提供者提供给了其他应用对数据库进行操作能力。

接下就是如何使用内容提供者提供的操作能力来访问这些数据了。在 Android 中使用 ContentResolver 类来完成。

第三步：创建一个新的程序实现数据共享

创建程序 SharedData_MyContacts，包名为 com.example.shareddata_mycontacts。

第四步：设计用户界面

我们将界面做成与通讯录类似的界面，activity_main 参考代码如下：

```
<?xml version="1.0" encoding="utf-8"?>
<androidx.constraintlayout.widget.ConstraintLayout  xmlns:android="http://schemas.android.com/apk/
res/android"
    xmlns:app="http://schemas.android.com/apk/res-auto"
    xmlns:tools="http://schemas.android.com/tools"
    android:layout_width="match_parent"
    android:layout_height="match_parent"
    tools:context=".MainActivity">

    <TextView
        android:id="@+id/tvname"
        android:layout_width="wrap_content"
        android:layout_height="wrap_content"
        android:layout_marginStart="20dp"
        android:layout_marginLeft="20dp"
        android:layout_marginTop="50dp"
        android:text="姓名："
        app:layout_constraintStart_toStartOf="parent"
        app:layout_constraintTop_toTopOf="parent" />

    <EditText
        android:id="@+id/et_name"
        android:layout_width="0dp"
        android:layout_height="wrap_content"
        android:layout_marginStart="8dp"
        android:layout_marginLeft="8dp"
        android:ems="10"
        android:inputType="textPersonName"
        app:layout_constraintBaseline_toBaselineOf="@+id/tvname"
        app:layout_constraintEnd_toEndOf="parent"
        app:layout_constraintStart_toEndOf="@+id/tvname" />

    <TextView
        android:id="@+id/tvphone"
        android:layout_width="wrap_content"
        android:layout_height="wrap_content"
        android:layout_marginTop="20dp"
        android:text="电话："
        app:layout_constraintStart_toStartOf="@+id/tvname"
        app:layout_constraintTop_toBottomOf="@+id/tvname" />

    <EditText
        android:id="@+id/et_phone"
        android:layout_width="0dp"
        android:layout_height="wrap_content"
        android:ems="10"
        android:inputType="phone"
        app:layout_constraintBottom_toBottomOf="@+id/tvphone"
        app:layout_constraintEnd_toEndOf="parent"
        app:layout_constraintStart_toStartOf="@+id/et_name"
        app:layout_constraintTop_toTopOf="@+id/tvphone"
        app:layout_constraintVertical_bias="0.346" />

    <Button
        android:id="@+id/btn_queryTest"
        android:layout_width="wrap_content"
```

```
        android:layout_height="wrap_content"
        android:layout_marginStart="13dp"
        android:layout_marginLeft="13dp"
        android:onClick="btn_query"
        android:text="查询"
        app:layout_constraintBottom_toBottomOf="@+id/btn_updTest"
        app:layout_constraintStart_toEndOf="@+id/btn_updTest" />

    <Button
        android:id="@+id/btn_delTest"
        android:layout_width="wrap_content"
        android:layout_height="wrap_content"
        android:layout_marginStart="12dp"
        android:layout_marginLeft="12dp"
        android:onClick="btn_delete"
        android:text="删除"
        app:layout_constraintBottom_toBottomOf="@+id/btn_addTest"
        app:layout_constraintStart_toEndOf="@+id/btn_addTest" />

    <Button
        android:id="@+id/btn_updTest"
        android:layout_width="wrap_content"
        android:layout_height="wrap_content"
        android:layout_marginStart="12dp"
        android:layout_marginLeft="12dp"
        android:onClick="btn_update"
        android:text="修改"
        app:layout_constraintBottom_toBottomOf="@+id/btn_delTest"
        app:layout_constraintStart_toEndOf="@+id/btn_delTest" />

    <Button
        android:id="@+id/btn_addTest"
        android:layout_width="wrap_content"
        android:layout_height="wrap_content"
        android:layout_marginStart="10dp"
        android:layout_marginLeft="10dp"
        android:layout_marginTop="24dp"
        android:onClick="btn_add"
        android:text="添加"
        app:layout_constraintStart_toStartOf="parent"
        app:layout_constraintTop_toBottomOf="@+id/tvphone" />

    <TextView
        android:id="@+id/tv_contacts"
        android:layout_width="0dp"
        android:layout_height="0dp"
        android:layout_margin="20dp"
        android:textSize="16sp"
        app:layout_constraintBottom_toBottomOf="parent"
        app:layout_constraintEnd_toEndOf="parent"
        app:layout_constraintStart_toStartOf="parent"
        app:layout_constraintTop_toBottomOf="@+id/btn_addTest" />

</androidx.constraintlayout.widget.ConstraintLayout>
```

设计用户界面与第 6 章通讯录项目“MyContacts”的界面类似，“添加”“删除”“修改”

和“查询”命令按钮都添加了 android:onClick 属性。在操作数据时需要实现对应的方法。

第五步：操作内容提供者中的数据

在程序 MyProviderTest 的 MainActivity 中实现 MyContacts 中数据的访问与操作，代码如下所示。

```
public class MainActivity extends AppCompatActivity {
    private EditText etName, etPhone;
    private TextView tvContacts;
    private Uri uri;
    private ContentResolver resolver;
    private ContentValues values;

    @Override
    protected void onCreate(Bundle savedInstanceState) {
        super.onCreate(savedInstanceState);
        setContentView(R.layout.activity_main);
        etName = findViewById(R.id.et_name);
        etPhone = findViewById(R.id.et_phone);
        tvContacts = findViewById(R.id.tv_contacts);
        uri = Uri.parse("content://com.example.mycontacts1/contancts");
        resolver = getContentResolver();
        values = new ContentValues();
    }

    public void btn_add(View view) {
        values.put("name", etName.getText().toString());
        values.put("phone", etPhone.getText().toString());
        resolver.insert(uri, values);
        Toast.makeText(this, "联系人已添加", Toast.LENGTH_LONG).show();
    }

    public void btn_delete(View view) {
        resolver.delete(uri, "name=?", new String[]{etName.getText().toString()});
        Toast.makeText(this, "联系人已删除", Toast.LENGTH_LONG).show();
    }

    public void btn_update(View view) {
        values.put("phone", etPhone.getText().toString());
        resolver.update(uri, values, "name=?", new String[]{etName.getText().toString()});
        Toast.makeText(this, "联系人电话已更新", Toast.LENGTH_LONG).show();

    }

    public void btn_query(View view) {
        Cursor cursor = resolver.query(uri, null, null, null, null);
        if (cursor.getCount() != 0) {
            tvContacts.setText("");
            while (cursor.moveToNext()) {
                tvContacts.append(cursor.getString(1) + ":" + cursor.getString(2) + "\n");
            }
        } else {
            Toast.makeText(this, "无联系人信息", Toast.LENGTH_LONG).show();
        }
    }
}
```

在 MainActivity 中的 onCreate()方法中对所有控件进行初始化，使用 Uri.parse()方法将一个内容提供者解析成 Uri 对象，然后使用 getContentResolver()获取 ContentResolver 实例 resolver。定义 ContentValues 对象用来存放要添加、删除、修改的数据。最后分别使用 resolver.insert()方法、resolver.delete()方法、resolver.update()方法和 resolver.query()方法对共享的数据进行增删改查。

第六步：运行程序

我们以修改数据为例运行程序，观察运行结果。先运行程序 MyContacts，添加如图 7-5（a）所示的通讯录信息，点击“查询”命令按钮看到如图 7-5（a）所示结果，接着运行程序 SharedData_MyContacts，修改如图 7-5（b）所示的内容，最后查看程序 MyContacts，点击“查询”结果如图 7-5（c）所示。从运行结果可以看出，在 SharedData_MyContacts 中修改的数据改变了 MyContacts 中的数据内容。对于数据的其他操作如添加、删除与数据修改类似。

职业素养拓展

【职业素养拓展7】

职业素养拓展内容详见文件“职业素养拓展”。

扫描二维码查看【职业素养拓展 7】

本章小结

- 内容提供者
  - 内容提供者概述
    - **内容提供者**是不同应用程序之间数据共享的一种接口机制，使用ContentProvider类将不同的应用程序的数据操作标准统一起来，这些标准在类中明确通过方法进行定义。其他应用程序在不知数据来源、路径的情况下，可以对共享数据进行增加、删除、修改和查询等操作
    - **应用场景：**
      1.访问其他应用中的现有内容提供者，如系统通讯录等
      2.在自己的应用中创建新的内容提供者，为其他应用共享数据
  - 共享系统应用数据
    - **实现过程：**
      第一步：在清单文件中申请授权
      第二步：使用getContentResolver()方法获取ContentResolver示例
      第三步：使用ContentResolver提供方法操作数据，如查询
      Cursor cursor=getContentResolver().query(ContactsContract.CommonDataKinds.Phone.*CONTENT_URI* ,null,null,null,null)
    - **创建内容提供者：**
      1.新建一个类继承自ContentProvider，重写insert ()、delete ()、update ()和query ()等的方法
      2.在清单文件中使用<provider>标记声明内容提供者
  - 程序间数据共享
    - 数据共享，至少涉及两个应用程序，一个是内容提供者的应用，另一个是内容访问者的应用。自定义程序间数据共享与使用系统应用提供的数据的区别在于要求在内容提供者应用中自定义内容提供者，在内容访问者中通过Uri对象和ContentResolver对象实现对数据的增加、删除、修改和查询等操作

第8章 服务的应用

服务 Service 是 Android 系统中的四大组件之一，它跟 Activity 的级别差不多，在每一个应用程序中都扮演着非常重要的角色，主要用于在后台处理一些耗时的逻辑，或者去执行某些需要长期运行的任务，比如下载文件、播放音乐等。本章将对服务的应用进行详细讲解。

学习目标：

1. 理解服务的应用；
2. 掌握服务启动的两种方法；
3. 能够熟练使用服务的应用。

8.1 服务概述

（1）服务简介

服务 Service 是一种可在后台长时间执行操作而不提供界面的应用组件，适合执行不需要和用户交互而且还需要长期运行的任务。服务可由其他应用组件启动，而且即使用户切换到其他应用，服务仍将在后台继续运行。此外，组件可通过绑定到服务与之进行交互，甚至是执行进程间通信(IPC)。例如，服务可在后台处理网络事务，执行文件 I/O 或与内容提供程序进行交互。

服务的创建

如要创建服务，必须创建继承自 Service 的子类或者使用它的现有子类，重写一些回调方法，并在清单文件中声明。根据服务启动方式的不同可以分为启动服务和绑定服务。

（2）服务的创建

当我们需要使用一个服务时，就需要创建服务，下面通过一个案例演示服务的创建，操作步骤如下。

第一步：新建项目

新建一个项目，命名为 Service，所在包名为 com.example.service。

第二步：创建服务

选择程序的包名，点击鼠标右键，选择“New”→“Service”→“Service”，弹出如图 8-1 所示的对话框。

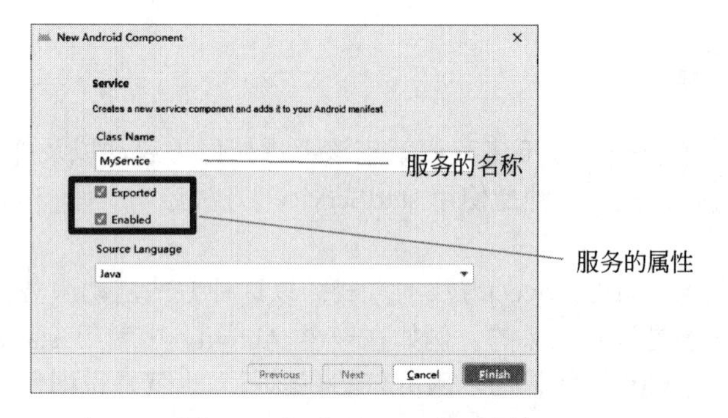

图 8-1　新建 Service 的对话框

在弹出的窗口中输入服务的名称，在此使用默认的服务名称“MyService”，注意有两个属性 Exported 和 Enabled 默认为选中状态。其中 Exported 属性表示是否允许除了当前程序之外的其他程序访问这个服务，Enabled 属性表示系统是否可以实例化此服务。选择“Finish”，在当前包中增加了一个 MyService 的类。打开 MyService，代码如下所示:

```
public class MyService extends Service {
    public MyService() {
    }

    @Override
    public IBinder onBind(Intent intent) {
        // TODO: Return the communication channel to the service.
        throw new UnsupportedOperationException("Not yet implemented");
    }
}
```

从代码中我们可以看到 MyService 继承自 Service 类,代码中包含一个构造方法 MyService(),重写了父类的 onBind()方法。onBind()方法是 Service 子类必须实现的方法,该方法返回一个 IBinder 对象，应用程序可以通过该对象与 Service 组件通信。

在学习 Activity 时我们提到 Activity 必须在 AndroidManifest.xml 文件中注册了才能用，对于服务也是一样的。我们打开 AndroidManifest.xml 文件可以看到如下代码:

```
<?xml version="1.0" encoding="utf-8"?>
<manifest xmlns:android="http://schemas.android.com/apk/res/android"
    package="com.example.service">

    <application
        ......>
        <service
            android:name=".MyService"
            android:enabled="true"
            android:exported="true"></service>
        ......
</manifest>
```

使用这种方法创建服务，Android Studio 会自动在清单文件 AndroidManifest 中增加<service>标记声明新建的服务。如果是通过新建继承 Service 类的方法创建服务，那么必须手动在 AndroidManifest 文件中添加<service>标记声明服务。

8.2 启动服务

启动服务的设计与实现

学习了如何创建一个服务，接下来我们就应该考虑如何去启动以及停止这个服务。服务的启动有两种方式：一种是使用 startService()方法启动；另一种是使用 bindService()方法启动。

启动服务是通过调用 startService()方法的方式，以这种方式启动服务会长期在后台运行，服务的状态与开启者的状态是没有关系的。例如在一个 Activity 中调用 startService()方法启动了一个服务，那么当这个 Activity 销毁以后，服务依然在运行，也就是说即使启动服务的组件已经被销毁，服务会依旧运行。

（1）创建启动服务

由另一个组件（如 Activity）通过调用 startService()方法启动服务，服务启动后，其生命周期即独立于启动它的组件，即使系统已销毁启动服务的组件，该服务仍可在后台无限期地运行。因此，服务应在其工作完成时通过调用 stopSelf()方法来自行停止运行，或者由另一个组件通过调用 stopService()方法来将其停止。

下面我们通过如图 8-2 所示效果实例学习使用这种方式如何来启动服务和关闭服务，具体步骤如下：

图 8-2　启动服务运行界面

第一步：打开项目 Service

打开上一节创建的项目 Service，在这个项目中我们已经创建了一个服务 MyService 类，以它为基础完成当前案例。

第二步：设计用户界面

我们要在 Activity 中启动服务和关闭服务，所以在 activity_main 布局文件中添加两个 Button，一个用来启动服务，设置 Button 的 android:text 属性为“启动服务”，android:onClick 属性为“startMyService”；另一个 Button 用来停止服务，设置 android:text 的属性为“停止服务”，android:onClick 的属性为“stopMyService”，后面需要在 MainActivity 中实现这两个方法。设计代码见布局文件【activity_main.xml】。

扫描二维码查看【8.2activity_main.xml】

8.2activity_main.xml

第三步：修改服务 MyService.Java 文件的内容

打开服务 MyService.java，重写父类的方法，MyService 修改后的代码如下所示：

```
public class MyService extends Service {
    public MyService() {    }

    @Override
    public IBinder onBind(Intent intent) {
        // TODO: Return the communication channel to the service.
        throw new UnsupportedOperationException("Not yet implemented");
    }

    @Override
    public void onCreate() {
        super.onCreate();
```

```
        Log.i("MyService","调用 MyService 中的 onCreate()方法");
    }

    @Override
    public int onStartCommand(Intent intent, int flags, int startId) {
        Log.i("MyService","调用 MyService 中的 onStartCommand()方法");
        return super.onStartCommand(intent, flags, startId);
    }

    @Override
    public void onDestroy() {
        super.onDestroy();
        Log.i("MyService","调用 MyService 中的 onDestroy()方法");
    }
}
```

在 MyService 服务中重写 onCreate()方法、onStartCommad()方法和 onDestroy()方法，在这三个方法中分别通过 Log.i()方法打印程序执行时调用到的方法。

第四步：编写界面交互代码

在 MainActivity 中实现单击按钮的点击事件，具体代码如下所示：

```
public class MainActivity extends AppCompatActivity {

    @Override
    protected void onCreate(Bundle savedInstanceState) {
        super.onCreate(savedInstanceState);
        setContentView(R.layout.activity_main);
    }

    public void startMyService(View view) {
        Intent intent = new Intent(this, MyService.class);
        startService(intent);
    }

    public void stopMyService(View view) {
        Intent intent = new Intent(this, MyService.class);
        stopService(intent);
    }
}
```

在以上代码中，“启动服务”命令按钮的单击事件在 startMyService()方法中实现,首先定义了一个 Intent 对象，指明当前活动 this 和目标服务 MyService，然后通过调用 startService()方法开启 Intent 对象；“停止服务”命令按钮的单击事件在 stopMyService()方法中实现，通过调用 stopService()方法关闭了 Intent 对象指定的服务 MyService。

第五步：运行程序

运行程序，运行效果如图 8-2 所示。单击界面上的“启动服务”命令按钮，将调用 startService()方法，可以在 Logcat 中看到如图 8-3 所示的效果，依次执行了

startService应用级
运行与总结

onCreate()方法和 onStartCommand()方法。

```
com.example.service I/MyService: 调用MyService中的onCreate()方法
com.example.service I/MyService: 调用MyService中的onStartCommand()方法
```

图 8-3　单击“启动服务”按钮后 Logcat

再次单击“启动服务”，可以看到如图 8-4 所示结果，onStartCommand()方法又一次执行了，也就是说第一次启动服务时调用了 onCreate()方法，在服务没有停止的情况下，再次执行启动服务操作，onCreate()方法将不再被调用，只是重复调用 onStartCommand()方法。

```
com.example.service I/MyService: 调用MyService中的onCreate()方法
com.example.service I/MyService: 调用MyService中的onStartCommand()方法
com.example.service I/MyService: 调用MyService中的onStartCommand()方法
```

图 8-4　再次单击“启动服务”按钮后 Logcat

单击界面上的“停止服务”命令按钮调用 stopService()方法关闭服务，效果如图 8-5 所示，即调用了 onDestroy()方法，这时服务被销毁了。

```
com.example.service I/MyService: 调用MyService中的onCreate()方法
com.example.service I/MyService: 调用MyService中的onStartCommand()方法
com.example.service I/MyService: 调用MyService中的onStartCommand()方法
com.example.service I/MyService: 调用MyService中的onDestroy()方法
```

图 8-5　单击“停止服务”按钮后 Logcat

（2）启动服务的生命周期

服务由其他组件调用 startService()方法创建时将无限期运行，必须通过调用 stopSelf()方法或者通过其他组件调用 stopService()方法停止此服务。服务停止后，系统会将其销毁。通过上一节的实例我们可以看到，通过使用 startService()方法启动服务时，服务的生命周期如图 8-6 所示。

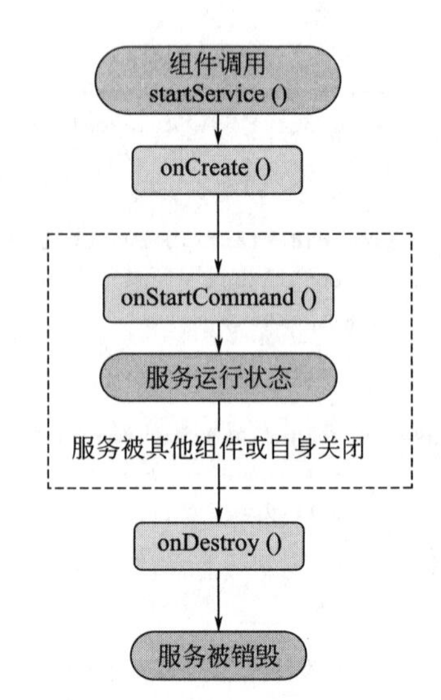

图 8-6　启动服务的生命周期

启动服务的生命周期过程的方法依次为 onCreate()方法、onStartCommand()方法和 onDestroy()方法。

- onCreate()方法：在第一次创建服务的时候系统将调用此方法。如果服务已经运行，则不会调用此方法，该方法只调用一次。
- onStartCommand()方法：当另一个组件通过调用 startService()方法请求启动服务时，系统将调用此方法。
- onDestroy()方法：是在关闭服务时执行的方法。

使用 startService()方法启动的服务，启动后就与启动它的组件不相关了，停止服务只能使用 stopService()方法或者是 stopSelf()方法。

8.3　绑定服务

（1）创建绑定服务

上一节的案例在 Activity 中调用 startService()方法启动了服务，服务虽然启动了，但是启动之后服务与启动它的 Activity 就没有什么关系了，即使 Activity 关闭了，服务依然在运行。那么如何让 Activity 与服务的关系更紧密，Activity 才能够参与服务运行的控制呢？这就可以使用绑定服务来实现了。

绑定服务是在其他组件调用 bindService()方法时创建的服务，其他组件通过 bindService()方法将服务与组件绑定，程序允许组件与服务交互，组件一旦退出或者调用 unbindService()方法解绑服务，服务就会被销毁。服务与组件的绑定就要用到我们之前在服务中重写的 onBind()方法。

下面通过如图 8-7 所示案例演示绑定服务和解绑服务，操作步骤如下。

第一步：新建项目

创建一个项目，命名为 BindService，指定程序的包名为 com.example.bindservice。

第二步：设计交互界面

在用户界面 activity_main 中放置 3 个命令按钮，其 android:text 属性分别为“绑定服务”“与服务交互”和“解除绑定”，三个命令按钮的 id 分别为：btn_bindService、btn_callMethod、btn_unbindService。

设计代码见布局文件【activity_main.xml】。

扫描二维码查看【8.3activity_main.xml】

图 8-7　BindService 运行界面

第三步：创建服务 MyService

创建一个继承于 Service 的类 MyService，注意如果手动创建的继承自 Service 的类，一定要在 AndroidManifest 文件中声明所建的服务，否则不能使用。MyService 的代码如下：

```
public class MyService extends Service {
    class MyBinder extends Binder {
        public void myMethod() {
            Log.i("MyService", "MyService 中自定义的方法 myMethod()方法");
        }
    }

    @Override
    public IBinder onBind(Intent intent) {
        Log.i("MyService", "调用 MyService 中的 onBind()方法");
        return new MyBinder();
    }

    @Override
    public void onCreate() {
        super.onCreate();
        Log.i("MyService", "调用 MyService 中的 onCreate()方法");
    }

    @Override
```

```
    public boolean onUnbind(Intent intent) {
        Log.i("MyService", "调用 MyService 中的 onUnbind()方法");
        return super.onUnbind(intent);
    }
}
```

在 MyService 类中定义一个类 MyBinder,并让它继承自 Binder，然后在这个类中又自定义了一个 myMethod()方法，在 onBind()方法中返回 MyBinder 的实例。重写了父类 onCreate()方法和 onUnbind()方法并各打印了一行日志。

第四步：在 MainActivity 中实现用户交互

BindService应用中Activify代码的实现

在 MainActivity 中实现“绑定服务”“与服务交互”和“解除绑定”的逻辑。实现绑定服务需要调用 bindService()方法来实现，bindService()方法的语法格式如下所示。

bindService(Intent service, ServiceConnection conn,int flags);

这个方法有三个参数。

第一个参数 service：用于指定要启动的 Service。

第二个参数 conn：用于调用者（服务绑定的组件）与 Service 之间的连接状态，当调用者与 Service 连接成功时，程序会回调该对象的 onServiceConnected(ComponentName name, IBinder service)方法。当调用者与 Service 断开时，程序会调用该对象的 onServiceDisconnected(ComponentName name)方法。

第三个参数 flags：是一个标志位，表示组件绑定服务时是否自动创建 Service(如果 Service 还未创建，该参数可设置为 0，是不自动创建 Service，也可设置为“BIND_AUTO_CREATE”，表示自动创建 Service)。

MainActivity.java 的代码如下所示：

```
public class MainActivity extends AppCompatActivity implements View.OnClickListener {

    private Button btn_bindService, btn_callMethod, btn_unbindService;
    private MyService.MyBinder myBinder;
    private ServiceConnection conn = new ServiceConnection() {
        @Override
        public void onServiceConnected(ComponentName name, IBinder service) {
            myBinder = (MyService.MyBinder) service;
            Log.i("MainActivity", "Activity 与服务绑定成功");
        }

        @Override
        public void onServiceDisconnected(ComponentName name) {
        }
    };

    @Override
    protected void onCreate(Bundle savedInstanceState) {
        super.onCreate(savedInstanceState);
        setContentView(R.layout.activity_main);
        btn_bindService = findViewById(R.id.btn_unbindService);
```

```
        btn_callMethod = findViewById(R.id.btn_callMethod);
        btn_unbindService = findViewById(R.id.btn_unbindService);
        btn_bindService.setOnClickListener(this);
        btn_callMethod.setOnClickListener(this);
        btn_unbindService.setOnClickListener(this);

    }

    @Override
    public void onClick(View v) {
        switch (v.getId()) {
            case R.id.btn_bindService://绑定服务
                Intent intent = new Intent(MainActivity.this, MyService.class);
                bindService(intent, conn, BIND_AUTO_CREATE);
                break;
            case R.id.btn_callMethod://调用服务中的自定义方法
                myBinder.myMethod();
                Log.i("MainActivity", "活动中成功调用了服务中的方法");
                break;
            case R.id.btn_unbindService://解绑服务
                unbindService(conn);
                break;
        }
    }
}
```

在上面的代码中，先定义了一个 MyService.MyBinder 对象 myBinder；创建了一个 ServiceConnection 对象 conn，设计当绑定服务成功时，程序会调用 onServiceConnected()方法，在该方法中获取传递过来的 IBinder 对象 service，实例化 myBinder 对象，然后打印 log 日志。

服务的创建与操作

在 MainActivity 的 onCreate()方法中对三个命令按钮初始化，并设置事件监听。

在“绑定服务”命令按钮中创建 Intent 对象，调用 bindService()方法，这个方法的三个参数中第一个参数 Intent 就是刚刚创建的 Intent 对象，第二个参数是 ServiceConnection 的实例 conn，第三个参数是 BIND_AUTO_CREATE，表示 Activity 和服务进行绑定后自动创建服务。

在“调用服务中的方法”命令按钮中调用 MyService.MyBinder 中自定义的方法 myMethod()，这样 Activity 就与服务相关联了。通过 Activity 中命令调用服务中定义的方法。

在“解绑服务”命令按钮中，使用 unbindService()方法解除绑定，这个方法的参数为 ServiceConnection 实例，也就是解除服务与 Activity 的连接。

第五步：运行程序

运行程序，分别选择图 8-7 界面中的“绑定服务”“调用服务中的方法”和“解绑服务”。

选择“绑定服务”命令按钮，Logcat 中的内容如图 8-8 所示。可以看到绑定服务先调用了服务中 onCreate()方法，再调用了 onBind()方法，最后显示了 onServiceConnected()方法中的打印日志“Activity 与服务绑定成功”。

```
com.example.bindservice I/MyService: 调用MyService中的onCreate()方法
com.example.bindservice I/MyService: 调用MyService中的onBind()方法
com.example.bindservice I/MainActiivty: Activity与服务绑定成功
```

图 8-8　选择“绑定服务”命令按钮 Logcat 内容

选择“调用服务中的方法”命令按钮，Logcat 中的内容如图 8-9 所示。显示在 Activity 中调用了服务中的自定义方法 myMethod()。

绑定解绑服务的运行与总结

```
com.example.bindservice I/MyService: 调用MyService中的onCreate()方法
com.example.bindservice I/MyService: 调用MyService中的onBind()方法
com.example.bindservice I/MainActiivty: Activity与服务绑定成功
com.example.bindservice I/MyService: 调用MyService中自定义的方法mymethod()方法
com.example.bindservice I/MainActiivty: 活动中成功调用了服务中的方法
```

图 8-9　选择“调用服务中的方法”命令按钮 Logcat 内容

选择“解绑服务”，Logcat 中的内容如图 8-10 所示。可以看到调用了服务中的 onUnbind()方法。

```
com.example.bindservice I/MyService: 调用MyService中的onCreate()方法
com.example.bindservice I/MyService: 调用MyService中的onBind()方法
com.example.bindservice I/MainActiivty: Activity与服务绑定成功
com.example.bindservice I/MyService: 调用MyService中自定义的方法mymethod()方法
com.example.bindservice I/MainActiivty: 活动中成功调用了服务中的方法
com.example.bindservice I/MyService: 调用MyService中的onUnbind()方法
```

图 8-10　选择“解绑服务”命令按钮 Logcat 内容

需要注意的是，解绑服务与启动它的组件的生命周期紧密关联，如果组件关闭，则服务与组件之间解绑。在 BindService 应用中，如果我们绑定了服务后，没有解绑服务，而直接退出了 Activity，则程序调用服务中定义的 onUnbind()方法。

(2) 绑定服务的生命周期

通过分析 BindService 实例的运行过程，我们可以看出使用 bindService()方法启动服务，如果之前没有创建过这个服务，则会调用 onCreate()方法，然后再调用 onBind()方法，获取 onBind()方法返回的 IBinder 对象的实例，这样就可以和组件进行通信了，只要组件和服务的连接没有断开，服务就会一直保持运行状态，当调用了 onBind()方法之后去调用 onUnbind()方法，服务与组件解绑，onDestroy()方法也会执行。绑定服务的生命周期如图 8-11 所示。

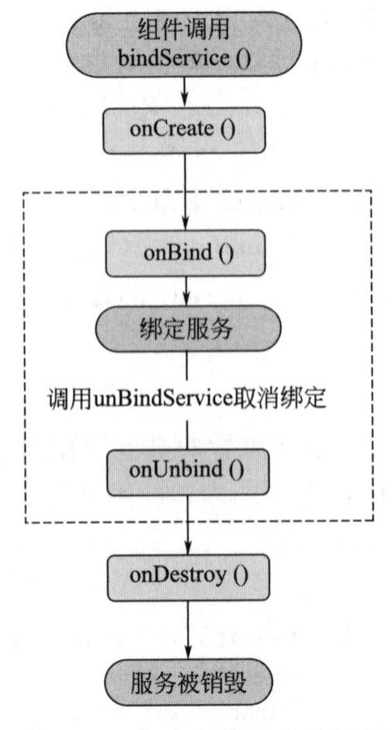

图 8-11　绑定服务的生命周期

8.4　后台播放音乐

我们时常会在后台播放音乐，下面我们通过一个简单实例实现后台播放音乐，设计程序界面如图 8-12 所示，单击“开始”将播放音乐，选择“停止”结束播放音乐。实现步骤如下。

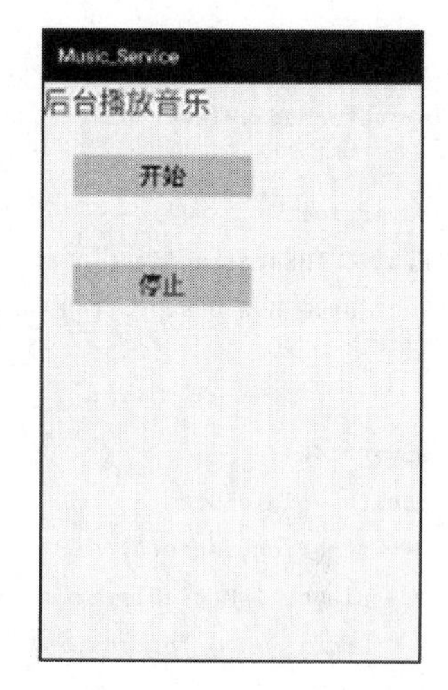

图 8-12　后台播放音乐

第一步：新建项目

新建一个项目，名为 Music_Service，指定包名为 com.example.music_service。

第二步：准备工作

在 Android 应用中播放音乐文件，音乐文件要存放在资源文件夹“raw”中，那么首先需要创建一个“raw”文件夹。创建“raw”文件夹的过程如下：选择程序中的 res 文件夹，单击右键选择“New”→“Android Resource Directory”，弹出如图 8-13 所示的“New Resource Directory”对话框，选择“Resource type”为“raw”，则“Directory name”自动填充为“raw”。将准备好的音乐文件 dawn.mp3 复制到此文件夹中。

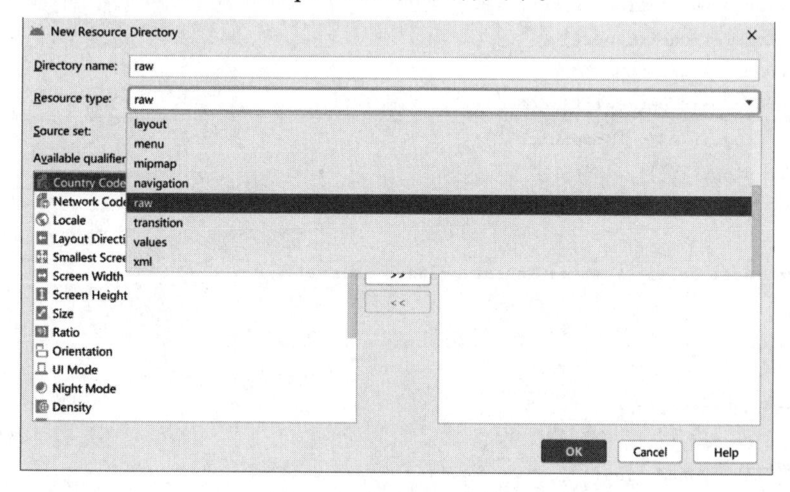

图 8-13　新建 raw 文件夹

第三步：设计用户界面

在用户界面 activity_main 中添加音乐播放的控制按钮，界面如图 8-12 所示。添加一个文本框和两个命令按钮，命令按钮“开始”的 id 为“btn_start”，命令按钮“停止”的 id 为“btn_stop”。设计代码见布局文件【activity_main.xml】。

扫描二维码查看【8.4activity_main.xml】

8.4activity_main.xml

第四步：创建服务 MusicService

选择当前应用程序的包名，单击右键选择“New”→“Service”→“Service”，创建名为

“MusicService”的服务，在服务中播放音乐和停止音乐。代码如下：

```
public class MusicService extends Service {
    private MediaPlayer player;

    @Override
    public IBinder onBind(Intent intent) {
        throw new UnsupportedOperationException("Not yet implemented");
    }

    @Override
    public void onCreate() {
        super.onCreate();
        player = MediaPlayer.create(getApplicationContext(), R.raw.dawn);
        Log.i("AA", "onCreate()");
    }

    @Override
    public int onStartCommand(Intent intent, int flags, int startId) {
        Log.i("AA", "onStartCommand()");
        player.start();
        player.setOnCompletionListener(new
MediaPlayer.OnCompletionListener() {
            @Override  //播放结束监听
            public void onCompletion(MediaPlayer mp) {
                Log.i("AA", "stopSelf()");
                stopSelf();
            }
        });
        return super.onStartCommand(intent, flags, startId);
    }

    @Override
    public void onDestroy() {
        super.onDestroy();
        Log.i("AA", "ondestroy()");
        player.stop();
        player = null;
    }
}
```

在服务 MusicService 中定义了一个 MediaPlayer 对象，在 onCreate()方法中使用 MediaPlayer.create()方法加载音乐文件，在 onStartCommand()方法中播放音乐，并且监听到音乐播放结束则调用 stopSelf()方法停止服务，在 onDestroy()方法中停止音乐播放。

第五步：用户界面交互的实现

在 MainActivity.java 实现音乐的“开始”和“停止”，代码如下：

```java
public class MainActivity extends AppCompatActivity implements View.OnClickListener {

    private Button btn_start, btn_stop;

    @Override
    protected void onCreate(Bundle savedInstanceState) {
        super.onCreate(savedInstanceState);
        setContentView(R.layout.activity_main);
        btn_start = findViewById(R.id.btn_start);
        btn_stop = findViewById(R.id.btn_stop);
        btn_start.setOnClickListener(this);
        btn_stop.setOnClickListener(this);
    }

    @Override
    public void onClick(View v) {
        Intent intent = new Intent(MainActivity.this, MusicService.class);
        switch (v.getId()) {
            case R.id.btn_start:
                startService(intent);
                break;
            case R.id.btn_stop:
                stopService(intent);
                break;
        }
    }
}
```

在“开始”和“停止”命令按钮中分别使用的是 startService()方法和 stopService()方法开启服务和停止服务。

第六步：运行程序

运行程序图如 8-12 所示，点击“开始”命令按钮，则音乐开始播放，即服务被启动，此时如果按模拟器“返回”键，音乐继续播放，直到音乐播放结束，则停止服务。如果在音乐播放结束之前需要关闭，则点击“停止”命令按钮，音乐结束，停止服务。注意观察 Logcat 的日志信息。

职业素养拓展

职业素养拓展内容详见文件“职业素养拓展”。

扫描二维码查看【职业素养拓展 8】

【职业素养拓展8】

本章小结

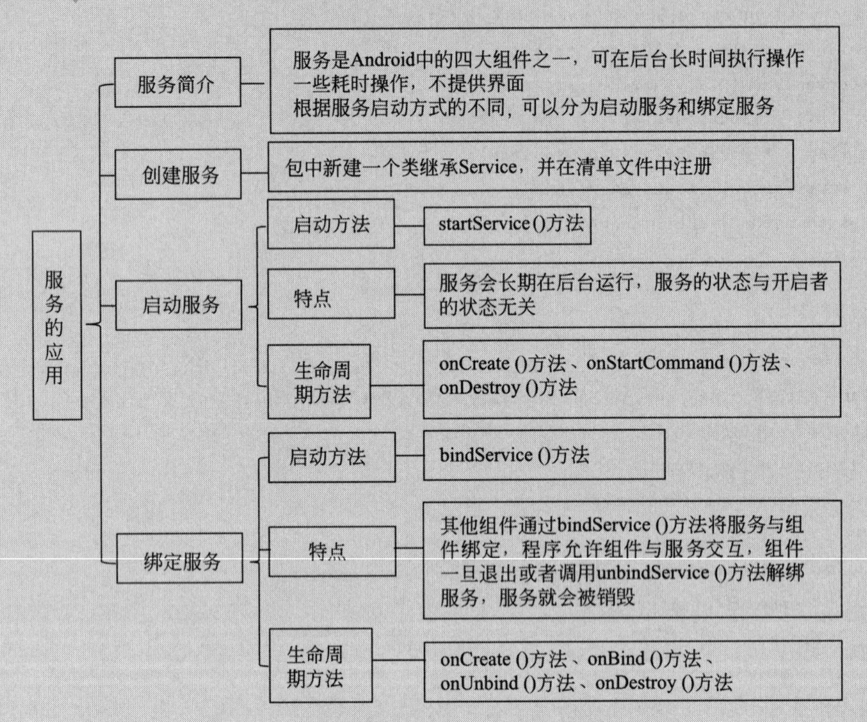
服务的应用
服务简介
服务是Android中的四大组件之一，可在后台长时间执行操作
一些耗时操作，不提供界面
根据服务启动方式的不同，可以分为启动服务和绑定服务
创建服务
包中新建一个类继承Service，并在清单文件中注册
启动服务
启动方法
startService ()方法
特点
服务会长期在后台运行，服务的状态与开启者的状态无关
生命周期方法
onCreate ()方法、onStartCommand ()方法、onDestroy ()方法
绑定服务
启动方法
bindService ()方法
特点
其他组件通过bindService ()方法将服务与组件绑定，程序允许组件与服务交互，组件一旦退出或者调用unbindService ()方法解绑服务，服务就会被销毁
生命周期方法
onCreate ()方法、onBind ()方法、onUnbind ()方法、onDestroy ()方法

第9章

广播的应用

广播接收者是 Android 系统当中的四大组件之一，是 Android 系统中的一种消息传递机制。当手机发生一些特定事件的时候，如电池的电量不足，网络状态发生变化，收到短信，等等，手机都会发出一个广播，而我们只需要在程序中去注册一个广播接收者，就能知道手机现在发生了什么样的事件。本章主要内容包括广播机制介绍，广播接收者的创建与使用，广播的分类及其应用，等等。

学习目标：

1. 了解 Android 中的广播机制；
2. 掌握广播接收者的用法；
3. 掌握标准广播和有序广播的发送与接收。

9.1 广播机制

广播的类型

（1）广播机制简介

在平时的生活中，我们时常会听到各类广播，比如在学校会听到校园广播，在公交车或是地铁上会听到报站广播等。在 Android 中，有一些操作完成以后，也会发送广播，这个广播与传统意义中的电台广播有些相似，之所以叫广播就是因为它只负责“说”而不管接收者“听不听”，也就是不管接收者如何处理。而且广播可以被不止一个应用程序接收，当然也可能不被任何应用程序接收。

Android 中的广播是操作系统中产生的各种各样的事件，例如手机电量不足会接收到一个广播，手机的网络状态发生变化也会收到一个广播，等等。Android 系统一旦内部产生了这些事件，就会向所有的广播接收者广播这些事件。对于 Android 中的每个应用程序来说都可以根据自己的需求对广播进行注册，然后使用广播接收者来监听处理接收到的广播事件。这样应用程序就只会接收到自己关心的广播内容，这些广播可能是 Android 系统发送的，也可能是其他应用程序发送。

（2）广播接收者

BroadcastReceiver 即广播接收者，是 Android 中接收广播消息并做出反应的组件，使用它可以实现对系统内部定义广播或者其他应用程序自定义广播的监听。在 Android 中的每个应用程序都可以对自己感兴趣的广播进行注册，这个程序也只会收到注册过的广播。应用程序实现广播监听需要在清单文件或者代码中先对广播

广播接收者简介

进行注册，然后创建一个继承自 BroadcastReceiver 的类，在这个类中重写 onReceive()方法，并在该方法中对广播事件进行处理。

（3）广播接收者的创建

广播接收者的创建

广播接收者的创建与 Activity 和 Service 的创建类似，可以通过选择弹出菜单创建，也可通过自定义类创建。具体操作：找到项目的包名，点击鼠标右键，选择"New"→"Other"→"Broadcast Receiver"，弹出如图 9-1 所示的"New Android Component"对话框，Class Name 使用默认的"MyReceiver"，点击"Finish"，完成广播接收者的创建。

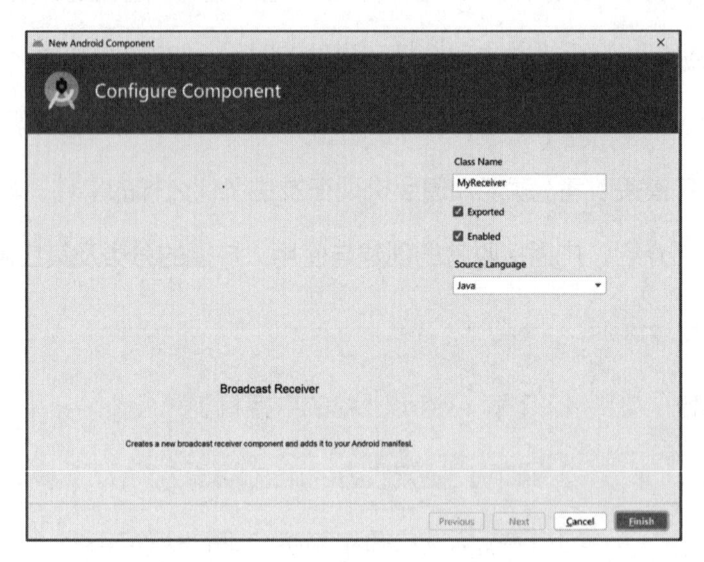

图 9-1 "New Android Component"对话框

观察程序所在包，新增了一个 MyReceiver.java 文件，打开此文件代码如下所示：

```
package com.example.myapplication;

import android.content.BroadcastReceiver;
import android.content.Context;
import android.content.Intent;

public class MyReceiver extends BroadcastReceiver {

    @Override
    public void onReceive(Context context, Intent intent) {
        // TODO: This method is called when the BroadcastReceiver is receiving
        // an Intent broadcast.
        throw new UnsupportedOperationException("Not yet implemented");
    }
}
```

新建的 MyReceiver 继承自 BroadcastReceiver，类中重写了 onReceive()方法。当广播接收者收到广播时执行 onReceive 中的逻辑，也就是当某个广播发生时，由广播接收者监听到然后给出响应。

也可以通过在应用程序所在包中新建一个 BroadcastReceiver 的子类，并重写 onReceive()方法来创建一个广播接收者。

创建了广播接收者后打开清单文件 AndroidManifest.xml，可以看到清单文件中自动增加了一对<receiver></receiver>标签，刚创建的广播接收者已经在清单文件中声明了。代码如下所示：

```
<application
    ……>
    <receiver
        android:name=".MyReceiver"
        android:enabled="true"
        android:exported="true" />
……
</application>
```

其中属性 enabled 表示广播接收者是否可以由系统实例化，exported 表示是否接收当前程序之外的广播。

这样我们的广播接收者就创建完成了。应用程序要监听什么广播，监听到后进行什么样的处理，这就需要完成广播接收者中的方法 onReceive()和注册要监听的广播，在后续的内容中我们通过监听系统广播和自定广播进行学习。

9.2　系统广播

系统会在发生各种系统事件时自动发送广播，例如当系统网络状态发生变化时，系统广播会被发送给所有同意接收相关事件的应用。

下面我们以如图 9-2 所示的监听网络状态为例，学习系统广播的使用。在本应用中当网络状态发生变换，发送网络状态变化的广播。实现过程如下：

第一步：新建项目

新建一个项目，名为 SYSBroadcastReceiver，包名为 com.example.sysbroadcastreceiver。

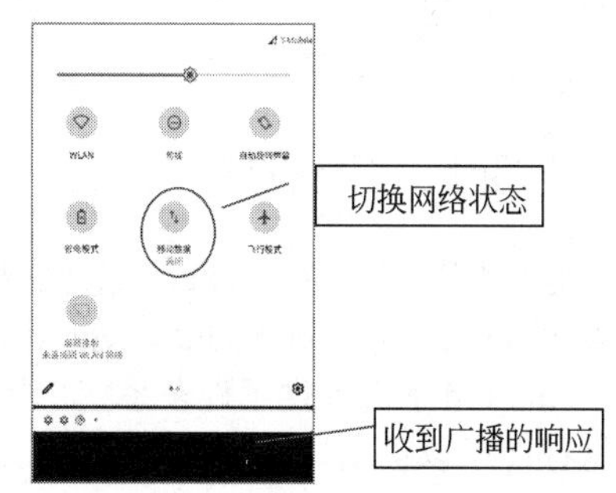

图 9-2　网络状态发生变化

第二步：创建广播接收者

找到项目的包名 com.example.sysbroadcastreceiver，点击鼠标右键，选择“New”→“Java Class”，类名为“myReceiver”。打开类 myReceiver，是其继承 BroadcastReceiver 类，并重写 onReceive()方法，修改代码如下所示：

```
public class MyReceiver extends BroadcastReceiver {
    @Override
    public void onReceive(Context context, Intent intent) {
        Toast.makeText(context, "网络状态发生变化", Toast.LENGTH_LONG).show();
    }
}
```

这样我们的广播接收者就创建完成了，当系统网络状态发生变化时，发送广播消息“网络状态发生变化”。如何监听系统的网络状态呢？这就需要将创建的广播接收者在 Activity 中对广播接收者进行注册。

第三步：注册广播

在 MainActivity 中注册广播，具体代码如下：

```
public class MainActivity extends AppCompatActivity {

    @Override
    protected void onCreate(Bundle savedInstanceState) {
        super.onCreate(savedInstanceState);
        setContentView(R.layout.activity_main);
        MyReceiver myReceiver=new MyReceiver();
        IntentFilter intentFilter=new IntentFilter();
        intentFilter.addAction("android.net.conn.CONNECTIVITY_CHANGE");
        registerReceiver(myReceiver,intentFilter);

    }
}
```

首先定义了一个广播接收者 MyReceiver 对象，然后创建一个 IntentFilter 对象并为其使用 addAction()方法添加动作为“android.net.conn.CONNECTIVITY_CHANGE”。“android.net.conn.CONNECTIVITY_CHANGE”是系统广播，它是“网络状态变化”的动作。最后使用 registerReceiver()方法注册广播。registerReceiver()方法格式如下所示：

```
registerReceiver(BroadcastReceiver receiver, IntentFilter filter);
```

此方法有两个参数：第一个为定义的广播接收者对象；第二个为 IntentFilter 对象。

使用 registerReceiver()方法注册广播接收者是动态注册，这种方式需要在代码中动态指定广播地址并注册，一般是在组件 Activity 或 Service 中注册一个广播。当用来注册广播的组件关掉后，广播也就失效了。

第四步：运行程序

运行程序，打开系统设置，当点击移动数据“ ”图标时，则会弹出“网络状态发生变化”提示，效果如图 9-2 所示。在 MainActivity 中注册了广播，在广播接收者中对发出的广播进行需要的响应。

Android 系统提供了许多系统广播供开发者使用，它们会随着系统的某些变化而被发送出去，应用程序可以借助这些广播并执行相应的处理。例如常见的有：

Intent.ACTION_BATTERY_CHANGED：充电状态，或者电池的电量发生变化。

Intent.ACTION_BATTERY_LOW：电池电量低。

Intent.ACTION_AIRPLANE_MODE_CHANGED：关闭或打开飞行模式。

Intent.ACTION_CLOSE_SYSTEM_DIALOGS：当屏幕超时进行锁屏时，当用户按下电源按钮，长按或短按(不管有没跳出话框)，进行锁屏。

Intent.ACTION_DATE_CHANGED：设备日期发生改变时。

Intent.ACTION_DEVICE_STORAGE_LOW：设备内存不足时发出的广播，此广播只能由系统使用，其他 App 不可用。

9.3 自定义广播

自定义广播

在 Android 中除了监听系统广播，用户还可以自定义广播并对其进行监听。用户自定义的广播主要可以分为两种类型：标准广播和有序广播。

标准广播是一种完全异步执行的广播，在广播发出之后，所有的广播接收者几乎都会在同一时刻接收到这条广播消息，因此它们之间没有任何先后顺序可言，这种广播的效率会比较高，但同时也意味着它是无法被拦截的。

有序广播是一种同步执行的广播，在广播发出之后，同一时刻只会有一个广播接收者能够收到这条广播消息，当这个广播接收者中的逻辑执行完毕后，广播才会继续传递。所以此时的广播接收者是有先后顺序的，优先级高的广播接收者就可以先收到广播消息，并且前面的广播接收者还可以截断正在传递的广播，这样后面的广播接收者就无法收到广播消息了。

9.3.1 标准广播

标准广播也称为无序广播，发送广播时，所有监听这个广播的广播接收者都会接收到这个广播，接收和执行的顺序不确定，标准广播效率比较高，但无法被拦截。在前面的学习中我们接收了系统的广播，下面我们通过一个自定义广播，实现如图 9-3 所示效果，学习标准广播的发送与接收。操作步骤如下：

图 9-3　发送标准广播

第一步：创建项目

新建一个项目，名为 Diy_BroadcastReceiver，程序所在的包名为 com.example.diy_broadcastreceiver。

第二步：用户界面

在 activity_main 中添加一个 Button 命令按钮，当单击命令按钮时发送广播。Button 控件的 text 属性设置为“发送自定义广播”，onClick 属性设置为“sendDIYBroadcast”，我们在 MainActivity 中实现 sendDIYBroadcast()方法。设计代码见布局文件【activity_main.xml】。

扫描二维码查看【9.3.1activity_main.xml】

9.3.1activity_main.xml

第三步：创建广播接收者

创建广播接收者 MyReceive，当接收到广播时，给出提示“收到了我自定义的广播”。选择当前包 com.example.diy_broadcastreceiver，单击右键“New”→“class”在弹出的提示框中输入“MyReceiver”创建一个新的类 MyReceiver.java，打开文件，修改代码使这个类继承 BroadcastReceiver，并重写 onReceive()方法，代码如下所示：

```
public class MyReceiver extends BroadcastReceiver {

    @Override
    public void onReceive(Context context, Intent intent) {
        Toast.makeText(context,"收到了我自定义的广播",Toast.LENGTH_LONG).show();
    }
}
```

第四步：清单文件中为广播接收者设置 action

在清单文件 AndroidManifest.xml 中使用<intent-filter>添加 action。

```
<?xml version="1.0" encoding="utf-8"?>
<manifest xmlns:android="http://schemas.android.com/apk/res/android"
    package="com.example.diy_broadcastreceiver">

    <application
        ……
        <receiver
            android:name=".MyReceiver"
            android:enabled="true"
            android:exported="true">
            <intent-filter>
                <action android:name="myself_broadcastReceiver" />
            </intent-filter>
        </receiver>
        ……
    </application>

</manifest>
```

代码在<intent-filter>中设置了 action 的值，这里要注意，这是自定义的过滤，在 MainActivity 中设置意图的 action 必须与清单文件中广播接收者中过滤的 action 一致。在上一节中我们是在 MainActivity 中创建 InterFilter 对象，使用 addAction()方法添加意图过滤，然后调用 registerReceiver()方法注册广播，属于动态注册广播。如果在清单文件中直接设置广播接收者<intent-filter>中的 action 则是静态注册，对于广播的注册，可以使用动态注册，也可使用静态注册。

第五步：用户交互界面中发送广播

打开 MainActivity.java，完成 sendDIYBroadcast()方法，具体代码如下所示：

```
public class MainActivity extends AppCompatActivity {

    @Override
    protected void onCreate(Bundle savedInstanceState) {
        super.onCreate(savedInstanceState);
        setContentView(R.layout.activity_main);
    }

    public void sendDIYBroadcast(View view) {
        Intent i = new Intent();
        i.setAction("myself_broadcastReceiver
");
        i.setPackage("com.example.diy_broadcastreceiver");
        sendBroadcast(i);
    }
}
```

在上面的代码中实现了 sendDIYBroadcast()方法，在方法中先定义了一个 Intent 对象，然后使用 setAction()方法设置 action 的内容，注意此处的 action 的内容要与清单文件中<receiver></receiver>标签中设置的 action 一致。再使用 setPackage()方法设置 Intent 对象所在包（当前程序的包 com.example.diy_broadcastreceiver），最后使用 sendBroadcast()方法发送广播。这样所有监听

“myself_broadcastReicer”这条广播的广播接收者就会收到消息，这样发送的广播就是标准广播。

第六步：运行程序

运行程序，单击“发送自定义广播”触发 sendDIYBroadcast()方法，广播接收者 MyRecive 接收到广播，提示“收到了我自定义的广播”，运行效果如图 9-3 所示。标准广播的工作流程如图 9-4 所示。

补充　标准广播是通过 Intent 对象发送给广播接收者的，如果在发送广播时需要携带数据，只要使用 Intent 对象的 putExtra()方法就可以携带数据发给广播接收者。

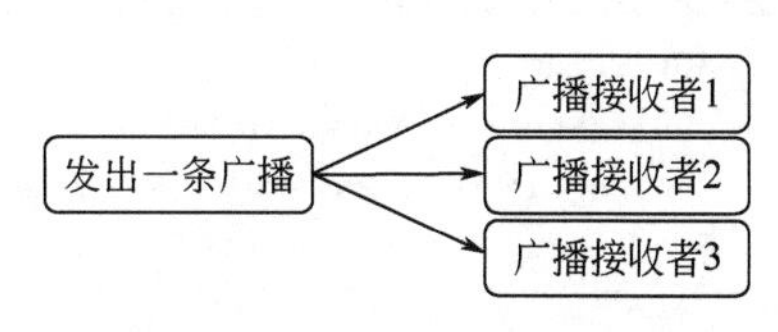

图 9-4　标准广播的工作流程

9.3.2　有序广播

标准广播在广播发出后所有广播接收者几乎是同时接收到这条广播的，而有序广播发出后，对于所有的广播接收者同一时刻只会有一个广播接收者能够收到这条广播，当这个广播接收者中的逻辑执行结束，广播才会继续传播，它的广播接收者是有顺序的。这个顺序通过设置优先级来实现，优先级高的广播接收者先收到广播，而且优先级高的广播也可以截断正在传播的广播，使得后面的广播接收者无法收到广播。

下面通过一个案例学习有序广播的发送和中断，操作步骤如下：

第一步：创建程序

新建一个程序，名为 Order_Broadcast，包名为 com.example.order_broadcast。

第二步：设计用户界面

在 activity_main 布局文件中添加一个命令按钮，用来发送广播。控件 Button 的 text 属性设置为“发送有序广播”，onClick 属性设置为“send_OrderBroadcast”，在 MainActivity 中需要实现这个方法。设计代码见布局文件【activity_main.xml】。

扫描二维码查看【9.3.2activity_main.xml】

9.3.2activity_main.xml

第三步：创建 3 个广播接收者

选择当前包 com.example.order_broadcast，单击右键“New”→“class”，在弹出的提示框中输入“MyReceiver1”创建一个新的类 MyReceiver1.java，打开文件，修改代码使这个类继承 BroadcastReceiver，并重写 onReceive()方法，在广播接收者中打印日志内容为“第一个广播接收者收到了广播”，代码如下所示：

```
public class MyReceiver1 extends BroadcastReceiver{
    @Override
    public void onReceive(Context context, Intent intent) {
        Log.i("MyReceiver1","第一个广播接收者收到了广播");
    }
}
```

使用相同的方法，分别创建第二个广播接收者 MyReceiver2 和第三个广播接收者 MyReceiver3，它们的代码与第一个广播接收者的类似，区别在 onReceiver()方法中打印的日志内容不同。

MyReceiver2 的 onReceiver()方法中代码如下：

```
public class MyReceiver2 extends BroadcastReceiver {
    @Override
    public void onReceive(Context context, Intent intent) {
        Log.i("MyReceiver2","第二个广播接收者收到了广播");
    }
}
```

MyReceiver3 的 onReceiver()方法中代码如下：

```
public class MyReceiver3 extends BroadcastReceiver {
    @Override
    public void onReceive(Context context, Intent intent) {
        Log.i("MyReceiver3","第三个广播接收者收到了广播");
    }
}
```

第四步：发送广播

在 MainActivity 中注册广播接收者和发送广播，代码如下所示：

```
public class MainActivity extends AppCompatActivity {

    @Override
    protected void onCreate(Bundle savedInstanceState) {
        super.onCreate(savedInstanceState);
        setContentView(R.layout.activity_main);
        //注册第一个广播接收者
        MyReceiver1 myReceiver1=new MyReceiver1();
        IntentFilter intentFilter1=new IntentFilter();
        intentFilter1.setPriority(100);
        intentFilter1.addAction("my_order_broadcast");
        registerReceiver(myReceiver1,intentFilter1);
        //注册第二个广播接收者
        MyReceiver2 myReceiver2=new MyReceiver2();
        IntentFilter intentFilter2=new IntentFilter();
        intentFilter2.setPriority(1000);
        intentFilter2.addAction("my_order_broadcast");
        registerReceiver(myReceiver2,intentFilter2);
        //注册第三个广播接收者
        MyReceiver3 myReceiver3=new MyReceiver3();
        IntentFilter intentFilter3=new IntentFilter();
        intentFilter3.setPriority(300);
        intentFilter3.addAction("my_order_broadcast");
        registerReceiver(myReceiver3,intentFilter3);
    }

    public void send_OrderBroadcast(View view) {
        Intent intent=new Intent();
```

```
        intent.setAction("my_order_broadcast");
        sendOrderedBroadcast(intent,null);
    }
}
```

上面的代码中，在 onCreate()方法中分别创建三个广播接收者实例，然后创建 IntentFilter 实例并使用 setPriority()方法设置广播接收者的优先级，通过 addAction()方法设置要过滤的 action，用 registerReceiver()方法注册广播接收者。

在 send_OrderBroadcast()方法中创建 Intent 对象，使用 setAction()方法设置广播事件的类型[这里必须与 onCreate()方法中 intentFilter 对象添加的 Action 一致]，最后使用 sendOrderedBroadcast()方法发送有序广播。发送广播方法的格式：

```
sendOrderedBroadcast(Intent intent,String receiverPermission)
```

第一个参数为广播的意图对象，第二个参数为广播接收者的权限，当这个参数设为 null 时表示任何广播接收者都可以接收该广播消息。

第五步：运行程序

程序运行效果如图 9-5 所示，点击界面上的“发送有序广播”，查看 Logcat，可以看到显示结果是第二个广播接收者先收到广播，然后是第三个广播接收者收到广播，最后是第一个广播接收者收到广播。这是因为在注册广播接收者时，第二个广播接收者的 Priority 的值为 1000，最大，优先级最高，然后是第三个广播接收者，它的 Priority 的值为 300，优先级较高，最后才是第一个广播接收者，它的 Priority 的值为 100，优先级最低。

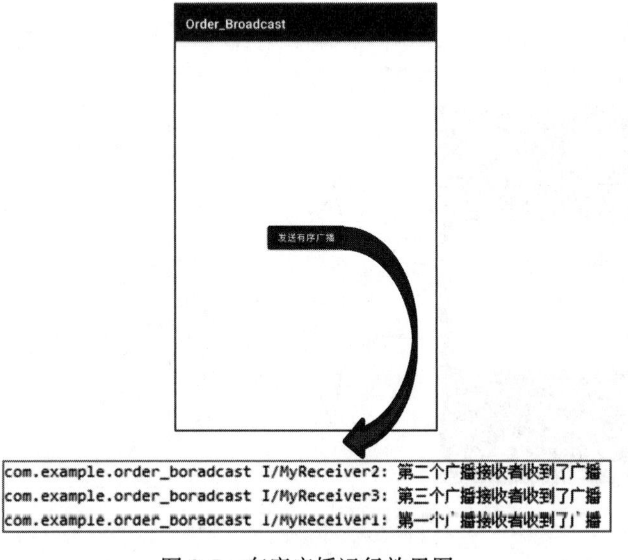

图 9-5 有序广播运行效果图

有序广播不但可以按广播接收者的优先级的高低决定接收广播的顺序，而且还可以由优先级高的广播接收者中断广播，不让广播继续发送。

第六步：中断有序广播

在上面程序的基础上修改第三个广播接收者 MyReceiver3，在打印日志的后面中断广播的继续发送，代码如下所示：

广播接收者
实例分析

```
public class MyReceiver3 extends BroadcastReceiver {

        @Override
        public void onReceive(Context context, Intent intent) {
            Log.i("MyReceiver3","第三个广播接收者收到了广播");
            abortBroadcast();
        }
    }
```

在 MyReceiver3 中添加了 abortBroadcast()方法，此方法的作用是拦截广播，当程序执行完此代码后，广播事件将会被终止，也就是广播不会再向下传递。

再次运行程序，点击“发送有序广播”，可以看到 Logcat 中的内容如图 9-6 所示，比当前优先级低的广播接收者没有接到广播。有序广播的工作流程如图 9-7 所示。

图 9-7　有序广播的工作流程

9.4　拦截电话

有了发送广播和接收广播的基础知识，下面我们实现一个拦截电话的应用，运行效果如图 9-8 所示，拨打指定号码“3232”则立刻中断。

图 9-8　拦截电话运行效果

实现拦截电话的操作步骤如下：

第一步：创建程序

创建一个新的程序，名为 Intercept_Call，包为 com.example.intercept_call。

第二步：创建广播接收者

选择当前包 com.example.intercept_call，单击右键“New”→“class”，在弹出的提示框中输

入“MyReceiver”，创建一个新的类 MyReceiver.java，打开文件，修改代码使这个类继承 BroadcastReceiver，并重写 onReceive()方法，在广播接收者中要求当拨出的号码是“3232”时，拦截电话。代码如下所示：

```
public class MyReceiver extends BroadcastReceiver {
    @Override
    public void onReceive(Context context, Intent intent) {
        String outPhone = getResultData();
        if (outPhone.equals("3232")) {
            setResultData(null);
        }
    }
}
```

在 MyReceiver 的 onReceive()方法中，使用 getResultData()方法获取外拨电话号码，如果号码为“3232”，则使用 setResultData()方法拦截。

第三步：清单文件中添加所需授权

在 AndroidManifest 文件中添加电话拦截授权。

```
<uses-permission android:name="android.permission.PROCESS_OUTGOING_CALLS"/>
```

第四步：在 MainActivity 中注册广播接收者

在 MainActivity 中进行动态检查和申请权限，并注册广播接收者，代码如下所示：

```
public class MainActivity extends AppCompatActivity {
    private MyReceiver myReceiver;

    @Override
    protected void onCreate(Bundle savedInstanceState) {
        super.onCreate(savedInstanceState);
        setContentView(R.layout.activity_main);
        if (Build.VERSION.SDK_INT >= 23) {
            int REQUEST_CODE_CONTACT = 101;
            String[] permissions = {Manifest.permission.PROCESS_OUTGOING_CALLS};
            for (String str : permissions) {  //验证是否有许可权限
               if (this.checkSelfPermission(str) != PackageManager.PERMISSION_GRANTED) { //申请权限
                    this.requestPermissions(permissions, REQUEST_CODE_CONTACT);
                }
            }
        }
        myReceiver = new MyReceiver();
        IntentFilter intentFilter = new IntentFilter();
        intentFilter.addAction("android.intent.action.NEW_OUTGOING_CALL");
        intentFilter.addAction("android.intent.action.PHONE_STATE");
        registerReceiver(myReceiver, intentFilter);
    }

    @Override
    protected void onDestroy() {
        super.onDestroy();
        unregisterReceiver(myReceiver);
    }
}
```

电话拦截权限“PROCESS_OUTGOING_CALLS”属于危险权限组，需要在代码中动态添加申请。在上面代码中判定 Build.VERSION.SDK_INT，然后定义了一个 String 类型数组 permissions，这个数组中存放的是需要动态检查和申请的权限，接下来验证是否有许可权限，如果没有授权，则进行授权。checkSelfPermission()方法检查应用是否具有某个危险权限，如果应用具有此权限，方法将返回 PackageManager.PERMISSION_GRANTED，并且应用可以继续操作。如果应用不具有此权限，方法将返回 PackageManager.PERMISSION_DENIED，且应用必须明确向用户要求权限。使用 requestPermissions()方法动态申请权限，调用后会弹出一个对话框提示用户授权所申请的权限。 最后重写 onDestroy()方法，调用 unregisterReceiver(myReceiver)方法注销广播接收者。

第五步：运行程序

运行程序，弹出“允许 Intercept_Call 访问您的手机通话记录吗?”点击“允许”。打开拨打电话应用拨号，如果拨打“3232”则拦截电话，拨号中断。

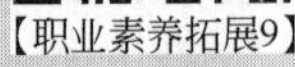

【职业素养拓展9】

职业素养拓展

职业素养拓展内容详见文件“职业素养拓展”。

扫描二维码查看【职业素养拓展 9】

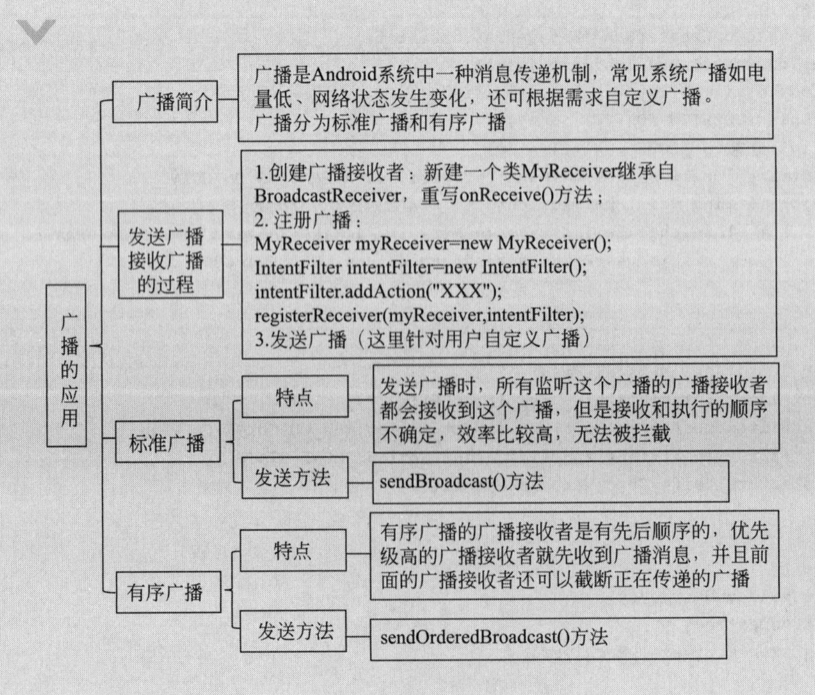

第 10 章

网络编程

使用手机上网接收信息，学习知识，浏览咨询等已经涉及我们生活、工作和学习的方方面面，这些都离不开网络技术的支持，网络编程也是移动应用开发中重要的组成部分。本章我们将介绍用于浏览器控件 WebView、HTTP 协议和 Volley 访问网络的方法，并通过查询天气情况的案例介绍网络数据的访问与处理方法。

学习目标：

1. 掌握 WebView 控件的使用；
2. 学会使用 HttpURLConnection 访问网络；
3. 掌握 JSON 数据结构及其解析方法；
4. 熟悉网络访问库 Volley 的使用方法。

10.1 WebView 的使用

10.1.1 使用 WebView 浏览网页

在 Android 系统之中，用户可以直接使用 WebView 控件显示网页的内容，或者将一些指定的 HTML 文件嵌入进来，下面我们使用 WebView 控件在界面上显示百度首页，运行效果如图 10-1。

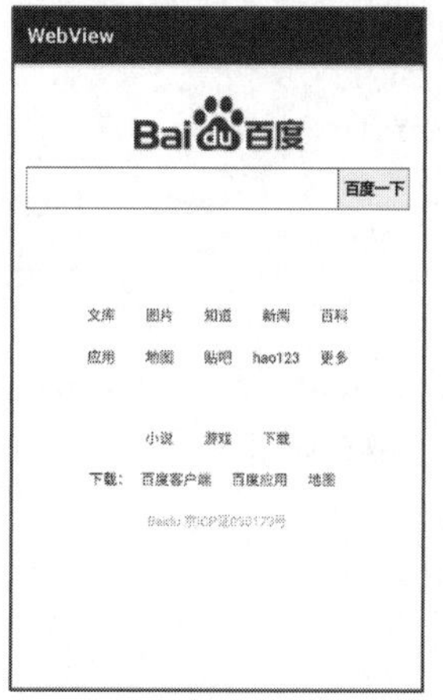

图 10-1　使用 WebView 加载网页

分析 在应用程序中浏览网页，要解决的问题是怎么链接网络，网页内容显示在什么容器中，怎么才能显示，带着问题来完成图 10-1 所示的效果，实现步骤如下所示。

第一步：新建项目

新建一个项目，名为 WebView，指定包为 com.example.webview。

第二步：设计用户界面

在 activity_main 中放置一个 WebView 控件，使其充满整个父容器。设置 id 为 webView。

第三步：申请网络访问权限

使用 WebView 加载网络网页，需要在清单文件中申请网络访问权限，在 AndroidManifest 文件中添加如下代码:

```
<uses-permission android:name="android.permission.INTERNET"/>
```

在<application>标记中设置属性:

```
android:usesCleartextTraffic="true"
```

设置 usesCleartextTraffic 为 true,指定 URL 使用 HTTP 协议也可以加载到 WebView 控件，如果不设置这个属性，对于 HTTP 协议不受影响。

第四步：在 WebView 中加载网页

打开 MainActivity，首先初始化控件，然后使用 loadUrl()方法加载网页。MainActivity 的代码如下所示:

```
public class MainActivity extends AppCompatActivity {
    private WebView webView;

    @Override
    protected void onCreate(Bundle savedInstanceState) {
        super.onCreate(savedInstanceState);
        setContentView(R.layout.activity_main);
        webView = findViewById(R.id.webView);
        webView.setWebViewClient(new WebViewClient());
        webView.loadUrl("https://www.baidu.com/");
    }
}
```

在上面的代码中主要使用了 WebView 的 setWebViewClient()和 loadUrl()两个方法。其中:

setWebViewClient()方法的作用是当我们从一个网页跳转到另外一个网页的时候目标网页仍然在 WebView 中显示，而不是打开系统浏览器。

loadUrl()方法：用于加载指定 URL 对应的网页。

第五步：运行程序

运行程序出现如图 10-1 所示的效果，继续浏览其他网页，也会在 WebView 控件中打开，但是当我们按返回按键时，应用程序直接就会关闭，针对这个问题，我们在 MainActivity 中重写 onKeyDown()方法：

```
@Override
public boolean onKeyDown(int keyCode, KeyEvent event) {
    if (keyCode == KeyEvent.KEYCODE_BACK && webView.canGoBack()) {
        // 返回上一页面
        webView.getSettings().setCacheMode(WebSettings.LOAD_NO_CACHE);
        webView.goBack();//执行后退操作，相当于浏览器上的后退按钮功能
        return true;
    }
    return super.onKeyDown(keyCode, event);
}
```

WebSettings 用于管理 WebView 状态配置，当 WebView 第一次被创建时，WebView 包含着一个默认的配置，这些默认的配置将通过 get 方法返回。这里使用 WebView 的 getSettings()方法获取 WebSettings 对象，调用 setCacheMode()方法设置缓存使用模式。setCacheMode()方法用于加载页面时检查缓存验证是否需要加载，如果不需要重新加载，将直接从缓存读取数据，默认值 LOAD_DEFAULT，这里使用 LOAD_NO_CACHE 表示不使用缓存，只从网络获取数据。goBack()方法用于执行后退操作，相当于浏览器上后退按钮的功能。

再次运行程序，在浏览网页过程中点击“返回”按钮，如果不是百度首页，则都会返回到上一个浏览过的页面。

10.1.2　WebView 中使用 JavaScript

在上一节中我们使用 WebView 控件加载了网页，在 Android 应用中，由于 WebView 控件加载的某些网页是通过 JavaScript 代码编写的，而 WebView 控件在默认情况下是不支持 JavaScript 代码的，要使 WebView 控件支持 JavaScript，可以通过 setJavaScriptEnabled()方法来设置，下面通过在 WebView 控件中加载一个本地的网页，调用 JavaScript 中的方法。程序运行效果如图 10-2 所示，实现过程如下。

图 10-2　WebView 中访问 JS 运行效果

第一步：创建程序

新建程序 WebView_JavaScript，包为 com.example.webview_javascript。

第二步：创建 assets 文件夹

本地网页放置在 assets 文件夹中，先需要创建一个 assets 文件夹。assets 文件夹用来存储 app 资源文件，包括文本文件、图像文件、网页文件、网页文件中引用的 js/ccs/jpg 等资源、音频视频文件。在 assets 目录下的文件在打包后会原封不动地保存在 apk 包中，不会被编译成二进制。

创建 assets 文件夹：选择“app”→“New”→“Folder”→“Assets Folder”，弹出“New Android Component”对话框，选择“Finish”,过程如图 10-3 所示。

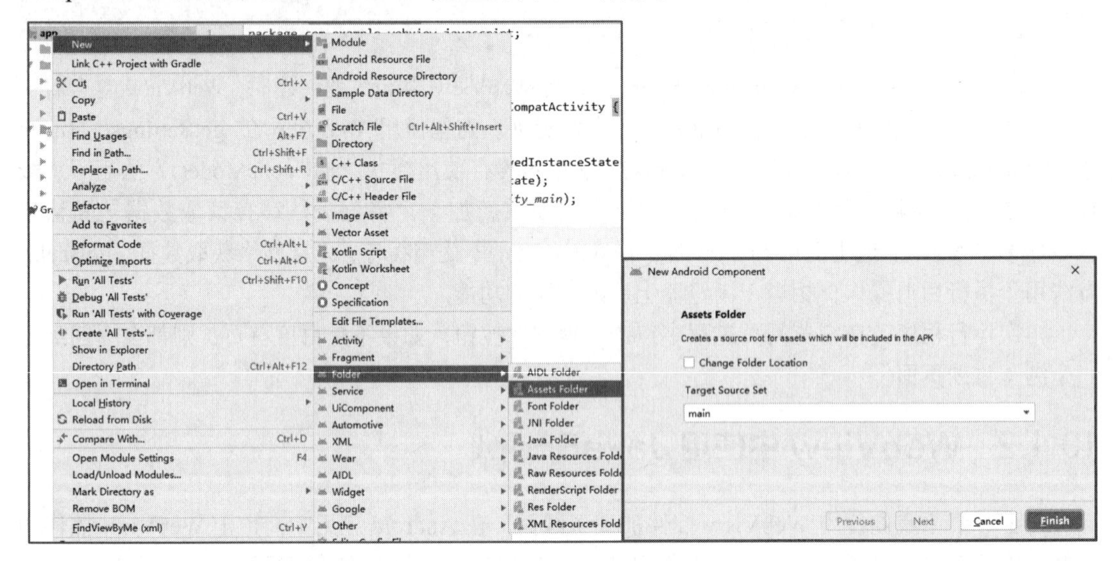

图 10-3　创建 assets 文件夹

第三步：创建网页文件

以加载包含 JavaScript 代码为例，我们先创建一个网页文件，选择“assets”文件夹，单击鼠标右键选择“New”→“File”,在弹出的“New File”对话框中输入“index.html”，创建一个名为 index 的网页文件，编写如下所示代码：

```
<html>
<head>
    <title>我的网页</title>
    <script language="JavaScript">
        function myAlert(){
        alert("调用了 JavaScript 中定义的方法");
        }
    </script>
</head>
<body>
   <h2>网页中调用 JavaScript</h2>
   <p>单击命令按钮调用 JavaScript 中的方法</p>
```

```
    <input type="button" value="调用 JavaScript 中的方法" onclick="myAlert()"/>
</body>
</html>
```

从上面的代码可以看出，在<script/>标记中定义一个方法 myAlert()，其作用是弹出一个对话框，提示“调用了 JavaScript 中定义的方法”。在网页文件的<body/>标记中，有一个标题、一个段落和一个命令按钮，点击命令按钮调用前面定义的方法 myAlert()。

第四步：界面设置

用户界面放置一个 WebView 控件，在默认布局下设置 WebView 控件使其充满整个屏幕。

第五步：用户交互的实现

在 MainActivity 中初始化控件，加载 assets 中的网页文件 index.html，具体代码如下所示。

```
public class MainActivity extends AppCompatActivity {

    private WebView webView;

    @Override
    protected void onCreate(Bundle savedInstanceState) {
        super.onCreate(savedInstanceState);
        setContentView(R.layout.activity_main);
        webView = findViewById(R.id.webview);
        webView.setWebChromeClient(new WebChromeClient());
        webView.getSettings().setJavaScriptEnabled(true);
        webView.loadUrl("file:///android_asset/index.html");
    }
}
```

这里使用 WebView 的 setWebChromeClient()方法实现网页的加载和网页中 JS 文件的访问。

setWebChromeClient()方法：主要用于处理解析、渲染网页等浏览器做的事情，如辅助 WebView 处理 JavaScript 的对话框、网站图标、网站 title、加载进度等。如果 WebView 要支持 JavaScript 脚本，需要设置 setJavaScriptEnabled()属性为 true。

我们在上一个案例中使用了 setWebViewClient()方法，在本案例中使用了 setWebChromeClient()方法。对于这两个方法，如果只是加载 HTML 网页，只需要用 WebViewClient 即可，但是在附加 JavaScript 的页面、调用 JavaScript 对话框的时候，或者功能较为复杂的内嵌操作的时候，建议使用 WebChromeClient。最后对于本地网页加载同样使用 loadUrl()方法，这里加载的是本地网页，需注意网页文件位置的写法。

在这里只是简单地调用了 JavaScript 中的 Alert()方法，如果要显示某一指定的格式，还得设定其格式，具体方法不是本案例的重点，感兴趣的同学可以自行完善。

第六步：运行程序

运行程序效果如图 10-2 所示，本地网页 index.html 直接加载在 WebView 控件中，单击网页中的“调用 JavaScript 中的方法”弹出由 JavaScript 调用的“alert”对话框。

10.2 网络访问

10.2.1 使用 HTTP 协议访问网络

HTTP 是 HyperText Transfer Protocol(超文本传输协议)的英文缩写，是一种应用层协议。它的工作原理就是客户端向服务器发出一条 HTTP 请求，服务器收到请求之后会返回一些数据给客户端，然后客户端再对这些数据进行解析和处理。在上一节中使用 WebView 控件访问百度其实就是我们向百度的服务器发送了一条 HTTP 请求，接着服务器分析出我们想要访问的百度首页，然后把该网页的 HTML 代码进行返回，WebView 再调用手机浏览器的内核对返回的 HTML 代码进行解析，最终将页面展示出来。

使用 WebView 展示百度首页的案例中，发送 HTTP 请求，接收服务器响应，解析返回数据，以及最终的页面展示，这些工作都是在后台处理的，我们不能直观地看出 HTTP 协议到底是如何工作的，下面通过一个案例手动发送 HTTP 请求，学习使用 HTTP 协议访问网络的过程。操作过程如下:

第一步：新建项目

新建项目，名为 Net_access_http,所在的包为 com.example.net_access_http。

第二步：用户界面设计

在 activity_main 布局文件中，添加两个命令按钮，一个 ScrollView 控件和一个 TextView 控件。其中一个命令按钮用于发送 HTTP 请求，对应的单击事件为 sendHTTP()，另一个命令按钮用于将网络访问返回的数据显示到 TextVeiw 控件中，在 MainActivity 中对应的单击事件为 setText() 方法。ScrollView 控件的作用相当于一个容器，其中放置一个 TextView 控件，TextView 用来显示服务器返回的数据，返回的数据如果整个屏幕不能容纳，有可能显示在屏幕外，使用 ScrollView 控件可以实现滚动查看屏幕外内容的功能，我们现在常用的 App 需要滚屏查看内容的都可以使用 ScrollView 控件。activity_main 代码如下所示:

```
<?xml version="1.0" encoding="utf-8"?>
<androidx.constraintlayout.widget.ConstraintLayout xmlns:android="http://schemas.android.com
/apk/res/android"
    xmlns:app="http://schemas.android.com/apk/res-auto"
    xmlns:tools="http://schemas.android.com/tools"
    android:layout_width="match_parent"
    android:layout_height="match_parent"
    tools:context=".MainActivity">

    <Button
        android:id="@+id/button"
        android:layout_width="150dp"
        android:layout_height="wrap_content"
```

```
        android:layout_marginStart="16dp"
        android:layout_marginTop="16dp"
        android:onClick="sendHTTP"
        android:text="发送 HTTP 请求"
        app:layout_constraintStart_toStartOf="parent"
        app:layout_constraintTop_toTopOf="parent" />

    <Button
        android:id="@+id/button2"
        android:layout_width="150dp"
        android:layout_height="wrap_content"
        android:layout_marginEnd="16dp"
        android:onClick="setText"
        android:text="返回的数据"
        app:layout_constraintBottom_toBottomOf="@+id/button"
        app:layout_constraintEnd_toEndOf="parent" />

    <ScrollView
        android:id="@+id/scrollView"
        android:layout_width="403dp"
        android:layout_height="580dp"
        android:layout_marginTop="28dp"
        app:layout_constraintEnd_toEndOf="parent"
        app:layout_constraintStart_toStartOf="parent"
        app:layout_constraintTop_toBottomOf="@+id/button">

        <TextView
            android:id="@+id/tvResponse"
            android:layout_width="match_parent"
            android:layout_height="match_parent"/>
    </ScrollView>

</androidx.constraintlayout.widget.ConstraintLayout>
```

第三步：实现 HTTP 访问网络

在 Android 上发送 HTTP 协议，我们使用的是 HttpURLConnection 的方法。使用这种方法首先需要获取到 HttpURLConnection 的实例，一般只需新建一个 URL 对象，并传入目标网络地址，然后调用一些 openConnection()方法即可。示例代码如下：

```
URL url=new URL("http://www.lzpcc.edu.cn/");
HttpURLConnection conn= (HttpURLConnection) url.openConnection();
```

在得到 HttpURLConnection 之后设置 HTTP 请求所使用的方法，常用的方法有 GET 和 POST。GET 把参数包含在 URL 字符串的后面，POST 方式的参数是放在 HTTP 请求中的。使用 setRequestMethod()方法设置请求方式：

```
conn.setRequestMethod("GET");
```

在使用 HttpURLConnection 对象访问网络时，需要设置超时时间，以防止连接被阻塞时无响应，影响用户体验。设置超时时间如

```
conn.setConnectTimeout(5000);
setConnectTimeout()方法的参数是毫秒数。
```

然后获取服务器返回的输入流:

```
InputStream in=conn.getInputStream();
```

最后使用 disconnect()方法关闭 http 连接:

```
conn.disconnect()。
```

基本理解使用 HttpURLConnection 方法网络的过程，下面在 MainActivity 中实现，核心代码如下所示。

```
public class MainActivity extends AppCompatActivity {
    private TextView tvResponse;
    private StringBuffer response;

    @Override
    protected void onCreate(Bundle savedInstanceState) {
        super.onCreate(savedInstanceState);
        setContentView(R.layout.activity_main);
        tvResponse = findViewById(R.id.tvResponse);
        response = new StringBuffer();
    }

    public void sendHTTP(View view) {
        new Thread(new Runnable() {
            @Override
            public void run() {
                try {
                    URL url = new URL("https://www.baidu.com/");
                    HttpURLConnection conn = (HttpURLConnection) url.openConnection();
                    conn.setRequestMethod("GET");
                    conn.setConnectTimeout(5000);
                    conn.setReadTimeout(5000);
                    InputStream in = conn.getInputStream();
                    BufferedReader reader = new BufferedReader(new InputStreamReader(in));

                    String line;
                    while ((line = reader.readLine()) != null) {
                        response.append(line);
                    }
                } catch (IOException e) {
                    e.printStackTrace();
                }
            }
```

```
        }).start();
    }

    public void setText(View view) {
        tvResponse.setText(response.toString());
    }
}
```

在 MainActivty 中定义了两个成员变量：tvResponse 为 TextVeiw 类型的变量，用于显示网络返回的数据；response 为 StringBuffer 类型的变量，用于保存从网络返回的数据，其结果最终显示在 TextView 控件上。

网络请求通常都属于耗时的操作，为防止主线程被阻塞住，需要开启一个子线程发起 HTTP 请求。我们在按钮的单击事件 sendHTTP()方法中，开启了一个子线程，然后在子线程里使用 HttpURLConnection 发出一条 HTTP 请求，请求的目标地址就是百度首页。接着使用 BufferedReader 对服务器返回的流进行读取，并将结果存储在 response 中。

实现命令按钮“返回的数据”对应的 setText()方法，设置 TextView 的 text 值为 response。

第四步：设置网络访问权限

要访问网络，需要在清单文件 AndroidManifest 中设置：

```
<uses-permission android:name="android.permission.INTERNET"/>
```

如果需要访问“http”开头的 URL，还需设置<application>的属性：

```
android:usesCleartextTraffic="true"
```

第五步：运行程序

运行程序，点击“发送 HTTP 请求”后，点击“返回的数据”将从百度首页获取的文本显示在下面的 TextView 中，结果如图 10-4 所示。

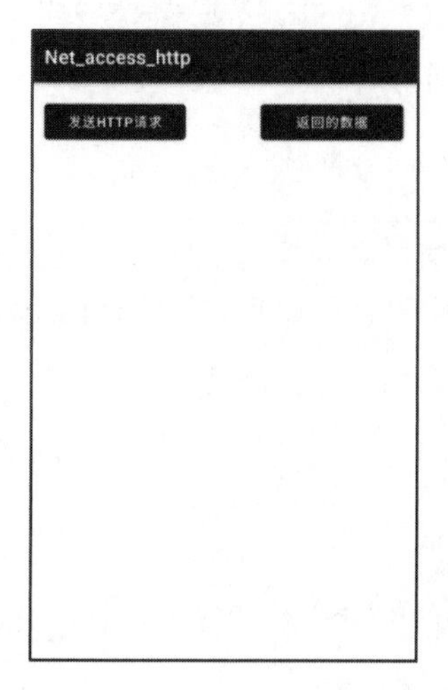

图 10-4　服务器的响应数据

10.2.2 使用 Volley 访问网络

我们使用 HttpURLConnection 可以实现访问网络，但是它的用法还是稍微有些复杂，如果不进行适当封装的话，很容易就会写出不少重复代码。于是一些 Android 网络通信框架也就应运而生，它们对 HTTP 通信细节进行了封装，我们只需要简单调用几行代码就可以完成通信操作了。在这里介绍一种对于广大初学来说最简单的网络访问框架 Volley。

Volley 是一个新的网络通信框架，它的特点是简单易用，适合数据量不大，但通信频繁的网络操作，并且 Volley 回调的时候是在主线程，可以直接操作 UI。

下面还是以访问百度首页为例，学习使用 Volley 访问网络。实现步骤如下所示。

第一步：创建项目

新建项目，名为 Volley_Simple,包名为 com.example.volley_simple。

第二步：添加 Volley

在使用 Volley 之前首先需要将 Volley 添加到项目中，最简单的方法是在应用程序的 build.gradle 文件中添加依赖 implementation"com.android.volley:volley:1.1.1"，然后点击“Sync Now”添加 Volley 库，如图 10-5 所示。

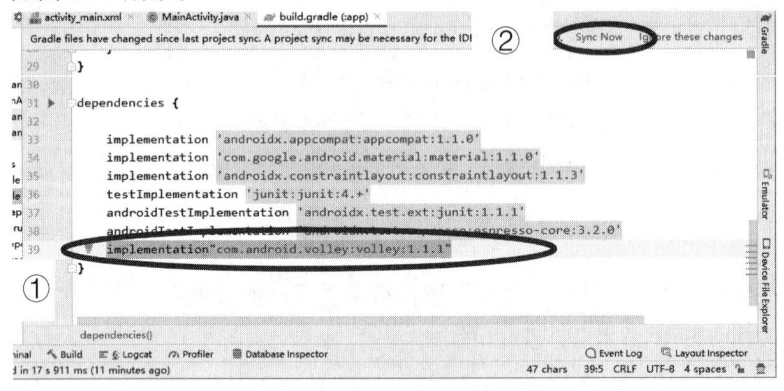

图 10-5 添加 Volley 库

第三步：设置用户界面

界面上有一个“volley 访问网络数据”的命令按钮对应的单击事件为 sendVolley()方法，有一个 ScrollView 控件和 id 为 tvResponse 的 TextView 控件。设计代码见布局文件【activity_main.xml】。

扫描二维码查看【10.2.2activity_main.xml】

10.2.2activity_main.xml

第四步：清单文件中申请网络访问权限

在清单文件中申请网络访问权限：

```
<uses-permission android:name="android.permission.INTERNET"/>
```

第五步：使用 Volley 访问百度首页

使用 Volley 访问网络首先使用 Volley.newRequestQueue 创建请求队列对象，接着创建请求对象 Request，最后使用请求队列的 add()方法将请求对象添加进请求队列。在 MainActivity 中实现百度首页访问的核心代码如下所示：

```
public class MainActivity extends AppCompatActivity {
    private TextView tvresult;
    @Override
    protected void onCreate(Bundle savedInstanceState) {
        super.onCreate(savedInstanceState);
        setContentView(R.layout.activity_main);
        tvresult = findViewById(R.id.tvResponse);
    }

    public void sendVolley(View view) {
        RequestQueue queue = Volley.newRequestQueue(this);
        String url = "https://www.baidu.com/";
        StringRequest strRequest = new StringRequest(Request.Method.GET, url, new Response.Listener
<String>() {
            @Override
            public void onResponse(String response) {
                tvresult.setText(response);
            }
        }, new Response.ErrorListener() {
            @Override
            public void onErrorResponse(VolleyError error) {
                tvresult.setText("Error");
            }
        });
        queue.add(strRequest);
    }
}
```

在上面的代码可以看到，使用 Volley.newRequestQueue(this)创建了一个 RequestQueue 对象，接着使用方法 StringRequest()方法创建 StringRequest 对象，StringRequest()方法格式如下所示：

StringRequest(int method, String url,Listener<String> listener, @Nullable ErrorListener errorListener);

此方法有四个参数：

第一个参数：表示使用什么方式请求网络，常用值为 Request.Method.GET 或者 Request.Method.POST。

第二个参数：是访问的网络地址。

第三个参数：请求成功的回调监听，在这个案例中 String response 为访问成功返回的数据，我们直接将它设置为 TextView 对象的值了。

第四个参数：请求错误的回调，也就是如果网络访问不成功时的处理在此处设置。

最后使用 queue.add(strRequest)将请求对象添加到请求队列中。简单的 Volley 的使用过程就完成了。

第六步：运行程序

运行程序，效果如图 10-6 所示。通过这个案例我们可以看到使用 Volley 访问网页内容主要有三步：

① 创建一个 RequestQueue 对象；

② 创建一个 StringRequest 对象；

③ 将 StringRequest 对象添加到 RequestQueue 里面。

Volley 除了使用 StringRequest，还有 JsonObjectRequest、JsonArrayRequest、ImageRequest，用法都与 StringRequest 类似，区别在于服务器返回的数据格式不同。

图 10-6　使用 Volley 发送网络请求

10.3　查看天气应用

查看天气应用访问的网络数据格式为 JSON，在实现网络数据访问与处理之前我们先了解一个 JSON 数据格式及其解析方法。

10.3.1　JSON 数据及解析

（1）JSON 数据

JSON(JavaScript Object Notation)是一种轻量级的数据交换格式，在网络传输中可减少网络传输数据流量从而加快传输速度，JSON 数据的文件后缀为.json。常用 JSON 数据有两种结构，一种是对象结构，一种是数组结构。

对象结构 json1 示例代码如下：

```
{"name":"zhangsan","age":20,"school":"shihuadaxue"}
```

JSON 数据使用键值对“key:value”的形式进行存储，每对键与值之间使用“：”分隔，键值对之间使用“，”分隔。由此可以看出在上面代码有三个键值对存储，分别存储一个学生的姓名、年龄和学校。

数组结构 json2 示例代码如下：

```
[{"name":"Tom","age":20,"school":"shihuadaxue"},
{"name":"Jack","age":21,"school":"pingliangzhiye"},
{"name":"Lisi","age":20,"school":"wuweizhiyuan"}]
```

从上面的代码可以看出，数组结构的JSON数据跟我们之前接触的数组类似，它包含多个对象结构的JSON数据。

了解了JSON数据结构，如果需要使用这些JOSN数据，就要对它进行解析。解析JSON数据有很多种方法，在这给大家介绍使用JSONObject与JSONArray和使用开源库GSON解析JSON数据的方法。

（2）使用JSONObject和JSONArray解析JSON数据

我们对json1使用JSONObject解析，代码如下：

```
JSONObject object=new JSONObject(json1);
String name=object.optString("name");
int age=object.optInt("age");
String school=object.optString("school");
```

解析对象结构的JSON数据，先创建一个JSONObject对象，然后通过optXxx()方法获取所含键（key）的值赋给对应类型的数据，optXxx()方法与getXxx()方法用法类似，可以是optString()方法、optInt()方法以及optBoolean()方法，等等，区别只是在于当获取的key不存在时，optXxx()方法会给出对应类型的默认值。

json2是一个数组，使用JSONArray来解析，代码如下：

```
JSONArray array = new JSONArray(json2);
for (int i = 0; i < array.length(); i++) {
    JSONObject object1 = array.optJSONObject(i);
    int age1 = object1.optInt("age");
    String name1 = object1.optString("name");
    String school1 = object1.optString("school");
}
```

对于数组形式的JSON数据，它是由多个相同结构的JSONObject数据组成，所以先定义一个JSONArray对象，然后将它的每个元素像JSONObject一样处理就可以了。

（3）使用GSON解析JSON数据

GSON是一个开源库，可以用来解析JSON数据，使用GSON解析JSON数据，首先需要在项目中添加GSON库文件。打开build.gradle文件添加如下代码：

```
    implementation 'com.google.code.gson:gson:2.8.6'
```

我们还是以json1和json2的解析为例来学习GSON的解析方法。先创建一个实体类Student，此类中的成员名称与JSON数据中的一致，Student类成员有name、age、school。Student.java代码如下所示：

```
public class Student {
    private String name;
    private int age;
    private String school;

    public String getName() {
```

```
        return name;
    }
    public void setName(String name) {
        this.name = name;
    }
    public int getAge() {
        return age;
    }
    public void setAge(int age) {
        this.age = age;
    }
    public String getSchool() {
        return school;
    }
    public void setSchool(String school) {
        this.school = school;
    }
}
```

使用 GSON 解析对象结构 JSON 数据：

```
Gson gson =new Gson();
Student student=gson.fromJson(json1,Student.class);
```

使用 GSON 的 fromJson()方法，将 JSON 数据解析为指定类型 Student 的对象。

使用 GSON 解析数组结构 JSON 数据：

```
Gson gson=new Gson();
Type listType=new TypeToken<List<Student>>(){}.getType();
List<Student> studentList=gson.fromJson(json2,listType);
```

（4）在 MainActivity 中解析 json1 和 json2

在 MainActivity 中使用 JSONObject 和 JSONArray 与 GSON 解析参考代码如下：

```
public class MainActivity extends AppCompatActivity {

    @Override
    protected void onCreate(Bundle savedInstanceState) {
        super.onCreate(savedInstanceState);
        setContentView(R.layout.activity_main);
        String json1="{\"name\":\"zhangsan\",\"age\":20,\"school\":\"shihuadaxue\"}";
        String  json2="[{\"name\":\"Tom\",\"age\":20,\"school\":\"shihuadaxue\"},\n"  +  "  {\"name
\":\"Jack\",\"age\":21,\"school\":\"pingliangzhiye\"},\n" + " {\"name\":\"Lisi\",\"age\":20,\"school\
":\"wuweizhiyuan\"}]\n";
        S//解析strjson
         try {
           JSONObject object=new JSONObject(json1);
           String name=object.optString("name");
           int age=object.optInt("age");
           String school=object.optString("school");
           Log.i("json1", name+"  "+age+"  "+school);
```

```
            JSONArray array = new JSONArray(json2);
            for (int i = 0; i < array.length(); i++) {
                JSONObject object1 = array.optJSONObject(i);
                int age1 = object1.getInt("age");
                String name1 = object1.optString("name");
                String school1 = object1.optString("school");
                Log.i("json2", name1 + "  " + age1 + "  " + school1);
            }
        } catch (JSONException e) {
            e.printStackTrace();
        }
        Gson gson =new Gson();
        Student student=gson.fromJson(json1,Student.class);
        Log.i("json1", student.getName()+"  "+student.getAge()+"  "+student.getSchool());
        Type listType=new TypeToken<List<Student>>(){}.getType();
        List<Student> studentList=gson.fromJson(json2,listType);
        for(Student student1:studentList){
          Log.i("json2", student1.getName()+"  "+student1.getAge()+"  "+student1.getSchool());
        }
    }
}
```

10.3.2　查看天气实例

在实际生活中很多人的手机中都会有关于天气预报的软件，通过查看天气预报安排各自的生活和工作。下面通过访问网络天气预报数据，对数据进行处理和展示。程序运行效果如图 10-7 所示。

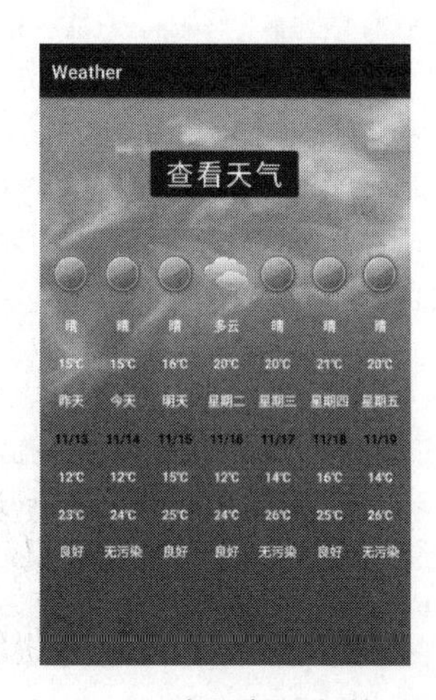

图 10-7　天气预报运行效果图

第一步：创建项目

新建一个项目，名为 Weather，包名为 com.example.weather。

第二步：准备工作

准备工作主要分两部分，一部分是导入项目要用到的图片，将准备好的背景图片 weather.jpg 和天气情况的图片 sun.jpg、clouds.jgp 复制到 drawable 文件夹中；另一部分是添加项目需要的库文件，这里需要加入 Volley 和 GSON。在 build.gradle 中添加库文件：

```
implementation"com.android.volley:volley:1.1.1"
implementation 'com.google.code.gson:gson:2.8.6'
```

第三步：用户界面设计

界面设置背景图片为 weather.jpg,在用户界面中放置一个命令按钮 button 和一个 RecyclerView，其 id 为 recycle，用户界面如图 10-8 所示。

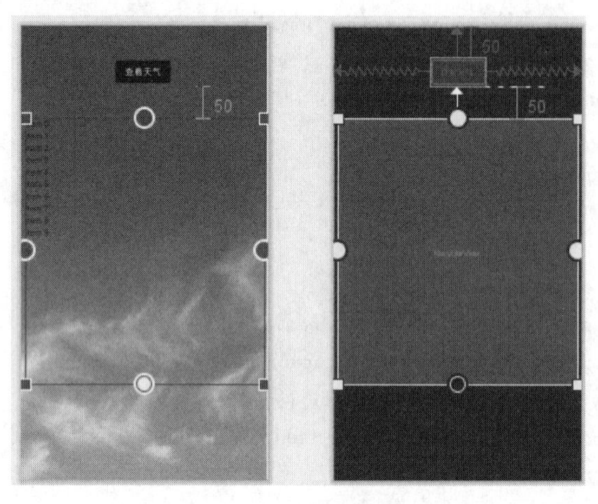

图 10-8　天气预报界面设计

第四步：创建天气类 Weather

要访问的网络数据放在本地服务器上，数据部分内容如下所示:


```
[{"minTemperature":"12℃","maxTemperature":  "23℃","temperature":"15℃","weather":"晴","air":"良好","label":"昨天","day":"11/13"},{"minTemperature":  "12℃","maxTemperature":"24℃","weather":  "晴","air":"无污染","temperature":"15℃","label":"今天","day":"11/14"},……]
```


详细的服务器数据，见文件【weather.json】。

扫描二维码查看【10.3.2weather.json】

10.3.2weather.json

创建一个 Weather 类，用来存放天气数据中的温度、天气、空气等信息，在当前项目的包中新建 Weather 类，代码如下所示。

```
public class Weather {
    private String maxTemperature;
    private String minTemperature;
    private String weather;
    private String air;
    private String temperature;
    private String label;
    private String day;
    private int image;

    public String getMaxTemperature() {
        return maxTemperature;
    }

    public String getMinTemperature() {
```

```
        return minTemperature;
    }

    public String getWeather() {
        return weather;
    }

    public String getAir() {
        return air;
    }

    public String getTemperature() {
        return temperature;
    }

    public String getLabel() {
        return label;
    }

    public String getDay() {
        return day;
    }
}
```

第五步：创建天气 item 界面

在 layout 文件夹中新建一个布局，名为 item.xml，在 item 上显示天气的图片、天气、温度、日期、最低温和最高温以及空气状况。

```
<?xml version="1.0" encoding="utf-8"?>
<LinearLayout xmlns:android="http://schemas.android.com/apk/res/android"
    android:layout_width="60dp"
    android:layout_height="match_parent"
    android:gravity="center_horizontal"
    android:orientation="vertical">

    <ImageView
        android:id="@+id/imgweather"
        android:layout_width="50dp"
        android:layout_height="50dp" />

    <TextView
        android:id="@+id/weather"
        android:layout_width="wrap_content"
        android:layout_height="wrap_content"
        android:layout_marginTop="20dp"
        android:text="天气"
```

```
        android:textColor="#ffffff"
        android:textStyle="bold" />

    <TextView
        android:id="@+id/temperature"
        android:layout_width="wrap_content"
        android:layout_height="wrap_content"
        android:layout_marginTop="20dp"
        android:text="温度"
        android:textColor="#ffffff"
        android:textStyle="bold" />

    <TextView
        android:id="@+id/lable"
        android:layout_width="wrap_content"
        android:layout_height="wrap_content"
        android:layout_marginTop="20dp"
        android:text="标签"
        android:textColor="#ffffff"
        android:textStyle="bold" />

    <TextView
        android:id="@+id/day"
        android:layout_width="wrap_content"
        android:layout_height="wrap_content"
        android:layout_marginTop="20dp"
        android:text="日期"
        android:textColor="#0000ff"
        android:textStyle="bold" />

    <TextView
        android:id="@+id/mintemperature"
        android:layout_width="wrap_content"
        android:layout_height="wrap_content"
        android:layout_marginTop="20dp"
        android:text="最低温度"
        android:textColor="#ffffff"
        android:textStyle="bold" />

    <TextView
        android:id="@+id/maxtemperature"
        android:layout_width="wrap_content"
        android:layout_height="wrap_content"
        android:layout_marginTop="20dp"
        android:text="最高温度"
        android:textColor="#ffffff"
        android:textStyle="bold" />
```

```
    <TextView
        android:id="@+id/air"
        android:layout_width="wrap_content"
        android:layout_height="wrap_content"
        android:layout_marginTop="20dp"
        android:text="空气质量"
        android:textColor="#ffffff"
        android:textStyle="bold" />
</LinearLayout>
```

第六步：定义适配器

为 RecyclerView 定义适配器，完成 item 界面布局的加载，item 界面上控件的初始化和控件与数据的绑定，代码如下所示:

```
public class MyAdapter extends RecyclerView.Adapter<MyAdapter.MyViewHolder> {
    Context context;
    List<Weather> weatherList;
    public MyAdapter( Context context,List<Weather> weatherList){
        this.context=context;
        this.weatherList=weatherList;
    }

    @NonNull
    @Override
    public MyViewHolder onCreateViewHolder(@NonNull ViewGroup parent, int viewType) {
        MyViewHolder holder = new MyViewHolder(LayoutInflater.from(context).inflate(R.layout.item,
parent, false));
        return holder;
    }

    @Override
    public void onBindViewHolder(@NonNull MyViewHolder holder, int position) {
        holder.weather.setText(weatherList.get(position).getWeather());
        if (weatherList.get(position).getWeather().equals("晴")) {
            holder.imageView.setImageResource(R.drawable.sun);
        } else if (weatherList.get(position).getWeather().equals("多云")) {
            holder.imageView.setImageResource(R.drawable.clouds);
        }
        holder.lable.setText(weatherList.get(position).getLabel());
        holder.maxtemperature.setText(weatherList.get(position).getMaxTemperature());
        holder.mintemperature.setText(weatherList.get(position).getMinTemperature());
        holder.air.setText(weatherList.get(position).getAir());
        holder.day.setText(weatherList.get(position).getDay());
        holder.temperature.setText(weatherList.get(position).getTemperature());
    }
```

```
    @Override
    public int getItemCount() {
        return weatherList.size();
    }

    public class MyViewHolder extends RecyclerView.ViewHolder {
        ImageView imageView;
        TextView weather, lable, day, mintemperature, maxtemperature, air, temperature;

        public MyViewHolder(@NonNull View itemView) {
            super(itemView);
            imageView = (ImageView) itemView.findViewById(R.id.imgweather);
            weather = (TextView) itemView.findViewById(R.id.weather);
            temperature = (TextView) itemView.findViewById(R.id.temperature);
            lable = (TextView) itemView.findViewById(R.id.lable);
            day = (TextView) itemView.findViewById(R.id.day);
            mintemperature = (TextView) itemView.findViewById(R.id.mintemperature);
            maxtemperature = (TextView) itemView.findViewById(R.id.maxtemperature);
            air = (TextView) itemView.findViewById(R.id.air);
        }
    }
}
```

为 RecyclerView 定义一个适配器 MyAdapter 类继承 RecyclerView.Adapter 类，并在该类定义一个构造方法，用于传递 Context 和天气数据。重写 onCreateViewHolder()方法用于加载 item 界面的布局文件，onBindViewHolder()方法用于将获取的数据设置到对应的控件上；使用 getItemCount()方法用于获取列表的总数。定义了一个内部类 MyViewHolder 继承 RecyclerView.ViewHolder 类，用于获取 item 界面上的控件。

第七步：实现查看天气

在 MainActivity 中初始化命令按钮和 RecyclerView，通过 Volley 访问网络数据，实现查看天气的功能，代码如下：

```
public class MainActivity extends AppCompatActivity {

    private RecyclerView recycler;
    private Button button;
    private List<Weather> weatherList;

    @Override
    protected void onCreate(Bundle savedInstanceState) {
        super.onCreate(savedInstanceState);
        setContentView(R.layout.activity_main);
        weatherList = new ArrayList<>();
        button = findViewById(R.id.button);
        recycler = findViewById(R.id.recycle);
        recycler.setLayoutManager(new GridLayoutManager(this, 7));
        button.setOnClickListener(new View.OnClickListener() {
            @Override
```

```
            public void onClick(View v) {
                weatherList = getData();
                MyAdapter adapter = new MyAdapter(MainActivity.this, weatherList);
                recycler.setAdapter(adapter);
            }
        });
    }

    private List<Weather> getData() {
        RequestQueue queue = Volley.newRequestQueue(this);
        String url = "http://10.0.2.2:8080/testdata/weatherList.json";
        JsonArrayRequest  jsonArrayRequest  =  new  JsonArrayRequest(url,  new  Response.Listener
<JSONArray>() {
            @Override
            public void onResponse(JSONArray response) {
                Gson gson = new Gson();
                Type listType = new TypeToken<List<Weather>>() {
                }.getType();
                weatherList = gson.fromJson(response.toString(), listType);
            }
        }, new Response.ErrorListener() {
            @Override
            public void onErrorResponse(VolleyError error) {
                Log.i("AA", "error");
            }
        });
        queue.add(jsonArrayRequest);
        return weatherList;
    }
}
```

在上面的代码中我们定义一个 getData()获取数据，数据指定访问的服务器地址是电脑本机。由于返回的数据是 JSONArray 类型，所以使用 JsonArrayRequest 请求数据，对于返回的 JSONArray response，使用 GSON 对其解析，将结果赋值给 weatherList。

第八步：清单文件中申请授权

我们访问的服务器，在这需要在 AndroidManifest 中添加 Intent 访问权限。

```
<uses-permission android:name="android.permission.INTERNET" />
```

第九步：运行程序

运行程序，单击界面“查看天气”命令按钮，显示如图 10-7 所示的运行效果。

职业素养拓展

职业素养拓展内容详见文件“职业素养拓展”。

扫描二维码查看【职业素养拓展 10】

【职业素养拓展10】

本章小结

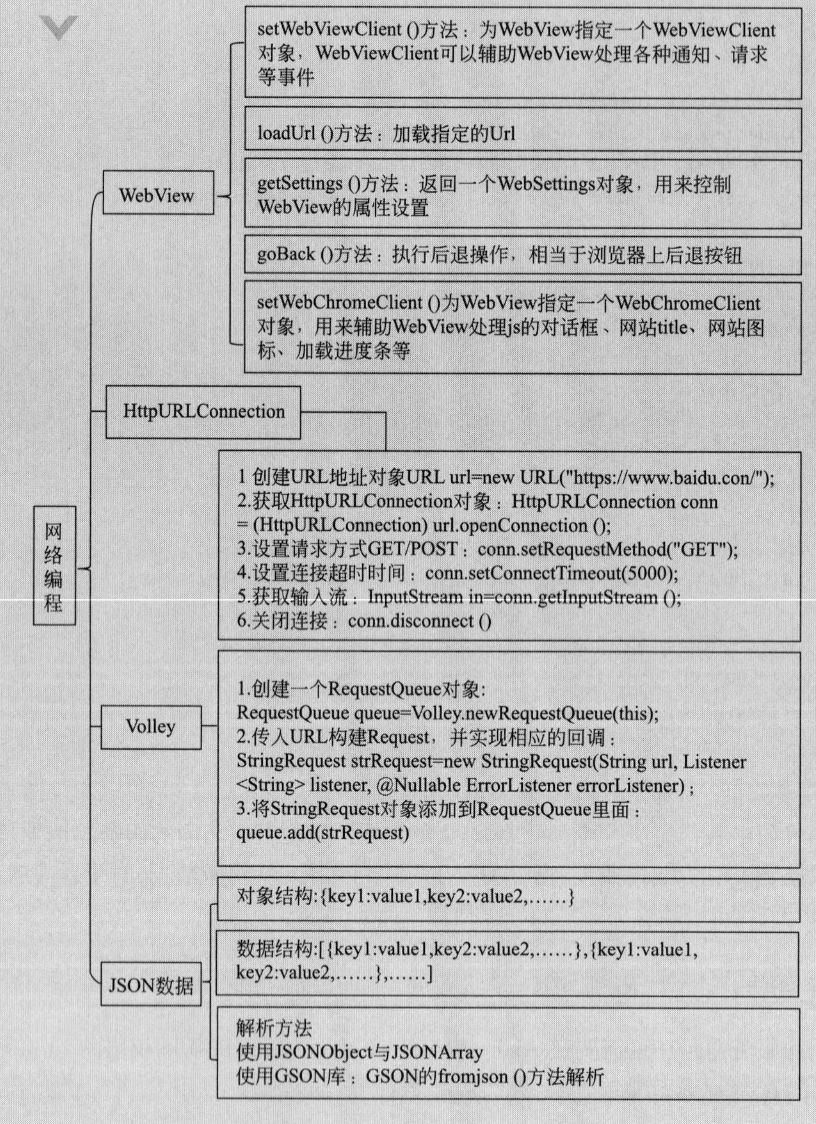
网络编程
WebView
setWebViewClient ()方法：为WebView指定一个WebViewClient对象，WebViewClient可以辅助WebView处理各种通知、请求等事件
loadUrl ()方法：加载指定的Url
getSettings ()方法：返回一个WebSettings对象，用来控制WebView的属性设置
goBack ()方法：执行后退操作，相当于浏览器上后退按钮
setWebChromeClient ()为WebView指定一个WebChromeClient对象，用来辅助WebView处理js的对话框、网站title、网站图标、加载进度条等
HttpURLConnection
1 创建URL地址对象URL url=new URL("https://www.baidu.con/");
2.获取HttpURLConnection对象：HttpURLConnection conn = (HttpURLConnection) url.openConnection ();
3.设置请求方式GET/POST：conn.setRequestMethod("GET");
4.设置连接超时时间：conn.setConnectTimeout(5000);
5.获取输入流：InputStream in=conn.getInputStream ();
6.关闭连接：conn.disconnect ()
Volley
1.创建一个RequestQueue对象:
RequestQueue queue=Volley.newRequestQueue(this);
2.传入URL构建Request，并实现相应的回调：
StringRequest strRequest=new StringRequest(String url, Listener <String> listener, @Nullable ErrorListener errorListener)；
3.将StringRequest对象添加到RequestQueue里面：
queue.add(strRequest)
JSON数据
对象结构:{key1:value1,key2:value2,……}
数据结构:[{key1:value1,key2:value2,……},{key1:value1, key2:value2,……},……]
解析方法
使用JSONObject与JSONArray
使用GSON库：GSON的fromjson ()方法解析

第 11 章

媒体应用技术

视频和音频的播放与录制是手机应用中常见的功能，Android 系统提供多媒体框架支持播放各种常见媒体类型，开发者可以轻松地将音频、视频和图片集成到应用中，也可以自行录制音频与视频。在本章我们将通过实例学习如何在手机应用中播放音频和视频，录制音频和视频；学习 MediaPlayer 类、VideoView 控件等内容。

学习目标：

1. 掌握使用 MediaPlayer 类播放音频文件的方法；
2. 掌握使用 VideoView 播放视频；
3. 学会如何录制音频和视频。

11.1 播放音频和视频

11.1.1 播放音频文件

Android 多媒体框架支持播放各种常见媒体类型，开发者可以轻松地将音频、视频和图片集成到应用中。可以使用 MediaPlayer 类播放存储在应用资源（原始资源）内的媒体文件、文件系统中的独立文件或者通过网络连接获得的数据流中的音频或视频。

下面我们通过如图 11-1 所示案例学习如何播放音频文件。

图 11-1 播放音频

分析 实现音频播放要解决的问题是由什么控制音频文件，音频文件放在哪里，如何获取，怎么开始播放，如何暂停和停止播放。Android 中播放音频文件可以通过 MediaPlayer 类来实现，MediaPlayer 类提供了一些常用的方法可以实现音频文件的播放。本应用要使用到的方法有以下几个：

setDataSource()方法：作用是设置要播放的音频文件的位置。

start()方法：作用为开始或者继续播放。

isPlaying()方法：是否正在播放，返回为 true 或者 false。

pause()方法：暂停播放。
stop()方法：停止播放。
release()方法：作用为释放 MediaPlayer 对象。
实现如图 11-1 所示的运行效果的操作过程如下所示。

第一步：新建项目

新建一个项目，名为 PlayAudio,包名为：com.example.playaudio。

第二步：准备所需资源文件

首先将准备好的图片文件 start.png、pause.png、stop.png 复制到 drawble 文件夹中。然后创建“raw”文件夹，创建过程为：选择 res 目录，单击右键选择“New Resource Directory”，在弹出的对话框中选择 Resource type 为“raw”，单击“OK”命令按钮，在 res 目录下建成一个“raw”文件夹，我们将准备好的音频文件 dawn.mp3 复制到 raw 文件夹中。

第三步：设计用户界面

用户界面放置一个 TextView 控件显示音频文件播放状态,初始值为“音乐文件准备好了……”；放置三个命令按钮分别表示“播放”“暂停”和“停止”，对应的单击事件通过设置 android:onClick 属性实现，分别为“playAudio”“pauseAudio”和“stopAudio”，布局文件 activty_main 的代码如下所示。

```
<?xml version="1.0" encoding="utf-8"?>
<androidx.constraintlayout.widget.ConstraintLayout  xmlns:android="http://schemas.android.com/apk/
res/android"
    xmlns:app="http://schemas.android.com/apk/res-auto"
    xmlns:tools="http://schemas.android.com/tools"
    android:layout_width="match_parent"
    android:layout_height="match_parent"
    tools:context=".MainActivity">
    <TextView
        android:id="@+id/textView"
        android:layout_width="411dp"
        android:layout_height="wrap_content"
        android:layout_marginTop="50dp"
        android:text="音乐文件准备好了……"
        android:textSize="20sp"
        app:layout_constraintTop_toTopOf="parent" />

    <Button
        android:id="@+id/button"
        android:layout_width="80dp"
        android:layout_height="80dp"
        android:layout_marginStart="16dp"
        android:layout_marginTop="50dp"
        android:background="@drawable/start"
          android:onClick="playAudio"
          app:layout_constraintStart_toStartOf="parent"
        app:layout_constraintTop_toBottomOf="@+id/textView" />
```

```
    <Button
        android:id="@+id/button2"
        android:layout_width="80dp"
        android:layout_height="80dp"
        android:layout_marginStart="69dp"
        android:background="@drawable/pause"
        android:onClick="pauseAudio"
        app:layout_constraintBottom_toBottomOf="@+id/button"
        app:layout_constraintStart_toEndOf="@+id/button" />

    <Button
        android:id="@+id/button3"
        android:layout_width="80dp"
        android:layout_height="80dp"
        android:layout_marginStart="70dp"
        android:layout_marginEnd="16dp"
        android:background="@drawable/stop"
        android:onClick="stopAudio"
        app:layout_constraintBottom_toBottomOf="@+id/button2"
        app:layout_constraintEnd_toEndOf="parent"
        app:layout_constraintStart_toEndOf="@+id/button2" />

</androidx.constraintlayout.widget.ConstraintLayout>
```

第四步：实现音乐播放

音频播放使用 MediaPlayer 类来实现，在 MainActivity 中实现音频文件播放的代码如下所示：

```
public class MainActivity extends AppCompatActivity {
    private MediaPlayer mediaPlayer;
    private TextView tv;
    @Override
    protected void onCreate(Bundle savedInstanceState) {
        super.onCreate(savedInstanceState);
        setContentView(R.layout.activity_main);
        tv = findViewById(R.id.textView);
        mediaPlayer = MediaPlayer.create(this, R.raw.dawn);
    }
    public void playAudio(View view) {
        if(!mediaPlayer.isPlaying()){
            mediaPlayer.start();
            tv.setText("正在播放音频文件……");
        }
    }

    public void pauseAudio(View view) {
        if(mediaPlayer.isPlaying()){
            mediaPlayer.pause();
            tv.setText("暂停播放音频文件……");
        }
    }
```

```
    public void stopAudio(View view) {
        if(mediaPlayer.isPlaying()){
            mediaPlayer.stop();
            tv.setText("停止播放音频文件……");
            mediaPlayer=MediaPlayer.create(this,R.raw.dawn);
        }
    }

    @Override
    protected void onDestroy() {
        super.onDestroy();
        if(mediaPlayer!=null){
            mediaPlayer.stop();
            mediaPlayer.release();
        }
    }
}
```

从上面的代码可以看到，在 onCreate()方法中对 TextView 控件 tv 进行初始化，使用 MediaPlayer.create()方法获取 MediaPlayer 实例，这里 create()方法是 MediaPlayer 的静态方法，有两个参数，其格式如下:

public static MediaPlayer create(Context context, int resid)

第一个参数为 Context 对象，第二个参数为要播放的文件，本应用中为 raw 文件夹中 dawn.mp3。

在实现播放、暂停和停止的操作中使用了 MediaPlayer 类的 start()方法开始播放,通过 isPlaying()方法判定音频文件是否在正在播放，如果正在播放可以使用 pause()方法或者 stop()方法暂停或停止播放。当 Activity 销毁时判断如果 mediaPlayer 对象不为空，先停止播放，然后调用 release()方法释放与 MediaPlayer 对象相关的资源。

补充 在这个应用中我们播放的是应用自带的音频文件，如果要播放网络音频文件，可以使用 setDataSource()方法，此方法的作用为设置多媒体数据来源。如:

mediaPlayer.setDataSource("http://www.XXX.com/aa.mp3");

当然如果要进行网络访问，必须在清单文件中添加网络访问权限。

注意: stop 状态不能直接调用 start()方法，要回到 prepared 状态[使用 prepare()或 prepareAsyn()]，才能调用 start()方法。

第五步：运行程序

运行程序，显示如图 11-1 所示界面。单击界面上的播放命令按钮，可以听到音频文件开始播放，并且界面上的 TextView 控件中的内容发生变为“正在播放音频文件……”。单击暂停命令按钮可以使音乐播放停止，TextView 控件中的内容发生变化。

11.1.2 播放视频文件

视频播放与音频播放的区别是需要提供图像输出的界面，在 Android 中提供了 VideoView 可以用来显示视频播放时的图像输出。

VideoView 控件可以作为视频播放的容器，而且它也提供了视频播放的方法。VideoView 播放视频用到的主要方法有:

setVideoURI()方法：加载要播放的视频。

start()方法：开始或者继续播放视频。

pause()方法：暂停播放视频。

resume()方法：将视频设置为从头开始播放。

stopPlayback()方法：停止播放视频。

VideoView 控件还可以与 MediaController 关联，通过 MediaController 的实例来控制视频的播放。MediaController 是 Android 封装的辅助控制器，带有暂停、播放、停止、进度条等控件。通过 VideoView 和 MediaController 可以很轻松地实现视频播放、停止、快进、快退等功能。

以图 11-2 所示的运行效果为例，学习使用 VideoView 控件实现视频播放。

分析　从图 11-2 可以看到，界面有一个视频播放容器，三个用来控制播放状态的命令按钮，而且在屏幕的底部有视频快进、后退、播放/暂停的控制选项。这里视频播放的容器使用 VideoView 控件，界面上的三个命令按钮中的控制视频的播放状态是由 VideoView 控件对应的相关方法来实现的，底部视频播放控制选项由 MediaController 类与 VideoView 控件绑定来实现。

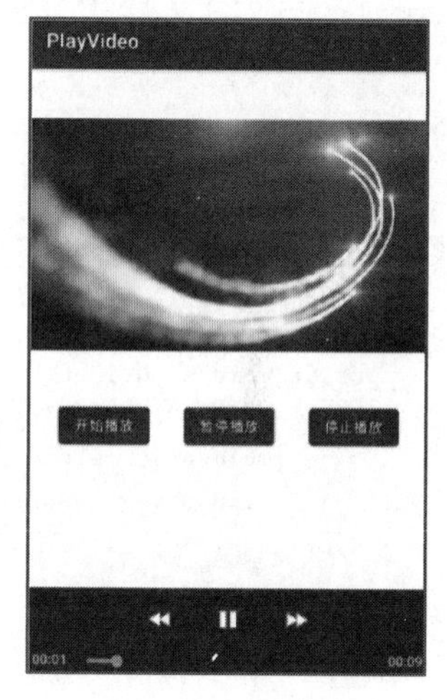

图 11-2　播放视频

使用 VideoView 播放视频的实现过程如下。

第一步：创建项目

新建一个项目，名为 PlayVideo，所在包为 com.example.playvideo。

第二步：准备视频资源

在当前包的 res 目录下，创建 raw 文件夹，将准备好的视频文件 video1.mp4 存入文件夹中。

第三步：设计用户界面

用户界面上放置一个 VideoView 控件，添加三个命令按钮分别为“开始播放”“暂停播放”和“停止播放”，对应 android:onClick 属性分别设置为“playVideo”“pauseVideo”和“stopVideo”。

第四步：实现视频播放

在 MianActivty 中实现视频播放，核心代码如下：

```
public class MainActivity extends AppCompatActivity {
    private VideoView videoView;

    @Override
    protected void onCreate(Bundle savedInstanceState) {
        super.onCreate(savedInstanceState);
        setContentView(R.layout.activity_main);
        videoView = findViewById(R.id.videoView);
        Uri mUri = Uri.parse("android.resource://" + getPackageName() + "/" + R.raw.video01);
        videoView.setVideoURI(mUri);
        MediaController mediaController = new MediaController(this);
```

```
        //VideoView 与MediaController 建立关联
        videoView.setMediaController(mediaController);
    }

    public void playVideo(View view) {
        if (!videoView.isPlaying()) {
            videoView.start();
        }
    }

    public void pauseVideo(View view) {
        if (videoView.isPlaying()) {
            videoView.pause();
        }
    }
    public void stopVideo(View view) {
        if (videoView.isPlaying()) {
            videoView.stopPlayback();//停止播放视频
            videoView.resume();//重新播放视频
        }
    }
}
```

从上面的代码中可以看出，在 MainActivity 的 onCreate()方法中初始化了 VideoView 控件，定义 Uri 对象获取视频文件，使用 VideoView 的 setVideoURI()方法加载要播放的视频。创建一个 MediaController 对象，再使用 setMediaController()方法将 VideoView 控件与 MediaController 对象关联。

然后在 playVideo()方法中使用 isPlaying()方法判定如果视频没有播放，则调用 start()方法播放视频；在 pauseVideo()方法中判定如果视频正在播放，则调用 pause()方法暂停；在 stopVideo()方法中判定如果视频在播放，则调用 stopPlayback()方法停止播放，使用 resume()方法设置重新播放。

第五步：运行程序

运行程序，效果如图 11-3 所示，开始运行时视频没有播放，将鼠标移动到 VideoView 控件单击控件，出现屏幕下方的视频播放控制界面，选择播放视频后视频开始播放。也可以通过“开始播放”“暂停播放”和“停止播放”命令按钮来控制视频的播放。

图 11-3　播放视频运行效果

11.2　录制音频和视频

11.2.1　录制音频

Android 多媒体框架支持捕获和编码各种常见的音频和视频格式。如果设备硬件支持，可以使用 MediaRecorder 从设备麦克风捕获音频、保存音频并使用 MediaPlayer 进行播放的应用。

以模拟手机录音为例，实现如图 11-4 所示运行界面功能。

分析　实现音频录制需要使用 MediaRecorder 类，首先创建并初始化 MediaRecorder 实例，然后使用 MediaRecorder 类提供的方法实现音频录制，用到的方法主要有以下几个：

setAudioSource()方法：设置音频源，参数为 MediaRecorder.AudioSource.MIC 说明声音源为麦克风。

setOutputFormat()方法：设置输出文件格式，输出格式有视频/音频，也有支持音频的，参数设置为 MediaRecorder.OutputFormat.MPEG_4 则设置输出文件格式为音频文件。

setOutputFile()方法：设置输出文件名，必须指定代表实际文件的文件描述符。

setAudioEncoder()方法：用来设置音频编码器。

prepare()方法：完成初始化，准备录音。

start()方法和 stop()方法：用来启动和停止录音功能。

release()方法：用来释放其资源，使用完 MediaRecorder 实例后，应尽快释放其资源。

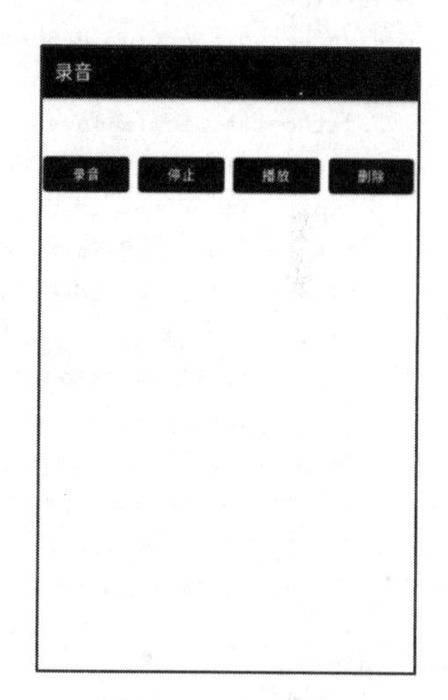

图 11-4　音频录制

实现录制音频的过程如下。

第一步：创建项目

新建一个项目，名为 Media_Recorder,包名为 com.example.media_recorder。

第二步：设计用户界面

在清单文件 AndroidManifest 中设置此应用的标题为“录音”，在用户界面中添加四个命令按钮“录音”“停止”“播放”和“删除”，界面运行效果如图 11-4 所示。

第三步：请求录制音频的权限

录制音频需要访问设备的音频输入。必须在应用的清单文件中添加以下权限标记：

```
<uses-permission android:name="android.permission.RECORD_AUDIO"/>
```

RECORD_AUDIO 是一项“危险”的权限，因为它可能会对用户的隐私构成威胁。从 Android6.0（API 级别 23）开始，使用危险权限的应用在运行时必须请求用户的批准，也就是还需要在程序中动态申请此权限。

第四步：实现音频录制以及播放

在 MainActivity 中实现音频的录制。

```
public class MainActivity extends AppCompatActivity implements View.OnClickListener {

    private Button btn1,btn2,btn3,btn4;
    private File recordAudioFile;
    private MediaRecorder mediaRecorder;
    private MediaPlayer mediaPlayer;

    @Override
    protected void onCreate(Bundle savedInstanceState) {
        super.onCreate(savedInstanceState);
        setContentView(R.layout.activity_main);
        btn1 = findViewById(R.id.button);
        btn2 = findViewById(R.id.button2);
        btn3 = findViewById(R.id.button3);
        btn4 = findViewById(R.id.button4);
        btn1.setOnClickListener(this);
        btn2.setOnClickListener(this);
        btn3.setOnClickListener(this);
        btn4.setOnClickListener(this);
        if (Build.VERSION.SDK_INT >= Build.VERSION_CODES.M) {
            String[] permissions = new String[]{Manifest.permission.RECORD_AUDIO};
            for (String permission : permissions) {
                if (ContextCompat.checkSelfPermission(this, permission) !=PackageManager.PERMISSION_
GRANTED) {
                    ActivityCompat.requestPermissions(this, permissions, 200);
                }
            }
        }
    }
    @Override
    public void onClick(View v) {
        switch (v.getId()){
            case R.id.button://录制音频
                try {
                    recordAudioFile = File.createTempFile("record", ".amr");
                } catch (IOException e) {
                    e.printStackTrace();
                }
                mediaRecorder = new MediaRecorder();
                mediaRecorder.setAudioSource(MediaRecorder.AudioSource.MIC);
                mediaRecorder.setOutputFormat(MediaRecorder.OutputFormat.MPEG_4);
                mediaRecorder.setAudioEncoder(MediaRecorder.AudioEncoder.DEFAULT);
                mediaRecorder.setOutputFile(recordAudioFile.getAbsolutePath());
                try {
                    mediaRecorder.prepare();
                } catch (IOException e) {
                    e.printStackTrace();
                }
                mediaRecorder.start();
                Toast.makeText(MainActivity.this,"录制开始",Toast.LENGTH_LONG).show();;
                break;
            case R.id.button2://停止录制
                if(mediaRecorder!=null){
```

```
                mediaRecorder.stop();
                mediaRecorder.release();
                mediaRecorder=null;
                Toast.makeText(MainActivity.this,"停止录制",Toast.LENGTH_LONG).show();
            }

            break;
        case R.id.button3://播放音频
            mediaPlayer=new MediaPlayer();
            try {
                mediaPlayer.setDataSource(recordAudioFile.getAbsolutePath());
                mediaPlayer.prepare();
            } catch (IOException e) {
                e.printStackTrace();
            }
            mediaPlayer.start();
            Toast.makeText(MainActivity.this,"正在播放录音",Toast.LENGTH_LONG).show();
            break;
        case R.id.button4://删除音频
            recordAudioFile.delete();
            Toast.makeText(MainActivity.this,"删除录音",Toast.LENGTH_LONG).show();
            break;
    }
  }
}
```

从上面的代码可以看到，在 MainActivity 的 onCreate()方法中对界面中的命令按钮进行初始化,并监听其单击事件，动态申请录制音频的权限。

在 onClick()方法中，根据所选择的命令按钮进行音频的录制、停止、播放与删除。

“录制”命令按钮：使用 File.createTempFile（）方法创建一个临时文件，接着创建 MediaRecorder 类的实例，使用 setAudioSource()方法设置音频源为麦克风，使用 setOutputFormat()方法设置输出文件格式 MPEG_4，使用 setAudioEncoder()方法设置音频编码器为 DEFAULT 默认的，调用 setOutputFile()方法设置输出文件名为刚创建的文件，使用 prepare()方法完成初始化，最后调用 start()方法启动录音功能。

“停止”命令按钮：先判定 MediaRecorder 实例，当其不为 null 时，调用 stop()方法停止录制，调用 release()方法释放 MediaRecorder 实例相关资源。

“播放”命令按钮：使用 MediaPlayer 来实现音频的播放，调用 MediaPlayer 的 setDataSource()方法设置播放的音频文件，使用 start()方法开始播放音频文件。

“删除”命令按钮：调用 delete()方法删除音频文件，并给出提示。

说明：对于音频的录制只是简单学习了 MediaRecorder 的使用，本应用的一些细节没有处理，例如播放或删除时音频文件是否存在，等等，读者可自行扩展完善。

第五步：运行程序

运行程序，弹出“是否允许‘录音’录制音频？”对话框，选择“允许”。出现如图 11-4 所示的运行界面，选择“录音”开始音频录制，点击“停止”停止音频录制，选择“播放”播放刚刚录制的音频，单击“删除”则删除了刚录制的音频文件。

11.2.2 视频的录制

图 11-5 录制视频运行效果

在很多应用中都会用到视频的录制，在很多 Android 设备中已经安装了一个用于录制视频的相机应用，本节内容通过一个案例学习 Android 应用中视频的录制，实现如图 11-5 所示的运行效果。

分析 实现视频的录制，主要解决两个问题，一个是如何实现录制视频，另一个是播放所录制的视频。对于第一个问题使用 startActivityForResult()方法启动系统摄像头录制视频，第二个问题把 onActivityResult()方法接收返回的视频用 VideoView 控件播放视频。具体视实现过程如下所示：

第一步：创建项目

新建一个项目，名为 Video_Recording,包名为 com. example.video_recording。

第二步：设计用户界面

在界面添加一个命令按钮，用于发送“录制视频”的命令，添加一个 VideoView 控件用来播放视频。

第三步：申请相机的权限

在清单文件 AndroidManifest 中添加申请相机的权限，代码如下：

```
<uses-permission android:name="android.permission.CAMERA"/>
```

第四步：实现录像功能

要实现录制视频后播放，需要调用相机应用录像，并将视频存在存储设备中，在 VideoView 控件中进行播放，这里就要用到 startActivityForResult()方法，然后重写 onActivityResult()方法播放视频。MainActivity 中的代码如下：

```
public class MainActivity extends AppCompatActivity {

    private VideoView videoView;
    private Button button;

    @Override
    protected void onCreate(Bundle savedInstanceState) {
        super.onCreate(savedInstanceState);
        setContentView(R.layout.activity_main);
        videoView = findViewById(R.id.videoView);
        button = findViewById(R.id.button);
```

```
        if (Build.VERSION.SDK_INT >= Build.VERSION_CODES.M) {
            String[] permissions = new String[]{Manifest.permission.CAMERA};
            for (String permission : permissions) {
                if  (ContextCompat.checkSelfPermission(MainActivity.this,  permission)  !=  Package-
Manager.PERMISSION_GRANTED) {
                    ActivityCompat.requestPermissions(MainActivity.this, permissions, 200);
                }
            }
        }
        button.setOnClickListener(new View.OnClickListener() {
            @Override
            public void onClick(View v) {
                Intent intent = new Intent(MediaStore.ACTION_VIDEO_CAPTURE);
                startActivityForResult(intent, 1);
            }
        });
    }

    @Override
    protected void onActivityResult(int requestCode, int resultCode, @Nullable Intent data) {
        super.onActivityResult(requestCode, resultCode, data);
        if(requestCode==1&&data!=null){
            Uri videoUri = data.getData();
            videoView.setVideoURI(videoUri);
            videoView.setMediaController(new MediaController(this));
            videoView.start();
        }
    }
}
```

在 MainActivity 的 onCreate()方法中对 Button 控件和 VideoView 控件进行初始化，然后动态申请使用相机的权限；在命令按钮的单击事件中定义 Intent 对象，其中 MediaStore.ACTION_VIDEO_CAPTURE 为打开相机录像的系统常量，调用 startActivityForResult()方法启动 Intent 对象，设置请求码为 1。

重写 onActivityResult()方法，当请求码为 1 且传回的 Intent 对象 data 不为空时，返回视频在存储设备中的 Uri，使用 setMediaController()方法绑定 VideoView 对象与 MediaController()对象，调用 VideoView 的 start()方法播放视频。

第五步：运行程序

运行程序，弹出“运行 Video_Recording 拍摄照片和录制视频吗？”，选择“允许”，出现如图 11-6 所示对话框，单击“录制视频”命令按钮，进入视频录制模式，视频录制完成后，返回并开始播放视频，运行效果如图 11-5 所示。

图 11-6　录制视频运行界面

11.3　仿微信头像设置

在很多应用中常会用到用户头像的设置，常见的设置用户头像可以通过拍照或者是从相册中选择照片来实现，本节将实现简单头像设置功能，运行效果如图 11-7 所示。

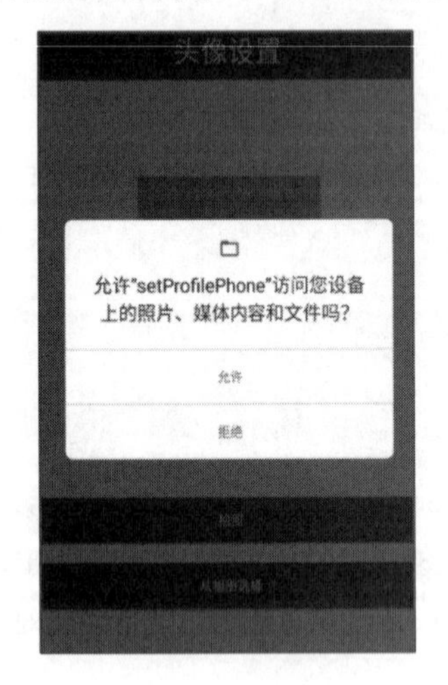

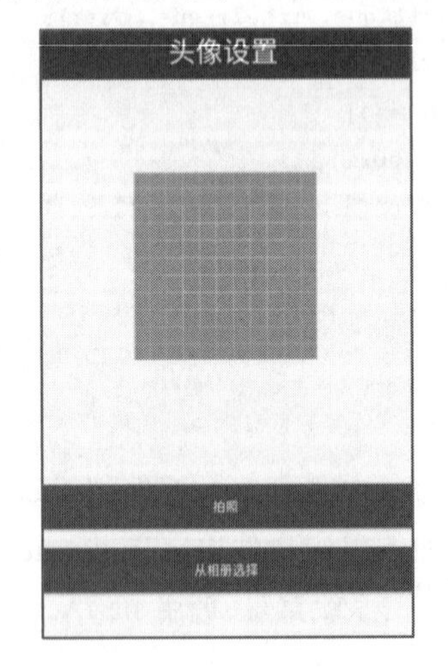

图 11-7　设置头像界面设计

分析　用户头像的设置主要通过两种途径，一种是在相册中选择已有的图片，另一种是使用相机拍照，将所拍照片设置为头像。也就是访问系统相册和调用系统相机的过程，具体实现过程如下。

第一步：创建项目

新建一个项目，名为 setProfilePhone,包名为 com.example.setprofilephone。

第二步：用户界面设计

用户界面设置比较简单，添加一个 TextView 控件显示“头像设置”文本，添加一个 ImageView

控件用来显示头像，添加两个命令按钮分别为“拍照”和“从相册选择”。布局如图 11-8 所示。

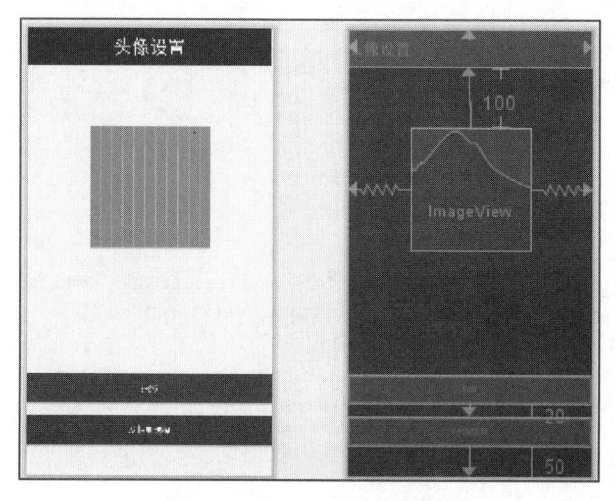

图 11-8　设置头像界面设计

第三步：在清单文件中申请权限

在 AndroidManifest 文件中添加权限：

```
<uses-permission android:name="android.permission.READ_EXTERNAL_STORAGE"/>
```

第四步：实现头像的设置

在 MainActivity 中实现通过拍照和相册选择设置用户头像，代码如下所示。

```
public class MainActivity extends AppCompatActivity {
    ImageView iv;
    Button btnTakePhoto, btnAlbum;
    Bitmap bitmap;
    private Uri imageUri;
    private String imagePath;

    @Override
    protected void onCreate(Bundle savedInstanceState) {
        super.onCreate(savedInstanceState);
        setContentView(R.layout.activity_main);
        getSupportActionBar().hide();
        iv = findViewById(R.id.imageView);
        btnTakePhoto = findViewById(R.id.button);
        btnAlbum = findViewById(R.id.button2);
        String[] permissions = {Manifest.permission.READ_EXTERNAL_STORAGE};
        ActivityCompat.requestPermissions(MainActivity.this, permissions, 101);
        //“拍照”
         btnTakePhoto.setOnClickListener(new View.OnClickListener() {
            @Override
            public void onClick(View v) {
                Intent intent = new Intent(MediaStore.ACTION_IMAGE_CAPTURE);
                startActivityForResult(intent, 1);
            }
        });
        //“从相册选择”
```

```
        btnAlbum.setOnClickListener(new View.OnClickListener() {
          @Override
          public void onClick(View v) {
              Intent  intent  =  new  Intent(Intent.ACTION_PICK,  MediaStore.Images.Media.EXTERNAL_
CONTENT_URI);
              startActivityForResult(intent, 2);
          }
      });
  }

  @Override
  protected void onActivityResult(int requestCode, int resultCode, @Nullable Intent data) {
      super.onActivityResult(requestCode, resultCode, data);
      if (requestCode == 1 && data != null) {// “拍照”
           Bundle bundle = data.getExtras();
          bitmap = (Bitmap) bundle.get("data");
          Bitmap resized = Bitmap.createScaledBitmap(bitmap, 300, 300, true);   //设置图片大小
            iv.setImageBitmap(resized);

      }else if (requestCode == 2 && data != null) {// “从相册选择”
           imageUri = data.getData();  //获取系统返回的照片的Uri
           ContentResolver contentResolver = getContentResolver();
          Cursor cursor = contentResolver.query(imageUri, null, null, null, null);
          if (cursor != null) {
              if (cursor.moveToFirst()) {
                  imagePath = cursor.getString(cursor.getColumnIndex(MediaStore.Images.Media.DATA));
                  bitmap = BitmapFactory.decodeFile(imagePath);
                  if (bitmap != null && iv != null) {
                      Bitmap resized = Bitmap.createScaledBitmap(bitmap, 300, 300, true);   //设置图
片大小
                          iv.setImageBitmap(resized);
                  }
              }
          }
          cursor.close();
      }
  }
}
```

在 MainActivity 的 onCreate()方法中首先初始化 ImageView 和 Button 控件，申请 READ_EXTERNAL_STORAGE 访问权限，接着实现“拍照”和“从相册选择”功能，最后在 onActivityResult()方法中设置用户头像。

“拍照”命令按钮的功能实现：创建 Intent 对象，使用 MediaStore.ACTION_IMAGE_CAPTURE 启动相机、拍摄照片并返回照片信息，调用 startActivityForResult()方法启动此意图，设置请求码 requestCode 为 1。

“从相册选择”命令按钮的功能实现：创建 Intent 对象，跳转到手机系统相册中，startActivityForResult()方法的请求码 requestCode 为 2。

重写 onActivityResult()方法，当请求码为 1 且传回来的 Intent 对象不为空时，获取传递回来的数据创建 Bitmap 对象，并且设置图像文件大小，将其显示在 ImageView 对象中，实现通过拍照设置头像；当请求码为 2 且传递过来的 Intent 对象不为空时，使用 getContentResolver()方法获取 ContentResolver 对象，从系统返回的照片的 Uri 中查询指定 Uri 对象对应的照片，将其显示在 ImageView 对象中。

第五步：运行程序

运行程序，效果如图 11-7 所示，先询问：允许“setProfilePhone”访问您设备上的照片、媒体内容和文件吗？单击“允许”后，选择界面上的“拍照”则开启系统相机进行拍照，拍照确认后，会将所拍照片显示在界面上的 ImageView 控件中，选择界面上的“从相册选择”会打开系统相册，从相册中选中照片后，所选照片会显示在 ImageView 控件中，实现用户头像的设置。

【职业素养拓展11】

职业素养拓展内容详见文件“职业素养拓展”。

扫描二维码查看【职业素养拓展 11】

本章小结

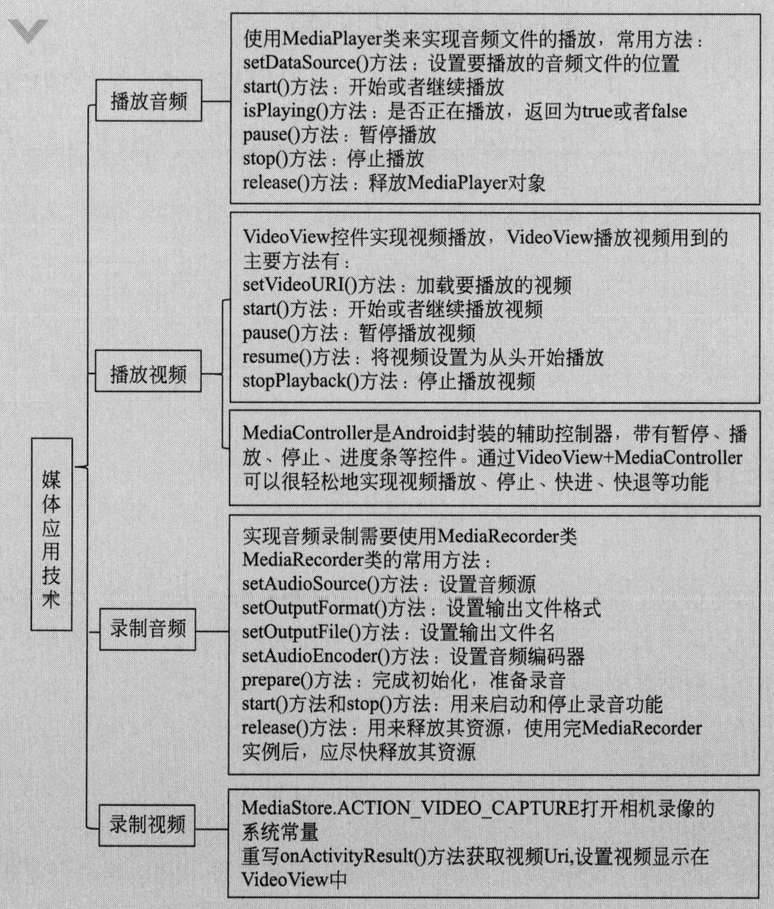

媒体应用技术
播放音频
使用MediaPlayer类来实现音频文件的播放，常用方法：
setDataSource()方法：设置要播放的音频文件的位置
start()方法：开始或者继续播放
isPlaying()方法：是否正在播放，返回为true或者false
pause()方法：暂停播放
stop()方法：停止播放
release()方法：释放MediaPlayer对象
播放视频
VideoView控件实现视频播放，VideoView播放视频用到的主要方法有：
setVideoURI()方法：加载要播放的视频
start()方法：开始或者继续播放视频
pause()方法：暂停播放视频
resume()方法：将视频设置为从头开始播放
stopPlayback()方法：停止播放视频
MediaController是Android封装的辅助控制器，带有暂停、播放、停止、进度条等控件。通过VideoView+MediaController可以很轻松地实现视频播放、停止、快进、快退等功能
录制音频
实现音频录制需要使用MediaRecorder类
MediaRecorder类的常用方法：
setAudioSource()方法：设置音频源
setOutputFormat()方法：设置输出文件格式
setOutputFile()方法：设置输出文件名
setAudioEncoder()方法：设置音频编码器
prepare()方法：完成初始化，准备录音
start()方法和stop()方法：用来启动和停止录音功能
release()方法：用来释放其资源，使用完MediaRecorder实例后，应尽快释放其资源
录制视频
MediaStore.ACTION_VIDEO_CAPTURE打开相机录像的系统常量
重写onActivityResult()方法获取视频Uri,设置视频显示在VideoView中

第12章

综合项目——智慧党建

本章主要是对 Android 开发基础知识的综合应用，通过智慧党建项目的实现，学习我们平常看到比较多的移动应用项目框架的设计与实现。项目应用知识点主要包括 Android 基础控件、常用资源、数据传递、数据库应用、多媒体技术和网络数据访问与应用等。

学习目标：

1. 了解智慧党建项目的功能与模块结构；
2. 掌握智慧党建首页的设计与实现；
3. 掌握 RecyclerView 控件及其适配器的使用；
4. 掌握网络资源的访问与应用；
5. 掌握数据存储与访问；
6. 学会调用系统资源；
7. 学会视频、音频的播放等。

12.1 项目概述

（1）项目概述

智慧党建是运用信息化新技术，整合各方资源，更有效地加强组织管理，提高服务群众水平，巩固党的执政基础的新平台、新模式、新形态。通过智慧党建系统建设，主要解决党建宣传、学习、管理、资源等方面的基础问题。

（2）开发环境

操作系统：Windows 系统。

开发工具：JDK8，Android studio v2020.3.1，Android 10.0。

（3）项目主要功能

智慧党建主要包括以下功能模块：党建活动轮播图显示，展示党建要闻、党建活动，等等；党建学习内容分类、学习内容展示，微课视频播放，等等；随手拍照上传的功能，可以将身边发现的先进事迹、存在的问题上传系统；建言献策功能可以提交问题以及建议给上级，查看留言以及对个人的信息进行维护，等等。

（4）项目功能概述

根据需要实现的主要功能，设计智慧党建的执行流程如图 12-1 所示。

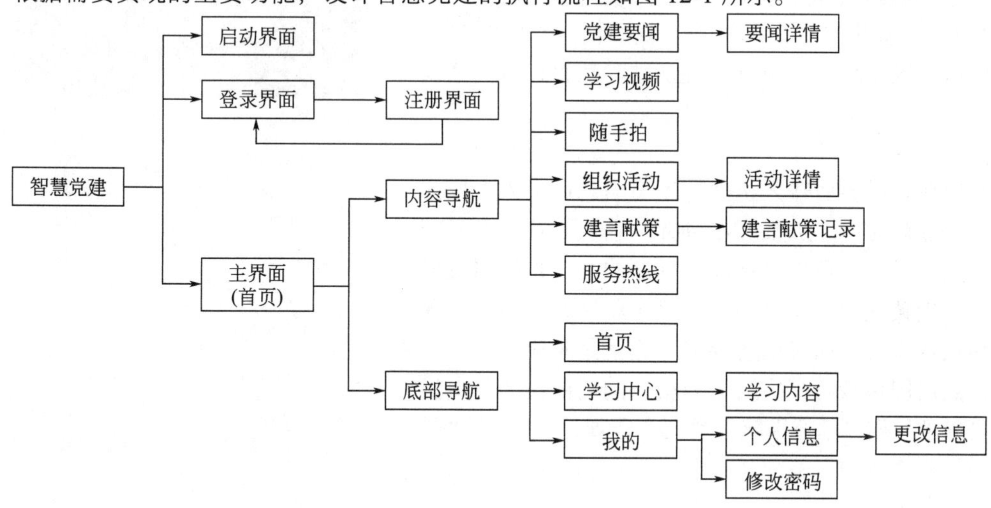

图 12-1　智慧党建执行流程图

12.2　项目运行效果展示

（1）启动界面

启动“智慧党建”打开启动界面，效果如图 12-2 所示，3s 后跳转进入登录界面。

（2）用户登录

在登录界面输入正确用户账号和密码信息登录系统，如果没有正确输入信息则会给出相应的提示。输入信息正确选择记住密码复选框，下次登录时直接显示用户登录信息，登录界面如图 12-3 所示。如果用户没有注册，则必须先注册再登录。

图 12-2　启动界面效果图

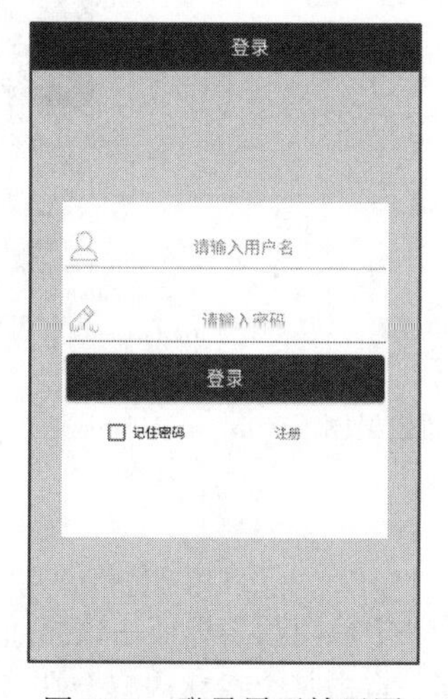

图 12-3　登录界面效果图

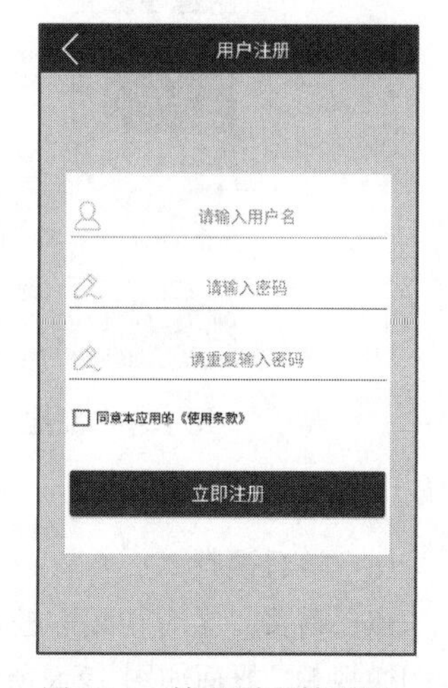

图 12-4　注册界面效果图

（3）用户注册

在登录界面选择“注册”进入注册界面，该界面要求输入用户名、密码和重复密码，如果密码和重复密码不一致，则给出提示，密码与重复密码一致，并且选中界面中的“同意本应用的《使用条约》”复选框，才可以注册。注册界面效果如图 12-4 所示。注册成功后直接返回登录界面。

（4）主界面（首页）

用户在登录界面输入正确的用户信息后进入主界面（即应用的首页），如图 12-5 所示。首页包括五个部分内容：从上至下最顶端是标题，标题的下方是轮播图显示党建的相关图片，轮播图下方是滚动字幕显示当前天气，滚动字幕的下面是内容导航，也就是本应用的主要功能，点击图标或者文字会进入相应的功能页。首页最下面是底部导航，底部导航包括三个选项：“首页”“学习中心”和“我的”。

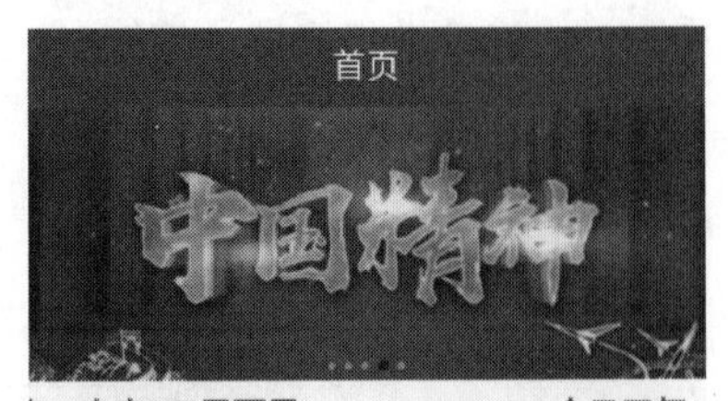

图 12-5　首页效果图

（5）党建要闻

点击首页“党建要闻”进入“党建要闻”界面，效果如图 12-6 所示，点击党建要闻某个列表项，进入党建要闻详情界面显示相关内容的详细信息，效果如图 12-7 所示。

图 12-6　党建要闻效果图

图 12-7　党建要闻详情效果图

（6）学习视频

点击首页“学习视频”进入“学习视频”界面，该界面显示微课视频、榜样精神以及专业课程的微视频，界面如图 12-8 所示，点击某个具体内容会在当前界面弹出视频播放对话框，播放对应的视频内容，效果如图 12-9 所示。

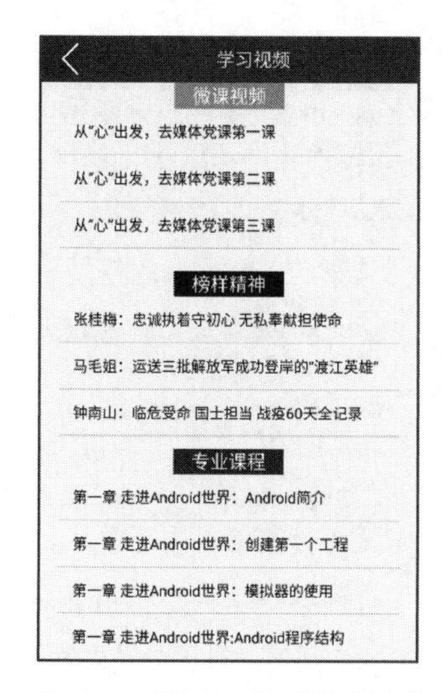

图 12-8 学习视频界面效果图

图 12-9 学习视频界面播放视频效果图

（7）随手拍

点击首页“随手拍”进入“随手拍”界面，在初始界面中填入主题和内容描述，点击图片上传下的图标，在界面的底部出现请选择图片的选项列表，可以从相册选择图片，也可以通过拍照得到图片，选择好的图片将显示在图片上传文字的下方，点击发布命令按钮，在按钮的下方显示随手拍的主题、时间、内容和图片、显示效果，随手拍的添加过程如图 12-10 所示，可以发布和显示多个随手拍内容。

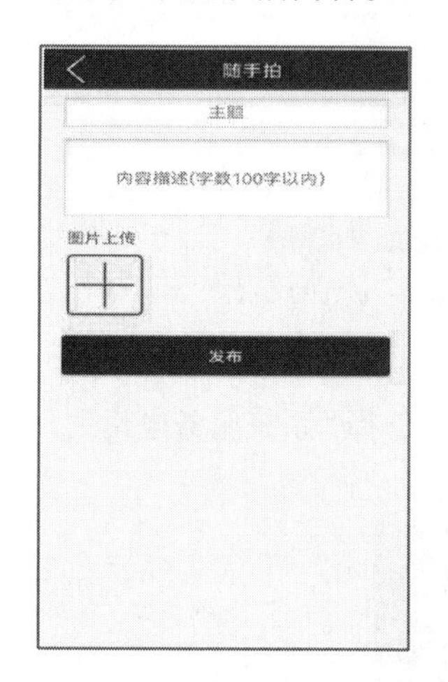

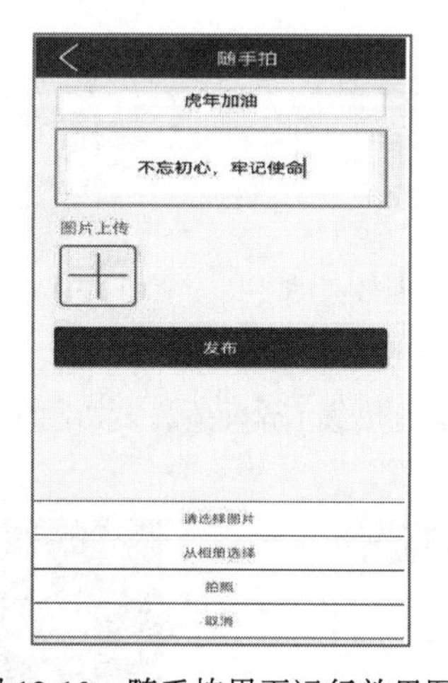

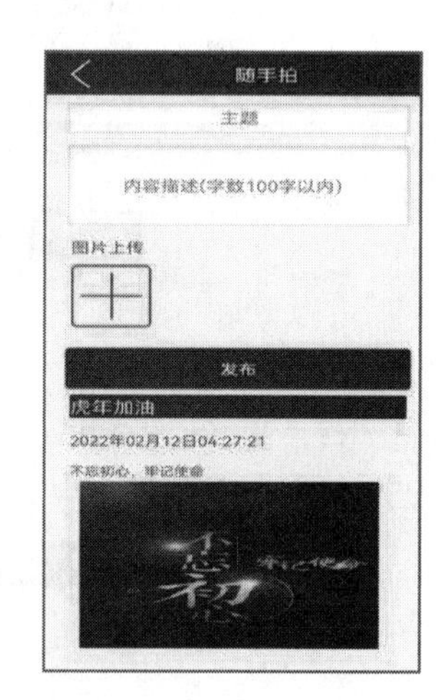

图 12-10 随手拍界面运行效果图

（8）建言献策

点击首页“建言献策”进入“建言献策”界面，效果如图 12-11 所示，填入主题、姓名、手机号、内容描述，点击“提交”命令按钮，进入“建言献策记录”界面，界面显示所有的提交了的建言献策内容，效果如图 12-12 所示。

（9）组织活动

点击首页“组织活动”界面，进入“组织活动”界面，效果如图 12-13 所示，点击每条活动，进入活动详情界面，效果如图 12-14 所示。

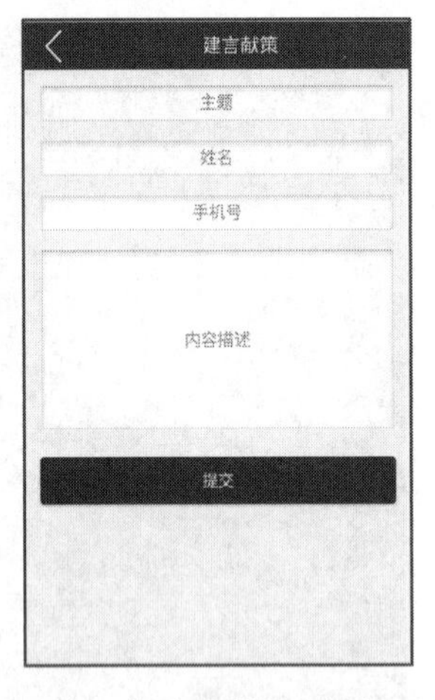

图 12-11　建言献策界面效果图

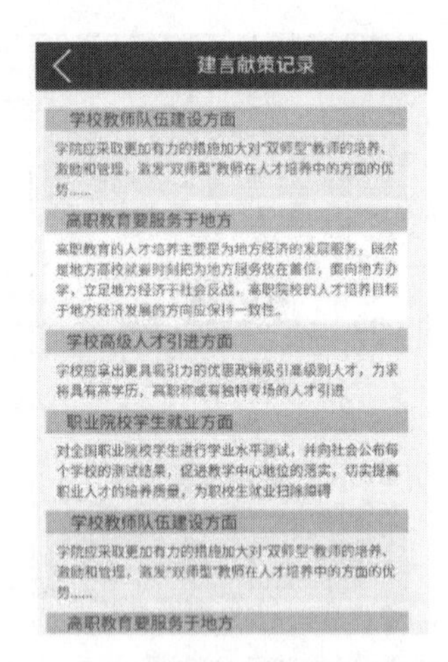

图 12-12　建言献策记录界面效果图

图 12-13　组织活动界面效果图

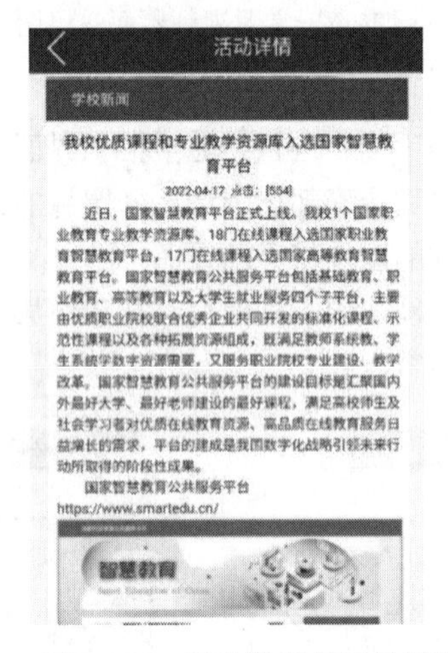

图 12-14　活动详情界面效果图

（10）服务热线

点击首页“服务热线”进入“服务热线”界面，效果如图 12-15 所示，选择“系统服务电话”，进入拨打电话界面。

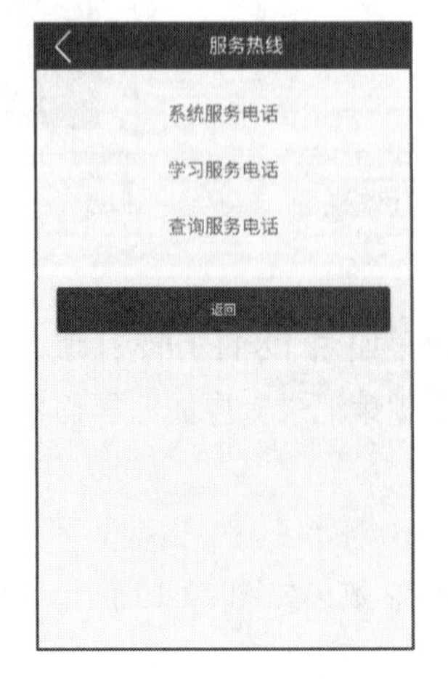

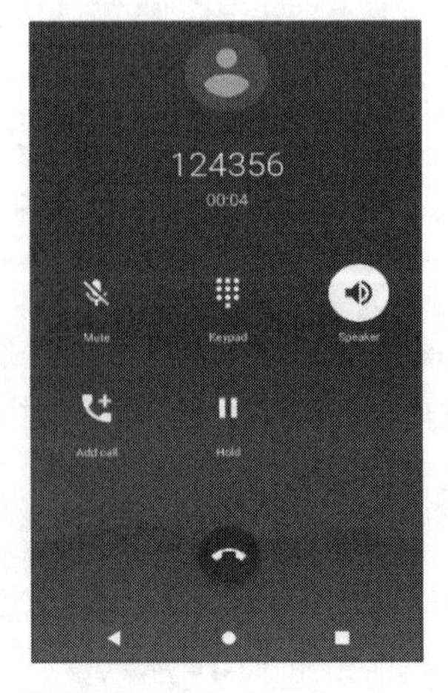

图 12-15　服务热线运行效果图

（11）学习中心

点击“首页”底部导航栏“学习中心”，界面切换到“学习中心”，在学习中心显示了“学—重要讲话”，“学—党规党章”和“学—榜样事迹”，效果如图 12-16 所示。选择某个学习内容下的列表项进入相应界面，可以选择“目录”或者“学习专栏”进行具体的学习。

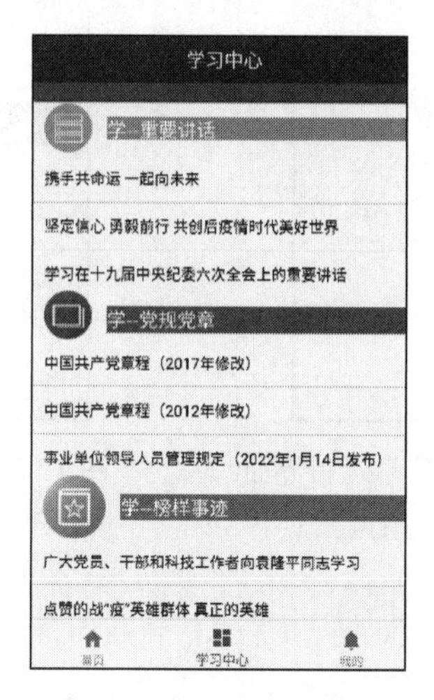

图 12-16 学习中心效果图

（12）我的

点击底部导航“我的”进入“个人中心”界面，效果如图 12-17 所示，在该界面显示当前用户的头像、用户名、“个人信息”和“修改密码”。选择“个人信息”后的图标“>”进入个人信息界面，效果如图 12-18 所示，选择“个性签名”后的“>”进入“修改个性签名”界面，效果如图 12-19 所示，修改后点击“保存”命令按钮自动返回“个人信息”界面，在个人信息界面将显示修改后的个性签名。“昵称”“性别”和“电话”与修改“个性签名”类似。在“个人信息”界面单击“修改”命令按钮个人信息保存到数据库中并返回到“个人中心”界面，在“个人中心”界面“修改密码”显示如图所示 12-20 所示的“修改密码”界面，输入对应内容，点击“确认”命令按钮，修改后的密码保存到用户信息数据中并返回到“个人中心”界面。

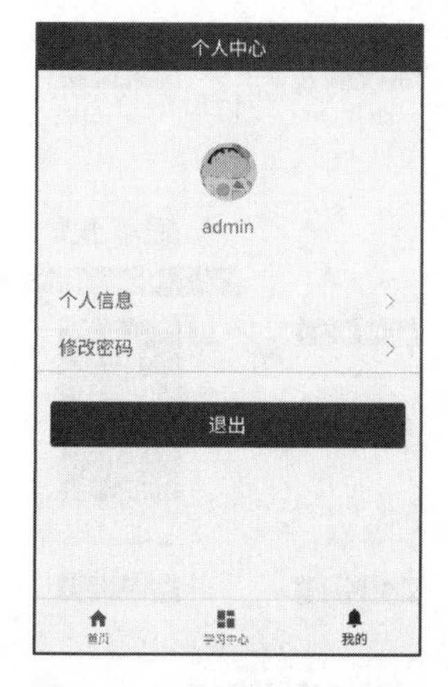

图 12-17 我的界面效果图

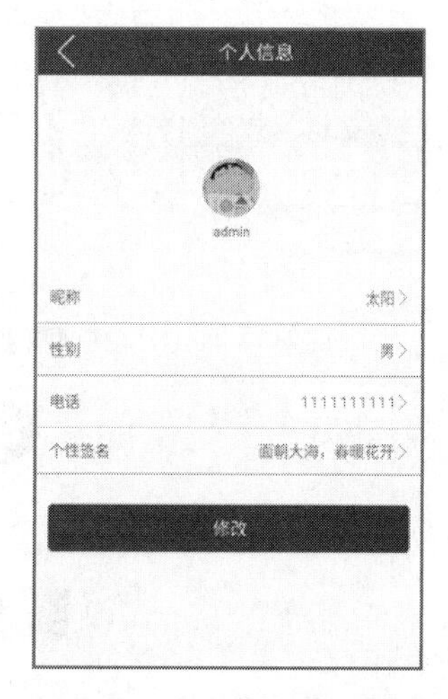

图 12-18 个人信息界面效果图

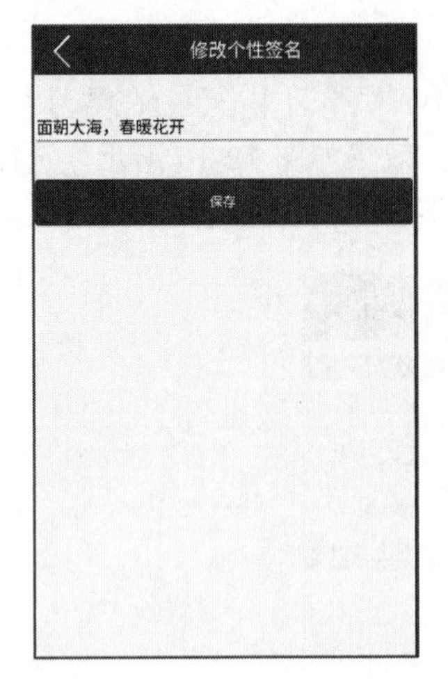

图 12-19 修改个人信息运行界面

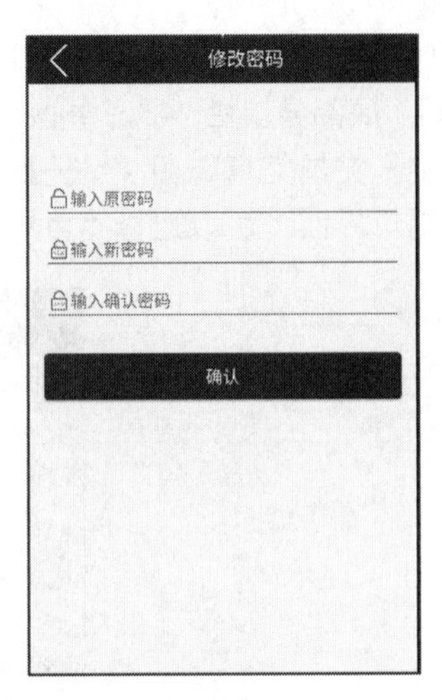

图 12-20 修改密码运行界面

12.3 项目的实现

12.3.1 创建项目

先准备项目开发需要的各类资源，如图片、图标、音频视频等，然后新建项目，设置项目公用的样式等，具体操作过程如下所示。

第一步：创建项目

选择新建项目，出现如图 12-21 所示界面，选择图中所示的“Bottom Navigation Activity”（使用 Bottom Navigation Activity 可以完成简单的底部导航栏功能），点击“Next”命令按钮，在弹出的对话框中设置项目名称为“ZhiHuiDangJian”，所在的包为 com.example.zhihuidangjian。

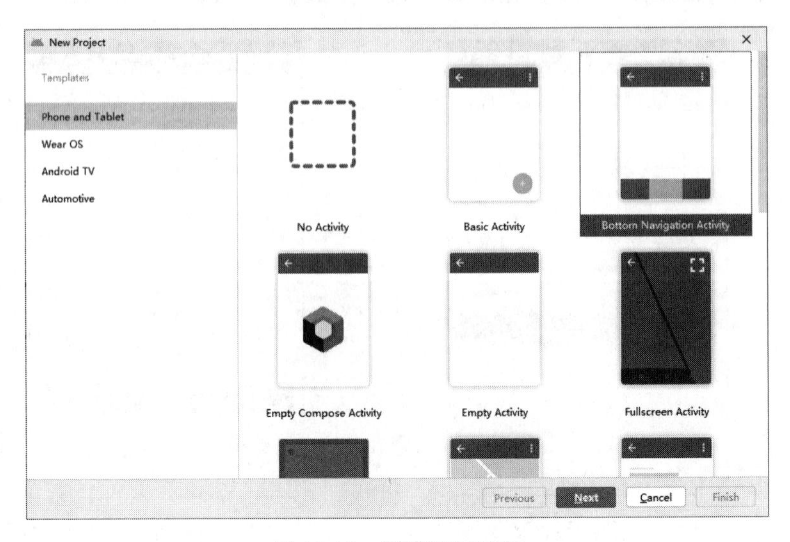

图 12-21 新建项目界面

第二步：准备所需资源

① 将图片复制到 res/drawable 文件下。

② 在 res 下创建 raw 文件夹，将项目中用的视频复制到 raw 文件夹中。

③ 创建资源文件 bg_btn。在 res/drawable 下新建一个文件 bg_btn 用来设置命令按钮的背景，选择“res”→“drawable”→“New Resource File”，弹出如图 12-22 所示的对话框，输入“File Name”和“Root element”后，点击“OK”。

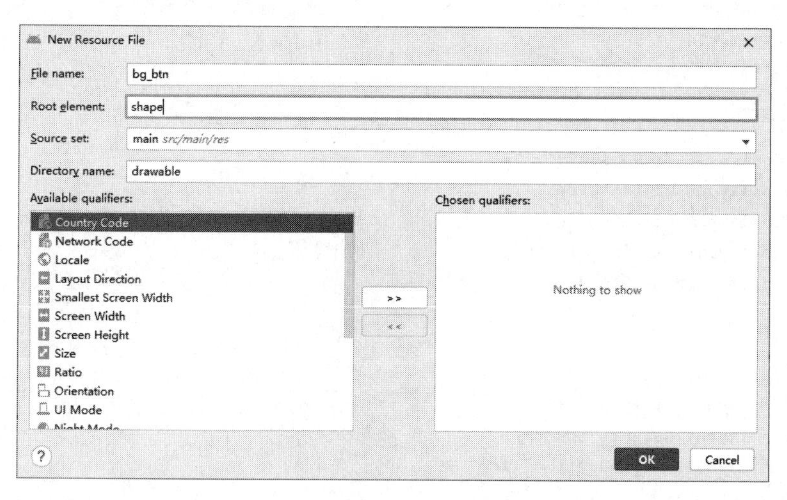

图 12-22　新建资源文件 bg_btn 对话框

完成 bg_btn.xml 文件创建，此文件主要用来设置项目中命令按钮显示为圆角，文件的 bg_btn.xml 代码如下所示：

```
<?xml version="1.0" encoding="UTF-8"?>
<shape xmlns:android="http://schemas.android.com/apk/res/android"
    android:shape="rectangle" >
    <solid android:color="#AC2C04" /> <!-- 填充的颜色 -->
    <corners
        android:bottomLeftRadius="4dp"
        android:bottomRightRadius="4dp"
        android:topLeftRadius="4dp"
        android:topRightRadius="4dp" />
    <stroke
        android:width="0px"
        android:color="#d9d9d9" />
</shape>
```

④ 创建样式 nobar。打开“res”→“themes”下的 themes.xml 文件，在该文件中增加一个自定义样式 nobar，隐藏 ActionBar。

```
<style name="nobar" parent="Theme.ZhiHuiDangJian">
    <item name="windowActionBar">false</item>
    <item name="windowNoTitle">true</item>
</style>
```

⑤ 创建标题栏布局。在智慧党建应用中，大部分界面中都有一个返回键和一个标题栏，为了

便于代码重复利用，这里将返回键和标题栏抽取出来单独放在一个布局文件中。

在 res/layout 下创建一个布局文件 activitytitle，用来设置 Activity 的标题，这个布局文件中有两个控件：一个 ImageView 控件，用来显示返回的图片；一个 TextView 控件，用来显示 Activity 的标题。activitytitle 代码如下所示：

```
<?xml version="1.0" encoding="utf-8"?>
<LinearLayout xmlns:android="http://schemas.android.com/apk/res/android"
    android:layout_width="match_parent"
    android:background="#AC2C04"
    android:layout_height="50dp">

    <ImageView
        android:id="@+id/iv_back"
        android:layout_width="35dp"
        android:layout_height="50dp"
        android:layout_marginLeft="16dp"
        android:src="@drawable/back"/>

    <TextView
        android:id="@+id/tvTitle"
        android:layout_width="match_parent"
        android:layout_height="50dp"
        android:layout_marginBottom="12dp"
        android:text="标题"
        android:gravity="center"
        android:textColor="#fff"
        android:textSize="20sp"/>
</LinearLayout>
```

第三步：修改清单文件

打开 AndroidManifest 文件，修改 application 标签下主题属性为下面所示的代码：

```
android:theme="@style/nobar"
```

这样程序在运行的时候，将不再显示系统自带的 ActionBar。

第四步：修改底部导航

修改底部导航需要修改 res/values 目录下的 strings 文件、res/menu 下的 bottom_nav_menu 文件和 res/navigation 下的 mobile_navigation 文件，并且对 MainAcitivity 中导航设置部分进行修改，具体操作如下所示。

① 打开 res/values 下的 strings.xml 文件修改字符串资源，修改后的代码如下所示：

```
<resources>
    <string name="app_name">智慧党建</string>
    <string name="title_home">首页</string>
    <string name="title_study">学习中心</string>
    <string name="title_personal">我的</string>
</resources>
```

② 打开 res/menu 下的 bottom_nav_menu.xml 文件进行修改，修改后的代码如下所示：

```
<?xml version="1.0" encoding="utf-8"?>
<menu xmlns:android="http://schemas.android.com/apk/res/android">
    <item
        android:id="@+id/navigation_home"
        android:icon="@drawable/ic_home_black_24dp"
        android:title="@string/title_home" />

    <item
        android:id="@+id/navigation_study"
        android:icon="@drawable/ic_dashboard_black_24dp"
        android:title="@string/title_study" />

    <item
        android:id="@+id/navigation_personal"
        android:icon="@drawable/ic_notifications_black_24dp"
        android:title="@string/title_personal" />

</menu>
```

③ 打开 res/navigation 下的 mobile_navigation.xml 文件进行修改，修改后的代码如下所示：

```
<?xml version="1.0" encoding="utf-8"?>
<menu xmlns:android="http://schemas.android.com/apk/res/android">

    <item
        android:id="@+id/navigation_home"
        android:icon="@drawable/ic_home_black_24dp"
        android:title="@string/title_home" />

    <item
        android:id="@+id/navigation_study"
        android:icon="@drawable/ic_dashboard_black_24dp"
        android:title="@string/title_study" />

    <item
        android:id="@+id/navigation_personal"
        android:icon="@drawable/ic_notifications_black_24dp"
        android:title="@string/title_personal" />

</menu>
```

④ 修改 MainActivity。根据 bottom_nav_menu 和 mobile_navigation 文件中 item 的 id 修改 MainActivity 中所对应的内容，修改后 MainActivity 的代码如下所示：

```
public class MainActivity extends AppCompatActivity {
    private ActivityMainBinding binding;
    @Override
    protected void onCreate(Bundle savedInstanceState) {
        super.onCreate(savedInstanceState);
        binding = ActivityMainBinding.inflate(getLayoutInflater());
        setContentView(binding.getRoot());
```

```
        BottomNavigationView navView = findViewById(R.id.nav_view);
        AppBarConfiguration appBarConfiguration = new AppBarConfiguration.Builder(
                R.id.navigation_home, R.id.navigation_study, R.id.navigation_personal)
                .build();
        NavController navController = Navigation.findNavController(this, R.id.nav_host_fragment_
activity_main);
        NavigationUI.setupWithNavController(binding.navView, navController);
    }
}
```

通过以上操作，包含底部导航的应用程序框架就完成了，接下来实现每个模块的功能。

12.3.2 实现项目启动界面

常见的应用都有启动界面，创建一个启动界面，界面显示一个图片，3s 后自动跳转到登录界面。操作过程如下所示。

第一步：新建 Activity

在当前包中新建一个 Activity，命名为 SplashActivity，对应布局文件为 activity_splash。

第二步：设计布局

打开 activity_splash 布局文件，为界面上布局添加背景图片 bg.jpg。

第三步：实现定时跳转

要求启动界面显示 3s 后，跳转到登录界面 LoginActivity（此 Activity 在后续操作中创建），实现代码如下所示：

```
public class SplashActivity extends AppCompatActivity {

    @Override
    protected void onCreate(Bundle savedInstanceState) {
        super.onCreate(savedInstanceState);
        setContentView(R.layout.activity_splash);
        Timer timer = new Timer();  //new 一个计时器
        timer.schedule(new TimerTask() {
            @Override
            public void run() {
                Intent intent = new Intent(SplashActivity.this,LoginActivity.class);
                startActivity(intent);
                finish();   //结束当前界面 (WelcomeActivity)
            }
        }, 3000);
    }
}
```

第四步：将 SplashActivity 设为启动界面

在清单文件 AndroidManifest 中将 SplashActivity 设置为启动 Activity。

项目的启动界面就设置好，运行程序直接显示 SplashActivity,3s 后跳转到登录界面。

12.3.3 实现登录和注册

登录界面中包括输入正确用户名、密码登录成功进入主界面和用户注册两个功能。实现这两个功能，操作过程如下。

第一步：新建 Activity

在当前包中创建两个 Activity，Activity 的名字分别为 LoginActivity 和 RegistActivity，对应的布局文件分别为 activity_login 和 activity_regist。

第二步：设计登录用户界面

用户登录界面顶部引入 activitytitle 用来自定义界面标题，显示标题“登录”。中间部分是两个输入框 EditText 控件，一个用来输入用户名，一个用来输入用户密码。一个命令按钮 Button 控件用来实现“登录”，一个表示“记住密码”的复选框 CheckBox 控件，还有一个 TextView 控件用来跳转到“注册”界面。设计代码见布局文件【activity_login.xml】。

扫描二维码查看【activity_login.xml】

activity_login.xml

第三步：编码实现用户登录

用户名和密码使用 SharedPreferences 方式存储在本地。登录界面中功能的实现首先对所用控件进行初始化，隐藏 activitytitle 布局中的“返回”图标，然后判定用户名、密码是否为空，如果不为空并且与从文件 myinfor 中取得的数据匹配，则跳转到主界面，如果用户名错误或不存在，选择“注册”跳转到注册界面注册后再登录。LoginActivity 实现的具体代码如下所示：

```
public class LoginActivity extends AppCompatActivity implements View.OnClickListener {

    private TextView tvtitle;
    private ImageView ivback;
    private EditText etusername;
    private EditText etuserpsw;
    private Button btnlogin;
    private CheckBox cbremeber;
    private TextView tvzhuce;
    private SharedPreferences sp;
    private SharedPreferences.Editor editor;
    private String myname;
    private String mypsw;
    private Boolean isremeber;

    @Override
    protected void onCreate(Bundle savedInstanceState) {
        super.onCreate(savedInstanceState);
        setContentView(R.layout.activity_login);
        init();//初始化控件
    }
```

```
private void init() {
    tvtitle = findViewById(R.id.tvTitle);
    tvtitle.setText("登录");
    ivback = findViewById(R.id.iv_back);
    ivback.setVisibility(View.INVISIBLE);
    etusername = findViewById(R.id.et_name);
    etuserpsw = findViewById(R.id.et_psw);
    btnlogin = findViewById(R.id.btnLogin);
    cbremeber = findViewById(R.id.checkBox);
    tvzhuce = findViewById(R.id.tv_regist);
    btnlogin.setOnClickListener(this);
    tvzhuce.setOnClickListener(this);
    sp = getSharedPreferences("myinfor", MODE_PRIVATE);
    editor = sp.edit();
    myname = sp.getString("name", "");
    mypsw = sp.getString("psw", "");
    isremeber = sp.getBoolean("isremberpsw", false);
    if (isremeber) {
        etusername.setText(myname);
        etuserpsw.setText(mypsw);
    }
}

@Override
public void onClick(View v) {
    Intent intent;
    switch (v.getId()) {
        case R.id.tv_regist:
            intent = new Intent(this, RegistActivity.class);
            startActivity(intent);
            break;
        case R.id.btnLogin:
            btnlogin();
            break;
    }
}

private void btnlogin() {
    String username = etusername.getText().toString().trim();
    String userpsw = etuserpsw.getText().toString().trim();
    if (TextUtils.isEmpty(username)) {
        Toast.makeText(this, "用户名不能为空", Toast.LENGTH_LONG).show();
        return;
    }
    if (TextUtils.isEmpty(userpsw)) {
        Toast.makeText(this, "密码不能为空", Toast.LENGTH_LONG).show();
        return;
    }
    if (!(username.equals(myname) && mypsw.equals(userpsw))) {
        Toast.makeText(this, "输入有误，请重试", Toast.LENGTH_LONG).show();
        return;
    } else {
        if (cbremeber.isChecked()) {
            SharedPreferences.Editor editor = sp.edit();
```

```
                editor.putBoolean("isremberpsw", true);
                editor.commit();
            } else {
                SharedPreferences.Editor editor = sp.edit();
                editor.putBoolean("isremberpsw", false);
                editor.commit();
            }
            Intent intent = new Intent(LoginActivity.this, MainActivity.class);
            startActivity(intent);
        }
    }
}
```

第四步：注册界面设计

用户注册界面顶部引入 activitytitle 用来自定义界面标题，中间部分有三个输入框 EditText 控件，分别用来输入用户名、用户密码和重复输入用户密码。一个复选框 CheckBox 控件用来表示是否“同意本应用的《使用条款》”，一个命令按钮 Button 控件执行“立即注册”操作。用户注册界面代码见布局文件【activity_regist.xml】。

扫描二维码查看【activity_regist.xml】

activity_regist.xml

第五步：注册功能的实现

在注册界面顶部显示标题“用户注册”，后退图标用于实现返回上一个运行界面即登录界面。然后判定用户名、用户密码和重复密码是否为空，并且要求密码和重复密码输入必须一致，如果都不为空，且密码和重复密码输入一致，并且选择了“同意本应用的《使用条款》”，则将用户信息存进“myinfor”文件，为用户登录提供信息，并显示“正在注册”对话框后返回登录界面。RegistActivity 的实现代码如下所示：

```
public class RegistActivity extends AppCompatActivity {
    private TextView tvtitle;
    private ImageView ivback;
    private EditText etuname;
    private EditText etrepsw;
    private EditText etrepsw2;
    private Button btnregist;
    private CheckBox cbregist;

    @Override
    protected void onCreate(Bundle savedInstanceState) {
        super.onCreate(savedInstanceState);
        setContentView(R.layout.activity_regist);
        init();
        btnregist.setOnClickListener(new View.OnClickListener() {
            @Override
            public void onClick(View v) {
                register();
            }
        });
        ivback.setOnClickListener(new View.OnClickListener() {
```

```
            @Override
            public void onClick(View v) {
                finish();
            }
        });
    }
    private void init() {
        tvtitle = findViewById(R.id.tvTitle);
        tvtitle.setText("用户注册");
        ivback = findViewById(R.id.iv_back);
        etuname = findViewById(R.id.et_regist_name);
        etrepsw = findViewById(R.id.et_psw);
        etrepsw2 = findViewById(R.id.et_regist_psw2);
        btnregist = findViewById(R.id.btnRegist);
        cbregist = findViewById(R.id.cb_regist);
    }
    private void register() {
        String username=etuname.getText().toString().trim();
        String userpsw=etrepsw.getText().toString().trim();
        String userpsw2=etrepsw2.getText().toString().trim();
        if(TextUtils.isEmpty(username)){
            Toast.makeText(this,"用户名不能为空",Toast.LENGTH_LONG).show();
            return;
        }
        if(TextUtils.isEmpty(userpsw)){
            Toast.makeText(this,"密码不能为空",Toast.LENGTH_LONG).show();
            return;
        }
        if(TextUtils.isEmpty(userpsw2)){
            Toast.makeText(this,"重复密码不能为空",Toast.LENGTH_LONG).show();
            return;
        }
        if(!userpsw.equals(userpsw2)){
            Toast.makeText(this,"密码与重复密码必须一致",Toast.LENGTH_LONG).show();
            return;
        }else if(!cbregist.isChecked()) {
            Toast.makeText(this,"是否同意本应用的《使用条款》？",Toast.LENGTH_LONG).show();
            return;
        }else{
            SharedPreferences sp=getSharedPreferences("myinfor",MODE_PRIVATE);
            SharedPreferences.Editor editor=sp.edit();
            editor.putString("name",username);
            editor.putString("psw",userpsw);
            editor.commit();
            final ProgressDialog pd = new ProgressDialog(this);
            pd.setMessage("正在注册……");
            pd.show();
            new Thread(new Runnable() {
                public void run() {
                    try {
                        Thread.sleep(3000);
                    } catch (InterruptedException e) {
                    }
                    pd.dismiss();
```

```
                }
            }).start();
            Intent intent=new Intent(getApplicationContext(),LoginActivity.class);
            startActivity(intent);
        }
    }
}
```

12.3.4 首页显示内容

根据运行效果可以看出首页上显示了轮播图、滚动字幕和内容导航。实现过程如下所示。

第一步：导入轮播图插件

轮播图的实现使用了第三方插件，打开 build.gradle 文件,在 dependencies 中添加如下代码：

```
    implementation 'com.youth.banner:banner:2.1.0'
```

第二步：设计首页布局

fragment_home.xml

在首页布局代码 fragment_home 中从上自下添加了一个 banner 控件实现轮播图，添加一个 TextView 控件显示滚动的天气情况，添加一个 RecyclerView 控件显示内容导航。首页布局代码 fragment_home 详细代码见【fragment_home.xml】

扫描二维码查看【fragment_home.xml】

注意　打开布局文件 activity_main.xml，删除布局 ConstraintLayout 的属性：

```
    android:paddingTop="?attr/actionBarSize"
```

解决 fragment_home 与 activity_main 之间留有空白的问题。

第三步：创建 ContentNav 类

选中当前包，在该包下创建一个 bean 包，在 bean 包下新建 ContentNav 类，该类定义首页内容导航中的图片和导航名。该类有两个成员变量，一个为 int 类型用来存储图片 id，一个为 String 类型用来存储内容导航的名称，类的定义如下所示：

```
public class ContentNav {
    private int imgId;
    private String contentName;

    public ContentNav(int imgId, String contentName) {
        this.imgId = imgId;
        this.contentName = contentName;
    }

    public int getImgId() {
        return imgId;
    }

    public String getContentName() {
        return contentName;
    }
}
```

第四步：搭建内容导航 item 布局

创建一个 item 布局文件 item_homenav 作为 RecyclerView 控件的布局，布局文件上有两个控件：一个是 ImageView 控件，用来放置内容导航的图标；一个 TextView 控件，用来显示导航的内容标题。代码详情见布局文件【item_homenav.xml】。

扫描二维码查看【item_homenav.xml】

item_homenav.xml

第五步：编写首页内容导航适配器

内容导航使用 RecyclerView 控件显示，因此需要创建一个数据适配器 HomeNavAdapter。在当前包下创建一个 adapter 包，在 adapter 包中新建一个 HomeNavAdapter 类，HomeNavAdapter 类定义如下所示：

```
public class HomeNavAdapter extends RecyclerView.Adapter<HomeNavAdapter.HomeViewHolder> {
    private List<ContentNav> contentnavLists;
    private Context context;

    public HomeNavAdapter(Context context, List<ContentNav> contentnavLists) {
        this.contentnavLists = contentnavLists;
        this.context = context;
    }

    @NonNull
    @Override
    public HomeViewHolder onCreateViewHolder(@NonNull ViewGroup parent, int viewType) {
        View view = LayoutInflater.from(context).inflate(R.layout.item_homenav, parent, false);
        HomeViewHolder holder = new HomeViewHolder(view);
        return holder;
    }

    @Override
    public void onBindViewHolder(@NonNull HomeNavAdapter.HomeViewHolder holder, int position) {
        ContentNav contentNav = contentnavLists.get(position);
        holder.imgnav.setImageResource(contentNav.getImgId());
        holder.tvnavName.setText(contentNav.getContentName());
        holder.itemView.setOnClickListener(new View.OnClickListener() {
            @Override//单击每个 item 时会打开一个 Activity
            public void onClick(View v) {
                Intent intent;//下面的所有 Activity 还没有创建，会显示出错
                switch (holder.getAdapterPosition()) {
                    case 0:
                        intent = new Intent(context, DangJianYaoWenActivity.class);
                        context.startActivity(intent);
                        break;
                    case 1:
                        intent = new Intent(context, XueXiShiPingActivity.class);
                        context.startActivity(intent);
                        break;
                    case 2:
                        intent = new Intent(context, SuiShouPaiActivity.class);
                        context.startActivity(intent);
```

```
                break;
            case 3:
                intent = new Intent(context, JianYanXianCeActivity.class);
                context.startActivity(intent);
                break;
            case 4:
                intent = new Intent(context, ZuZhiHouDongActivity.class);
                context.startActivity(intent);
                break;
            case 5:
                intent = new Intent(context, FuWuRexianActivity.class);
                context.startActivity(intent);
                break;
          }
        }
    });
  }

  @Override
  public int getItemCount() {
      return contentnavLists.size();
  }

  public class HomeViewHolder extends RecyclerView.ViewHolder {
      private ImageView imgnav;
      private TextView tvnavName;

      public HomeViewHolder(@NonNull View itemView) {
          super(itemView);
          imgnav = itemView.findViewById(R.id.imgnav);
          tvnavName = itemView.findViewById(R.id.tvNavName);
      }
  }
}
```

说明 在上面的代码中对每个 item 单击会实现界面跳转，目标 Activity 还没有创建，调试时可以先注释起来，在后续陆续创建好对应的 Activity 去掉注释就可以了。

第六步：滚动天气情况

这里使用一个 TextView 控件显示滚动的文字,在布局文件中注意设置 TextView 控件的属性：

```
android:marqueeRepeatLimit="marquee_forever"
```

在功能代码中设置如下操作：

```
TextView tvtianqi=view.findViewById(R.id.tv_tianqi);
tvtianqi.setMovementMethod(LinkMovementMethod.getInstance());
tvtianqi.setEllipsize(TextUtils.TruncateAt.MARQUEE);
tvtianqi.setSingleLine(true);
tvtianqi.setSelected(true);
tvtianqi.setFocusable(true);
tvtianqi.setFocusableInTouchMode(true);
```

第七步：实现首页显示功能

首页需要显示的内容的准备工作做好了，下面就是实现 HomeFragment，HomeFragment 的实现主要包括三部分：第一部分是轮播图；第二部分是文字滚动；第三部分是内容导航。具体代码如下所示：

```
public class HomeFragment extends Fragment {
    private Banner banner;
    private List<Integer> list;
    private View view;
    private List<DangJianYaowen> yaowenList;
    private List<ContentNav> contentNavList;
    private RecyclerView rc1;

    public View onCreateView(@NonNull LayoutInflater inflater,
                        ViewGroup container, Bundle savedInstanceState) {
        view = inflater.inflate(R.layout.fragment_home, container, false);
        return view;
    }

    @Override
    public void onActivityCreated(@Nullable Bundle savedInstanceState) {
        super.onActivityCreated(savedInstanceState);
        bannerinit();//轮播图
        gundongwenzi();//文字滚动效果
        initnav();//内容导航
    }

    private void bannerinit() {
        banner = view.findViewById(R.id.banner);
        list = new ArrayList();
        list.add(R.drawable.lunbo1);
        list.add(R.drawable.lunbo2);
        list.add(R.drawable.lunbo3);
        list.add(R.drawable.lunbo4);
        list.add(R.drawable.lunbo5);
        banner.setAdapter(new BannerImageAdapter<Integer>(list) {
            @Override
            public void onBindView(BannerImageHolder holder, Integer data, int position, int size) {
                holder.imageView.setImageResource(data);
            }
        });
        banner.isAutoLoop(true);
        banner.setIndicator(new CircleIndicator(getContext()));
        banner.start();
        banner.setOnBannerListener(new OnBannerListener() {//单击轮播图跳转到相应的界面
            @Override
            public void OnBannerClick(Object data, int position) {
                //参数：轮播绑定的数据，轮播图片的索引，position 从 0 开始
                Intent intent;
                if (position == 1) {
                    intent = new Intent(getContext(), LoginActivity.class);
                    startActivity(intent);
```

```
            }
        }
    });
}
private void gundongwenzi() {
    TextView tvtianqi = view.findViewById(R.id.tv_tianqi);
    tvtianqi.setMovementMethod(LinkMovementMethod.getInstance());
    tvtianqi.setEllipsize(TextUtils.TruncateAt.MARQUEE);
    tvtianqi.setSingleLine(true);
    tvtianqi.setSelected(true);
    tvtianqi.setFocusable(true);
    tvtianqi.setFocusableInTouchMode(true);
}
private void initnav() {
  int[]imgs={R.drawable.dongtai,R.drawable.studyall,R.drawable.camera,R.drawable.jianyanxiance,
R.drawable.zhengce, R.drawable.call};
    String[] contentName = {"党建要闻", "微课视频", "随手拍", "建言献策", "组织活动", "服务热线"};
    contentNavList = new ArrayList<>();
    for (int i = 0; i < imgs.length; i++) {
        ContentNav contentNav = new ContentNav(imgs[i], contentName[i]);
        contentNavList.add(contentNav);
    }
    rc1 = view.findViewById(R.id.rc_nav);
    rc1.setLayoutManager(new GridLayoutManager(getActivity(), 2));
    HomeNavAdapter homeNavAdapter = new HomeNavAdapter(getActivity(), contentNavList);
    rc1.setAdapter(homeNavAdapter);
}

@Override
public void onDestroyView() {
    super.onDestroyView();
}
}
```

在上面的代码中 onCreateView()方法指定 HomeFragment 对应的布局文件为 fragment_home，然后在 onActivityCreated()方法中实现显示轮播图、文字滚动和内容导航，分别调用 bannerinit()、gundongwenzi()和 initnav()自定义方法。

到现在为止，项目的启动界面、注册登录以及首页显示功能就实现了，接下来完成各内容导航模块的功能。

12.3.5　党建要闻展示

党建要闻界面包括两部内容：一个是标题栏，使用已经定义好的 item 布局文件 item_dangjina.xml；另一个是显示党建要闻列表的 RecyclerView 控件。这里在 RecyclerView 控件上显示的是网络数据（本地服务器），需要申请网络访问权限，定义对应 JavaBean 和数据适配器。具体实现过程如下：

第一步：创建 Activity

在当前包下创建一个 Activity，命名为 DangJianYaoWenActivity，对应的布局界面为 activity_ dang_jian_yao_wen.xml;

第二步：设计党建要闻界面布局

党建要闻界面只包括两部分内容，在界面的顶端引入标题栏布局 activitytitle，标题栏的下面添加一个 RecyclerView 控件用来显示党建要闻条目。布局文件详细代码见文件【activity_dang_jian_yao_wen.xml】。

扫描二维码查看【activity_dang_jian_yao_wen.xml】

activity_dang_jian_yao_wen.xml

第三步：导入相关插件

在实际应用开发中很多时候需要从服务器中取数据，在党建要闻界面中要实现访问服务器上的数据。数据详情见文件【dangjiandata.json】。扫描二维码查看【dangjiandata.json】

dangjiandata.json

党建要闻图片存储在服务器上，扫描二维码下载【党建要闻图片】

党建要闻图片

访问网络数据需要加入 Volley 和 GSON,在 build.gradle 中添加库文件：

```
implementation"com.android.volley:volley:1.1.1"
implementation 'com.google.code.gson:gson:2.8.6'
```

访问网络图片需要加 glide,在 build.gradle 中添加库文件代码如下：

```
implementation 'com.github.bumptech.glide:glide:4.11.0'
annotationProcessor 'com.github.bumptech.glide:compiler:4.11.0'
```

第四步：申请网络访问权限

我们访问的服务器，在这需要在清单文件 AndroidManifest 中添加 Intent 访问权限：

```
<uses-permission android:name="android.permission.INTERNET" />
```

为清单文件 AndroidManifest 的 application 标签添加如下属性：

```
android:usesCleartextTraffic="true"
```

第五步：创建 DangJianYaowen 类

选中当前包，在该包下创建一个 bean 包，在 bean 包下新建 DangJianYaowen 类，该类有三个 String 类型的成员变量，img 用来存储网络图片的地址，title 用来存放党建要闻的标题，content 用来存放党建要闻的内容，DangJianYaowen 类定义代码如下所示：

```
public class DangJianYaowen {
    String img;
    String title;
    String content;

    public DangJianYaowen(String img, String title, String content) {
        this.img = img;
        this.title = title;
        this.content = content;
    }
```

```
    public String getImg() {
        return img;
    }

    public String getTitle() {
        return title;
    }
    public String getContent() {
        return content;
    }
}
```

第六步：搭建党建要闻 item 布局

创建一个 item 布局文件 item_dangjian 作为 RecyclerView 控件的布局，布局文件上有四个控件：一个是 ImageView 控件，用来放置党建要闻的图片；一个是 TextView 控件，用来显示党建要闻的标题；还有一个 TextView 控件，用来显示党建要闻的内容；最后在布局的下方添加一个 View 控件，设置目 item 分割线。item_dangjian.xml 布局代码如下所示：

```
<?xml version="1.0" encoding="utf-8"?>
<androidx.constraintlayout.widget.ConstraintLayout xmlns:android="http://schemas.android.com/apk/
res/android"
    xmlns:app="http://schemas.android.com/apk/res-auto"
    xmlns:tools="http://schemas.android.com/tools"
    android:layout_width="match_parent"
    android:layout_height="wrap_content">

    <ImageView
        android:id="@+id/iv_dagnjian_img"
        android:layout_width="120dp"
        android:layout_height="80dp"
        android:layout_marginStart="8dp"
        android:layout_marginTop="8dp"
        app:layout_constraintStart_toStartOf="parent"
        app:layout_constraintTop_toTopOf="parent"
        app:srcCompat="@drawable/zuzhihoudong1" />

    <TextView
        android:id="@+id/tv_dangjian_title"
        android:layout_width="240dp"
        android:layout_height="wrap_content"
        android:layout_marginStart="8dp"
        android:layout_marginEnd="16dp"
        android:ellipsize="end"
        android:maxLines="1"
        android:text="TextView"
        android:textSize="18sp"
        app:layout_constraintBottom_toTopOf="@+id/tv_dangjian_content"
        app:layout_constraintEnd_toEndOf="parent"
        app:layout_constraintStart_toEndOf="@+id/iv_dagnjian_img"
```

```
        app:layout_constraintTop_toTopOf="@+id/iv_dagnjian_img" />

    <TextView
        android:id="@+id/tv_dangjian_content"
        android:layout_width="240dp"
        android:layout_height="0dp"
        android:ellipsize="end"
        android:maxLines="3"
        android:text="TextView"
        android:textSize="12sp"
        app:layout_constraintBottom_toBottomOf="@+id/iv_dagnjian_img"
        app:layout_constraintEnd_toEndOf="@+id/tv_dangjian_title"
        app:layout_constraintStart_toStartOf="@+id/tv_dangjian_title"
        app:layout_constraintTop_toBottomOf="@+id/tv_dangjian_title" />

    <View
        android:layout_width="match_parent"
        android:layout_height="1dp"
        android:layout_marginTop="8dp"
        android:layout_marginBottom="5dp"
        android:background="#eee"
        app:layout_constraintBottom_toBottomOf="parent"
        app:layout_constraintEnd_toEndOf="parent"
        app:layout_constraintStart_toEndOf="@id/tv_dangjian_content"
        app:layout_constraintTop_toBottomOf="@+id/iv_dagnjian_img" />
</androidx.constraintlayout.widget.ConstraintLayout>
```

第七步：创建党建要闻适配器

党建要闻使用 RecyclerView 控件显示，需要创建一个数据适配器 DangJianYaowenAdapter。在当前包下的 adapter 包中新建一个 DangJianYaowenAdapter 类，DangJianYaowenAdapter 类定义代码如下所示：

```
public class DangJianYaowenAdapter extends RecyclerView.Adapter<DangJianYaowenAdapter.DangJian-
YaowenViewHoder> {
    private Context context;
    private List<DangJianYaowen> lists;

    public DangJianYaowenAdapter(Context context, List<DangJianYaowen> lists) {
        this.context = context;
        this.lists = lists;
    }

    @NonNull
    @Override
    public DangJianYaowenAdapter.DangJianYaowenViewHoder onCreateViewHolder(@NonNull ViewGroup
parent, int viewType) {
        View view = LayoutInflater.from(context).inflate(R.layout.item_dangjian, parent, false);
        DangJianYaowenAdapter.DangJianYaowenViewHoder holder=new DangJianYaowenAdapter. DangJian-
YaowenViewHoder(view);
        return holder;
    }
```

```
    @Override
    public void onBindViewHolder(@NonNull DangJianYaowenViewHoder holder, int position) {
        DangJianYaowen yaowen = lists.get(position);
        Glide.with(context).load(yaowen.getImg()).into(holder.iv_yaowem);
        holder.tv_yaowemtitle.setText(yaowen.getTitle());
        holder.tv_yaowemcontent.setText(yaowen.getContent());
        holder.itemView.setOnClickListener(new View.OnClickListener() {
            @Override
            public void onClick(View v) {
                Intent intent = new Intent(context, DangJianXiangQingActivity.class);
                switch (holder.getAdapterPosition()) {
                    case 0:
                        intent.putExtra("x", 1);
                        break;
                    case 1:
                        intent.putExtra("x", 2);
                        break;
                    case 2:
                        intent.putExtra("x", 3);
                        break;
                    case 3:
                        intent.putExtra("x", 4);
                        break;
                    case 4:
                        intent.putExtra("x", 1);
                        break;
                    case 5:
                        intent.putExtra("x", 2);
                        break;
                }
                context.startActivity(intent);
            }
        });
    }

    @Override
    public int getItemCount() {
        return lists.size();
    }

    public class DangJianYaowenViewHoder extends RecyclerView.ViewHolder {
        private ImageView iv_yaowem;
        private TextView tv_yaowemtitle;
        private TextView tv_yaowemcontent;

        public DangJianYaowenViewHoder(@NonNull View itemView) {
            super(itemView);
            iv_yaowem = itemView.findViewById(R.id.iv_dagnjian_img);
            tv_yaowemtitle = itemView.findViewById(R.id.tv_dangjian_title);
            tv_yaowemcontent = itemView.findViewById(R.id.tv_dangjian_content);
        }
    }
}
```

在适配器的定义中使用 Glide.*with*(context).load(yaowen.getImg()).into(holder.iv_yaowem)用来实现 ImageView 网络图片的加载。

第八步：实现党建要闻显示功能

党建要闻显示的数据来自服务器，通过 Volloy 实现网络访问，使用 Gson 解析返回结果。

```
public class DangJianYaoWenActivity extends AppCompatActivity {
    private TextView tvtitle;
    private ImageView ivback;
    private RecyclerView redangjian;
    private List<DangJianYaowen> yaowenLists;
    MHandler handler;
    private List<DangJianYaowen> listsl;
    @Override
    protected void onCreate(Bundle savedInstanceState) {
        super.onCreate(savedInstanceState);
        setContentView(R.layout.activity_dang_jian_yao_wen);
        init();
        getData();
    }

    private void init() {
        tvtitle = findViewById(R.id.tvTitle);
        tvtitle.setText("党建要闻");
        ivback = findViewById(R.id.iv_back);
        redangjian = findViewById(R.id.rc_dangjianyaowen);
        yaowenLists = new ArrayList<>();
        handler=new MHandler();
        ivback.setOnClickListener(new View.OnClickListener() {
            @Override
            public void onClick(View v) {
                finish();
            }
        });
    }
    private void getData() {
        listsl=new ArrayList<>();
        RequestQueue queue = Volley.newRequestQueue(this);
        String url = "http://10.0.2.2:8080/dangjiandata/dangjiandata.json";
        JsonArrayRequest jsonArrayRequest = new JsonArrayRequest(url, new Response.Listener<JSON-
Array>() {
            @Override
            public void onResponse(JSONArray response) {
                Gson gson = new Gson();
                Type listType = new TypeToken<List<DangJianYaowen>>() {
                }.getType();
                listsl = gson.fromJson(response.toString(), listType);
                Message msg =new Message();
                msg.what=1;
                msg.obj = listsl ;
                handler.sendMessage(msg);
            }
```

```
        }, new Response.ErrorListener() {
            @Override
            public void onErrorResponse(VolleyError error) {
                Log.i("AA", "error" + error.getMessage());
            }
        });
        queue.add(jsonArrayRequest);
    }

    class MHandler extends Handler {
        @Override
        public void dispatchMessage(@NonNull Message msg) {
            super.dispatchMessage(msg);
            if (msg.obj != null) {
                yaowenLists = (List<DangJianYaowen>) msg.obj;
                redangjian.setLayoutManager(new LinearLayoutManager(DangJianYaoWenActivity.this));
                DangJianYaowenAdapter myAdapter = new DangJianYaowenAdapter(DangJianYaoWenActivity.
this, yaowenLists);
                redangjian.setAdapter(myAdapter);
            }
        }
    }
}
```

在上面的代码中，对于每个 item 对应的党建要闻，选择不同的 item 都会显示在党建详情 DangJianXiangQingActivity（在后续操作中创建）中，所以使用 Intent 对象传递了参数 x,在党建详情中根据传递的 x 的值，确定显示具体的内容。

12.3.6　显示党建要闻详情

在上一节中，实现了党建要闻条目的显示，对于每个条目的详细内容展示，在本节中实现。操作过程如下所示：

第一步：创建 Activity

在当前包下创建一个 Activity，命名为 DangJianXiangQingActivity，对应的布局界面为 activity_dang_jian_xiang_qing.xml。

第二步：设计党建要闻界面布局

党建要闻详情界面只包括两部分内容，在界面的顶端引入标题栏布局 activitytitle，标题栏的下面添加一个 ScrollView 布局，在该布局中有一个 WebView 控件。布局文件详细代码见文件【activity_dang_jian_xiang_qing.xml】。

扫描二维码查看【activity_dang_jian_xiang_qing.xml】

activity_dang_jian_xiang_qing.xml

第三步：显示党建要闻详情内容

在 DangJianXiangQingActivity 中根据 DangJianYaoWenActivity 通过 Intent 传递过来的参数，决定显示什么内容，DangJianXiangQingActivity 具体实现代码如下所示：

```
public class DangJianXiangQingActivity extends AppCompatActivity {
    private TextView tvtitle;
    private ImageView ivback;
    private WebView webView;
    int x=1;
    @Override
    protected void onCreate(Bundle savedInstanceState) {
        super.onCreate(savedInstanceState);
        setContentView(R.layout.activity_dang_jian_xiang_qing);
        init();
    }

    private void init() {
        tvtitle = findViewById(R.id.tvTitle);
        tvtitle.setText("党建要闻详情");
        ivback=findViewById(R.id.iv_back);
        ivback.setOnClickListener(new View.OnClickListener() {
            @Override
            public void onClick(View v) {
                finish();
            }
        });
        webView = findViewById(R.id.webView);
        webView.setWebViewClient(new WebViewClient());
        webView.getSettings().setJavaScriptEnabled(true);
        Intent intent=getIntent();
        x=intent.getIntExtra("x",1);
        extracted(x);
    }

    private void extracted(int x) {
        if (x == 1) {
            webView.loadUrl("http://dangjian.people.com.cn/n1/2022/0128/c117092-32342008.html");
        }
        if (x == 2) {
            webView.loadUrl("http://dangjian.people.com.cn/n1/2022/0128/c117092-32341933.html");
        }
        if (x == 3) {
            webView.loadUrl("http://dangjian.people.com.cn/n1/2022/0128/c117092-32341874.html");
        }
        if (x == 4) {
            webView.loadUrl("http://dangjian.people.com.cn/n1/2022/0127/c117092-32341470.html");
        }

        if (x == 5) {
            webView.loadUrl("http://dangjian.people.com.cn/n1/2022/0127/c117092-32341274.html");
        }
        if (x == 6) {
            webView.loadUrl("http://dangjian.people.com.cn/n1/2022/0127/c117092-32340889.html");
        }
        if (x == 7) {
            webView.loadUrl("http://dangjian.people.com.cn/n1/2022/0127/c117092-32340854.html");
        }
```

```
        if (x == 8) {
            webView.loadUrl("http://dangjian.people.com.cn/n1/2022/0127/c117092-32340850.html");}
        if (x == 9) {
            webView.loadUrl("http://dangjian.people.com.cn/n1/2022/0127/c117092-32341274.html");
        }
        if (x == 10) {
            webView.loadUrl("http://dangjian.people.com.cn/n1/2022/0127/c117092-32340889.html");
        }
    }
}
```

在上面的代码中，x 是上一个 Activity 通过 Intent 传递过来的值，根据 x 的值决定 webView 控件加载的页面。

12.3.7　实现播放学习视频

进入学习视频界面，选择需要播放的视频，界面弹出一个播放对话框，完成视频的播放。实现此功能的步骤如下所示：

第一步：创建 Activity

在当前包下创建一个 Activity，命名为 XueXiShiPingActivity，对应的布局界面为 activity_xue_xi_shi_ping.xml。

第二步：设计微课视频界面布局

微课视频界面只包括两部分内容，在界面的顶端引入标题栏布局 activitytitle，标题栏的下面使用 TextView 显示视频分类，各个分类的下面使用 ListView 控件显示视频的标题。布局文件详细代码见文件【activity_xue_xi_shi_pin.xml】。

扫描二维码查看【activity_xue_xi_shi_pin.xml】

activity_xue_shi_pin.xml

第三步：创建播放视频的对话框的布局

创建一个布局文件 video_dialog.xml,视频播放控件是使用 SurfaceView 控件来实现的，对话框布局中只有一个 SurfaceView 控件，布局文件的详细代码见文件【video_dialog.xml】。

扫描二维码查看【video_dialog.xml】

video_dialog.xml

第四步：设置对话框样式

我们希望视频在播放的时候不能出现对话框标题等样式，打开 res/values 下的 themes 文件，在其中自定义 style 名称为 MyDialog，定义代码如下所示：

```
<!--对话框样式-->
<style name="MyDialog" parent="android:Theme.Dialog">
    <item name="android:windowFrame">@null</item><!--是否有边框-->
    <item name="android:windowNoTitle">true</item><!--是否有标题-->
    <item name="android:windowIsFloating">true</item><!--是否是浮动的-->
</style>
```

第五步：创建播放视频的对话框

创建播放视频的对话框，定义一个类 VideoDialog 继承自 Dialog，在该类创建一个构造方法参数为 Contenxt 和 theme，通过 theme 引用自定义样式，重写 findViewById()方法获取对话框中的控件，定义 getDialogView()方法获取对话框 view。VideoDialog.java 的详细代码如下所示：

```
public class VideoDialog extends Dialog {
    Context context;            //上下文
    private View dialogView;    //对话框控件

    public VideoDialog(Context context, int theme) {
        super(context, theme);
        this.context = context;
        LayoutInflater inflater = LayoutInflater.from(context);//加载对话框布局
        dialogView = inflater.inflate(R.layout.video_dialog, null);
    }

    @Override
    protected void onCreate(Bundle savedInstanceState) {
        super.onCreate(savedInstanceState);
        this.setContentView(dialogView);
    }

    @Override
    public View findViewById(int id) {//重写findViewById 方法获取对话框中控件
            return super.findViewById(id);
    }

    public View getDialogView() {
        return dialogView; //获得对话框view
    }
}
```

第六步：实现学习视频的播放功能

在 XueXiShiPingActivity 中，使用 TextView 控件和 ListView 控件显示视频分类和各类视频的列表项，然后调用视频播放对话框播放视频，XueXiShiPingActivity 功能具体实现代码如下所示：

```
public class XueXiShiPingActivity extends AppCompatActivity {
    private TextView tvtitle;
    private ImageView ivback;
    private SurfaceView surfaceView;              //显示图像的控件
    private MediaPlayer mediaPlayer;              //播放器
    private SurfaceHolder surfaceHolder;          //用于控制surfaceView
    private VideoDialog videoDialog;              //自定义视频播放对话框
    private ListView listView1;
    private ListView listView2;
    private ListView listView3;
    private View view1;
    private String[] strdangshi={"从“心”出发，去媒体党课第一课","从“心”出发，去媒体党课第二课","从“心”
```

```
出发，去媒体党课第三课"};
    private String[] strbangy = new String[]{"张桂梅：忠诚执着守初心 无私奉献担使命", "马毛姐：运送三批解放军成功登岸的“渡江英雄”", "钟南山：临危受命 国士担当 战疫60天全记录"};
    private String[] strzhuanye = new String[]{"第一章 走进 Android 世界：Android 简介", "第一章 走进 Android 世界：创建第一个工程", "第一章 走进 Android 世界：模拟器的使用","第一章 走进 Android 世界:Android 程序结构 "};

    @Override
    protected void onCreate(Bundle savedInstanceState) {
        super.onCreate(savedInstanceState);
        setContentView(R.layout.activity_xue_xi_shi_ping);
        init();
        initData(listView1, strdangshi);
        initData(listView2, strbangy);
        initData(listView3, strzhuanye);
    }

    private void init() {
        tvtitle = findViewById(R.id.tvTitle);
        tvtitle.setText("学习视频");
        ivback = findViewById(R.id.iv_back);
        listView1 = findViewById(R.id.list_dangshi);
        listView2 = findViewById(R.id.list_bangy);
        listView3 = findViewById(R.id.list_zhuangye);
        ivback.setOnClickListener(new View.OnClickListener() {
            @Override
            public void onClick(View v) {
                finish();
            }
        });
    }

    private void initData(ListView listView, String[] strArr) {
        listView.setAdapter(new  ArrayAdapter<String>(this,  android.R.layout.simple_list_item_1,
strArr));
        listView.setOnItemClickListener(new
AdapterView.OnItemClickListener() {
            @Override
            public void onItemClick(AdapterView<?> parent, View view, int position, long id) {
                videoDialog = new VideoDialog( XueXiShiPingActivity.this, R.style.MyDialog);//新建对话框
                    videoDialog.setCanceledOnTouchOutside(true);
                view1 = videoDialog.getDialogView();
                surfaceView = view1.findViewById(R.id.surfaceView);
                switch (position) {
                    case 0:
                        playVideo(R.raw.video01);              //调用播放视频方法
                            break;
                    case 1:
                        playVideo(R.raw.video02);
                        break;
```

```
                case 2:
                    playVideo(R.raw.video01);
                    break;
            }
            videoDialog.show();          //显示对话框
        }
    });
}

private void playVideo(int x) {
    mediaPlayer = MediaPlayer.create(this, x);
    surfaceHolder = surfaceView.getHolder();//获取 surfaceHolder
     surfaceHolder.setKeepScreenOn(true); // 设置播放时打开屏幕
     mediaPlayer.setAudioStreamType(AudioManager.STREAM_MUSIC);//设置音频类型
     try {
       mediaPlayer.prepare(); //准备
     } catch (IllegalStateException e1) {
       e1.printStackTrace();
    } catch (IOException e1) {
       e1.printStackTrace();
    }
    //等待 surfaceHolder 初始化完成才能执行
mPlayer.setDisplay(surfaceHolder)
    mediaPlayer.setOnPreparedListener(new
MediaPlayer.OnPreparedListener() {
        @Override
        public void onPrepared(MediaPlayer mp) {
            mediaPlayer.setDisplay(surfaceHolder);// 把视频画面输出到 SurfaceView
            mediaPlayer.start();          //播放视频
        }
    });
    //视频播放完成后的操作
    mediaPlayer.setOnCompletionListener(new
MediaPlayer.OnCompletionListener() {
        @Override
        public void onCompletion(MediaPlayer mp) {
            if (mediaPlayer != null)
                mediaPlayer.release();//重置 mediaplayer 等待下一次播放
            if (videoDialog.isShowing())
                videoDialog.dismiss(); //关闭对话框
        }
    });
}

@Override
protected void onDestroy() {
    super.onDestroy();
    if (mediaPlayer != null) { // 释放资源
        mediaPlayer.release();
    }
}
}
```

在上面的代码中定义了三个方法：init()方法用来初始化控件，initData()方法用来为每类视频添加内容项和对应的播放视频内容，playVideo()方法用来播放视频。由于篇幅问题这里 initData()方法将三类视频播放内容没有进行区别。

12.3.8 实现随手拍功能

在常见的应用中都有随手拍的功能，本节课实现智慧党建中的随手拍功能，实现步骤如下所示：

第一步：创建 Activity

在当前包下创建一个 Activity，命名为 SuiShouPaiActivity，对应的布局界面为 activity_sui_shou_pai.xml。

第二步：设计随手拍界面布局

随手拍界面包括四部分内容：第一部分在界面的顶端引入标题栏布局 activitytitle；第二部分随手拍的主题使用 EditText 控件、随手拍内容输入框 EditText 控件、图片上传的 ImageView 控件和命令按钮 Button 控件“发布”；第三部分是一个 RecyclerView 控件，用来显示上传的随手拍记录；第四部分是用来选择上传的图片是从相册选择还是通过拍照获得，这部分只有点击了上传图片图标才会显示。随手拍布局文件详细代码见文件【activity_sui_shou_pai.xml】。

扫描二维码查看【activity_xue_xi_shi_pin.xml】

activity_xue_xi_shi_pin.xml

第三步：创建 SuiShouPai 类

选中当前包下的 bean 包，在 bean 包下新建 SuiShouPai 类，该类有四个成员变量，三个 String 类型成员变量分别用来存储随手拍的主题 title、内容 content、时间 time 和 Bitmap 类型的所拍的图片。SuiShouPai 类的定义如下所示：

```
public class SuiShouPai {
    String title;
    String content;
    String time;
    Bitmap tupian;

    public SuiShouPai(String title, String content, String time, Bitmap tupian) {
        this.title = title;
        this.content = content;
        this.time = time;
        this.tupian = tupian;
    }

    public String getTitle() {
        return title;
    }

    public String getContent() {
        return content;
    }
```

```
    public String getTime() {
        return time;
    }

    public Bitmap getTupian() {
        return tupian;
    }
}
```

第四步：搭建随手拍显示 item 布局

创建一个 item 布局文件 item_suishoupai 作为 RecyclerView 控件的布局，布局文件上有三个 TextVeiw 控件用来显示随手拍的主题、内容和时间；一个 ImageView 控件显示随手拍的图片，一个 View 控件用作分割线。代码详情见布局文件【item_suishoupai.xml】

扫描二维码查看【item_suishoupai.xml】

item_suishoupai.xml

第五步：编写随手拍适配器

随手拍记录通过 RecyclerView 控件显示，因此需要创建一个数据适配器 SuiShouPaiAdapter 对 RecyclerView 控件进行数据适配。在当前包下的 adapter 包新建一个 SuiShouPaiAdapter 类，SuiShouPaiAdapter 类定义如下所示：

```
public class SuiShouPaiAdapter extends RecyclerView.Adapter<SuiShouPaiAdapter.ViewHolder> {
    private List<SuiShouPai> suiShouPais;
    private Context context;

    public SuiShouPaiAdapter(Context context,List<SuiShouPai> suiShouPais ) {
        this.suiShouPais = suiShouPais;
        this.context = context;
    }

    @NonNull
    @Override
    public ViewHolder onCreateViewHolder(@NonNull ViewGroup parent, int viewType) {
        View view = LayoutInflater.from(context).inflate(R.layout.item_suishoupai, parent, false);
        ViewHolder holder = new SuiShouPaiAdapter.ViewHolder(view);
        return holder;
    }

    @Override
    public void onBindViewHolder(@NonNull ViewHolder holder, int position) {
        SuiShouPai shouPai=suiShouPais.get(position);
        holder.suishoupaiTitle.setText(shouPai.getTitle());
        holder.suishoupaiContent.setText(shouPai.getContent());
        holder.suishoupaiTime.setText(shouPai.getTime());
        holder.suishoupaiImage.setImageBitmap(shouPai.getTupian());
        shouPai.toString();
```

```
    }

    @Override
    public int getItemCount() {
        return suiShouPais.size();
    }

    public class ViewHolder extends RecyclerView.ViewHolder {
        TextView suishoupaiTitle;
        TextView suishoupaiContent;
        TextView suishoupaiTime;
        ImageView suishoupaiImage;

        public ViewHolder(@NonNull View itemView) {
            super(itemView);
            suishoupaiTitle=itemView.findViewById(R.id.item_suishoupai_title);
            suishoupaiContent=itemView.findViewById(R.id.item_suishoupai_content);
            suishoupaiTime=itemView.findViewById(R.id.item_suishoupai_time);
            suishoupaiImage=itemView.findViewById(R.id.item_suishoupai_image);
        }
    }
}
```

第六步：创建数据库

随手拍的内容存储在本地数据库，这里使用 SQLite 数据库实现数据存储。在当前包中新建一个 data 包，在 data 包中创建一个名为的 SQLiteHelper 类，该类的定义代码如下所示：

```
public class SQLiteHelper extends SQLiteOpenHelper {
    public SQLiteHelper(@Nullable Context context) {
        super(context,"dangjian.db", null,1);
    }

    @Override
    public void onCreate(SQLiteDatabase db) {
        db.execSQL("create table suipai_infor(_id integer Primary key autoincrement,title varchar
(30),content text,tupian BLOB,time text)");
    }

    @Override
    public void onUpgrade(SQLiteDatabase db, int oldVersion, int newVersion) {
    }
}
```

第七步：在清单文件中申请授权

读取数据时就需要有访问数据的权限，在清单文件 AndroidManifest 中添加授权申请：

<uses-permission android:name="android.permission.READ_EXTERNAL_STORAGE" />

在 application 标签下添加如下属性:

```
android:requestLegacyExternalStorage="true"
```

第八步：实现随手拍功能

在界面中输入随手拍的主题和内容，选择上传图片，出现选项是上传拍照还是从相册中选择，如果从相册中选择，则会访问系统相册，将选中的图片返回到界面；如果选择拍照，则会调用系统相机进行拍照，将所拍照片返回界面。单击“发布”命令按钮，将输入的主题、内容、图片和获取的系统时间存进数据库，然后显示在“发布”命令按钮下面的 RecyclerView 控件中。SuiShouPaiActivity 的具体代码如下所示:

```
public class SuiShouPaiActivity extends AppCompatActivity implements View.OnClickListener {
    private TextView tvtitle;
    private ImageView ivback;
    private ImageView ivtupian;
    private LinearLayout ll;
    private TextView tvquxiao;
    private TextView tvpaizhao;
    private TextView tvxiangce;
    private Bitmap bitmap;
    private Uri imageUri;
    private String imagePath;
    private Button btn_tijiao;
    private EditText ettitle;
    private EditText etcontent;
    private Bitmap resized;
    private List<SuiShouPai> suiShouPais;
    private RecyclerView rc1;
    private SQLiteHelper helper;
    private SQLiteDatabase db;
    private String suishoupai_title;
    private String suishoupai_content;
    @Override
    protected void onCreate(Bundle savedInstanceState) {
        super.onCreate(savedInstanceState);
        setContentView(R.layout.activity_sui_shou_pai);
        init();
    }
    private void init() {
        tvtitle = findViewById(R.id.tvTitle);
        tvtitle.setText("随手拍");
        ettitle = findViewById(R.id.et_suishoupai_title);
        etcontent = findViewById(R.id.et_suishoupai_content);
        ivback = findViewById(R.id.iv_back);
        ivtupian = findViewById(R.id.iv_tupian);
        btn_tijiao = findViewById(R.id.bt_suishoupai_tijiao);
        ll = findViewById(R.id.ll);
```

```
        tvquxiao = findViewById(R.id.tv_quxiao);
        tvpaizhao = findViewById(R.id.tv_paizhao);
        tvxiangce = findViewById(R.id.tv_xiangce);
        ivback.setOnClickListener(this);
        ivtupian.setOnClickListener(this);
        tvpaizhao.setOnClickListener(this);
        tvxiangce.setOnClickListener(this);
        tvquxiao.setOnClickListener(this);
        btn_tijiao.setOnClickListener(this);
        rc1 = findViewById(R.id.recycle_suishoupai);
        rc1.setLayoutManager(new LinearLayoutManager(this));
        String[] permissions = {Manifest.permission.READ_EXTERNAL_STORAGE};
        ActivityCompat.requestPermissions(this, permissions, 101);
        recycleInit(db);
    }

    private void recycleInit(SQLiteDatabase db) {
        helper = new SQLiteHelper(this);
        db = helper.getWritableDatabase();
        suiShouPais = new ArrayList<SuiShouPai>();
        Cursor c = db.query("suipai_infor", null, null, null, null, null, null);
        while (c.moveToNext()) {
            String title = c.getString(1);
            String content = c.getString(2);
            byte[] in = c.getBlob(c.getColumnIndex("tupian"));
            Bitmap bit = BitmapFactory.decodeByteArray(in, 0, in.length);
            String time = c.getString(4);
            SuiShouPai suiShouPai = new SuiShouPai(title, content, time, bit);
            suiShouPais.add(suiShouPai);
        }
        if (suiShouPais.size() > 0) {
            SuiShouPaiAdapter adapter = new SuiShouPaiAdapter(this, suiShouPais);
            rc1.setAdapter(adapter);
        }
        c.close();
        db.close();
    }

    @Override
    public void onClick(View v) {
        Intent intent;
        switch (v.getId()) {
            case R.id.iv_back:
                finish();
                break;
            case R.id.iv_tupian:
                ll.setVisibility(View.VISIBLE);
                break;
            case R.id.tv_paizhao:
```

```
            intent = new Intent(MediaStore.ACTION_IMAGE_CAPTURE);
            startActivityForResult(intent, 1);
            ll.setVisibility(View.GONE);
            break;
        case R.id.tv_xiangce:
            intent = new Intent(Intent.ACTION_PICK,
MediaStore.Images.Media.EXTERNAL_CONTENT_ URI);
            startActivityForResult(intent, 2);
            ll.setVisibility(View.GONE);
            break;
        case R.id.tv_quxiao:
            ll.setVisibility(View.GONE);
            break;
        case R.id.bt_suishoupai_tijiao:
            if (ettitle.getText().toString().equals("")) {
                suishoupai_title = "无题";
            } else {
                suishoupai_title = ettitle.getText().toString().trim();
            }
            if (etcontent.getText().toString().equals("")) {
                suishoupai_content = "";
            } else {
                suishoupai_content =
etcontent.getText().toString().trim();
            }
            SimpleDateFormat simpleDateFormat = new
SimpleDateFormat("yyyy年MM月dd日 hh:mm:ss");
            Date date = new Date(System.currentTimeMillis());
            String suishoupai_time = simpleDateFormat.format(date);
            ContentValues values = new ContentValues();
            final ByteArrayOutputStream os = new
ByteArrayOutputStream();
            if(resized!=null){
                resized.compress(Bitmap.CompressFormat.PNG, 100, os);
                values.put("title", suishoupai_title);
                values.put("content", suishoupai_content);
                values.put("time", suishoupai_time);
                values.put("tupian", os.toByteArray());
                helper = new SQLiteHelper(this);
                db = helper.getWritableDatabase();
                db.insert("suipai_infor", null, values);
                ettitle.setText("");
                etcontent.setText("");
                ivtupian.setImageResource(R.drawable.shangchuan);
                recycleInit(db);
            }else{
                Toast.makeText(this,"无发布内容",Toast.LENGTH_LONG).show();
            }
            db.close();
            break;
    }
```

```
    }
    @Override
    protected void onActivityResult(int requestCode, int resultCode, @Nullable Intent data) {
        super.onActivityResult(requestCode, resultCode, data);
        if (requestCode == 1 && data != null) {// "拍照"
            Bundle bundle = data.getExtras();
            bitmap = (Bitmap) bundle.get("data");
            resized = Bitmap.createScaledBitmap(bitmap, 300, 300, true);
            ivtupian.setImageBitmap(resized);
        } else if (requestCode == 2 && data != null) {// "从相册选择"
            imageUri = data.getData();  //获取系统返回的照片的Uri
            ContentResolver contentResolver = getContentResolver();
            Cursor cursor = contentResolver.query(imageUri, null, null, null, null);
            if (cursor != null) {
                if (cursor.moveToFirst()) {
                    imagePath =
cursor.getString(cursor.getColumnIndex(MediaStore.Images.Media.DATA));
                    bitmap = BitmapFactory.decodeFile(imagePath);
                    if (bitmap != null && ivtupian != null) {
                        resized = Bitmap.createScaledBitmap(bitmap, 300, 300, true);
                        ivtupian.setImageBitmap(resized);
                    }
                }
            }
            cursor.close();
        }
    }
}
```

12.3.9　其他功能实现说明

本系统中的建言献策的实现与随手拍类似，只是对主题文字、内容文字等的处理不同，数据存储和建言献策显示可以仿照随手拍。系统功能组织活动可以参看党建要闻，服务热线是对打电话的应用。

底部导航中的“学习中心”，显示学习分类和学习内容，可以结合学习视频中的学习分类、学习列表显示和党建要闻详情显示来实现。

底部导航中的“我的”主要功能是个人信息维护和修改密码。个人的信息存储在 SQLite 数据库中，在个人信息界面可以修改昵称、性别、电话，核心知识是 Activity 间的数据回传和 SQLite 中数据的存储，修改密码主要是数据的存储。

由于篇幅有限就不一一详细列举了，如果需要可以下载课程资源中源码参考，也可作为扩展练习巩固前面几节的学习内容。

【职业素养拓展12】

职业素养拓展

职业素养拓展内容详见文件“职业素养拓展”。

扫描二维码查看【职业素养拓展 12】

本章小结

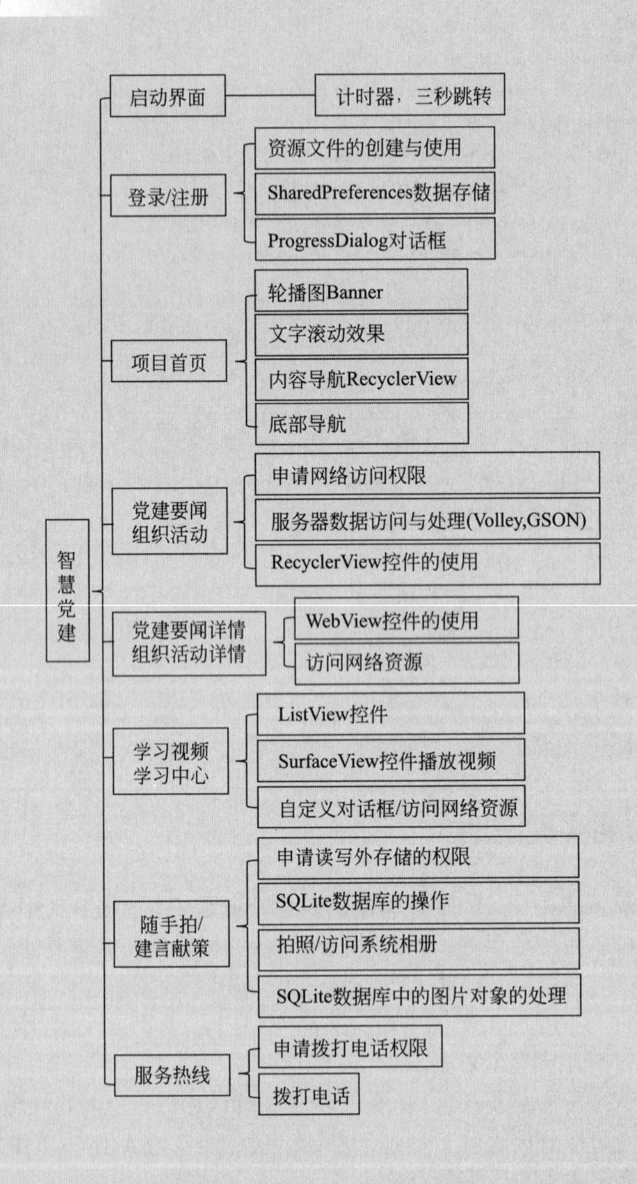
智慧党建
启动界面
计时器，三秒跳转
登录/注册
资源文件的创建与使用
SharedPreferences数据存储
ProgressDialog对话框
项目首页
轮播图Banner
文字滚动效果
内容导航RecyclerView
底部导航
党建要闻
组织活动
申请网络访问权限
服务器数据访问与处理(Volley,GSON)
RecyclerView控件的使用
党建要闻详情
组织活动详情
WebView控件的使用
访问网络资源
学习视频
学习中心
ListView控件
SurfaceView控件播放视频
自定义对话框/访问网络资源
随手拍/
建言献策
申请读写外存储的权限
SQLite数据库的操作
拍照/访问系统相册
SQLite数据库中的图片对象的处理
服务热线
申请拨打电话权限
拨打电话

参考文献

[1] Dawn Griffiths，David Griffiths. Head First Android 开发 .2 版. 乔莹，刘海洋，等译. 北京：中国电力出版社，2018.
[2] 明日科技. Android 精彩编程 200 例. 长春. 吉林大学出版社，2017.
[3] 黑马程序员. Android 移动应用基础教程（Android Studio）.2 版. 北京：中国铁道出版社有限公司，2019.
[4] 郭霖著. 第一行代码 Android .2 版. 北京：人民邮电出版社，2016.
[5] 唐亮，杜秋阳. 用微课学 Android 开发基础. 北京：高等教育出版社，2016.
[6] 罗文. Android 应用开发教程.北京：机械工业出版社，2013.